I0787960

Silicon-on-Insulator Technology and Devices 14

Editors:

Y. Omura
Kansai University
Osaka, Japan

S. Cristoloveanu
IMEP-INPG-Grenoble
Grenoble, France

F. Gámiz
University of Granada
Granada, Spain

B-Y. Nguyen
SOITEC USA
Bernin, France

Sponsoring Division:

 Electronics and Photonics

Published by
The Electrochemical Society

65 South Main Street, Building D
Pennington, NJ 08534-2839, USA
tel 609 737 1902
fax 609 737 2743
www.electrochem.org

ecstransactions ™

Vol. 19 No. 4

Copyright 2009 by The Electrochemical Society.
All rights reserved.

This book has been registered with Copyright Clearance Center.
For further information, please contact the Copyright Clearance Center,
Salem, Massachusetts.

Published by:

The Electrochemical Society
65 South Main Street
Pennington, New Jersey 08534-2839, USA

Telephone 609.737.1902
Fax 609.737.2743
e-mail: ecs@electrochem.org
Web: www.electrochem.org

ISSN 1938-6737 (online)
ISSN 1938-5862 (print)

Printed in the United States of America.

Preface

The Fourteenth International Symposium on Silicon-On-Insulator Technology and Devices was part of the 215th Meeting of the Electrochemical Society, held on May 24 - May 29, 2009, in San Francisco, California. The Electronics and Photonics Division of the Electrochemical Society sponsored the Symposium.

About 50 years have passed since SOS wafers became available. SOI-based integrated circuits are now well commercialized and many new device technologies based on advanced materials have been proposed. SOI activists had been hoping for the emergence of many new applications.

The Symposium did not disappoint with its 12 invited reviews, including 2 plenary talks, and 33 contributed papers. The Symposium has preserved its international nature, with 4 papers from North America, 22 from Europe, 16 from Asia, and 3 from South America. Many of the papers include collaborations between researchers from institutions on different continents.

This issue of *ECS Transactions* contains 41 papers presented at the conference. I would like to thank all the authors for submitting excellent manuscripts and submitting them on time despite the tight schedule. Most of the papers in these proceedings follow the same sequence as the symposium presentations in order to make it convenient to follow the text at the conference site.

It is my great pleasure to acknowledge the symposium co-organizers – Sorin Cristoloveanu, Francisco Gamiz, and Bich-Yen Nguyen – for their active involvement and efforts in spite of their heavy workload. Many thanks to the ECS staff: Paul Urso, John Lewis, Amir Zaman, Stephanie Plassa, and Timothy Fest for their support and advice. On behalf of the organizers, I like to express our great appreciation of the invited speakers who set the tone for the meeting and I thank all symposium speakers and co-authors for coming to San Francisco to share their recent exciting data and their technical expertise. Their contributions made the conference a success. The organizers would also like to thank the Electronics and Photonics Division for their technical support.

Yasuhisa Omura
Osaka, Japan
March 2009

iii

ECS Transactions, Volume 19, Issue 4
Silicon-on-Insulator Technology and Devices 14

Table of Contents

Preface   *iii*

Chapter 1
Plenary Talks

SOI Technology Driving the 21st Century Ubiquitous Electronics   3
   G. K. Celler

Impact of Confinement of Semiconductor and Band Engineering on Future   15
Device Performance
   V. Sverdlov, O. Baumgartner, T. Windbacher, F. Schanovsky and S. Selberherr

Chapter 2
Materials and Characterization I

Copper Decoration Combined with Preferential Etching for Delineation of Crystal   29
Defects in SOI and sSOI Materials
   H. Idrisi, M. Pejic, F. Uzelli and B. Kolbesen

Simulation and Characterization of the Strain Induced by an Original "Embedded   37
Buried Nitride" Technique
   S. Baudot, F. Andrieu, M. Kostrzewa, J. Widiez, J. Barbe, O. Faynot,
   Y. Lamrani, C. Vizioz, L. Brevard, H. Denis, V. Delaye, F. Rieutord and
   J. Eymery

Chapter 3
Device Technology I

Advanced FinFET Devices for Sub-32nm Technology Nodes: Characteristics and   45
Integration Challenges *
   A. Veloso, N. Collaert, A. De Keersgieter, L. Witters, R. Rooyackers,
   M. J. van Dal, R. Duffy, B. J. Pawlak, R. J. Lander, T. Hoffmann,
   S. Biesemans and M. Jurczak

v

Ultra Compact FDSOI Transistors including Strain and Orientation: Processing and Performance *   55
C. Fenouillet-Beranger, L. Pham Nguyen, P. Perreau, S. Denorme, F. Andrieu, O. Faynot, L. Tosti, L. Brevard, C. Buj, O. Weber, C. Gallon, V. Fiori, F. Boeuf, S. Cristoloveanu and T. Skotnicki

Impact of New Approach to Improve MOSFETs Performance with Ultrathin Gate Insulator   65
W. Cheng, A. Teramoto and T. Ohmi

Process and Characterization of Hybrid-Oriented Complementary Single-Crystal Si Thin-Film Transistors on A Plastic Substrate   71
H. Pang, G. K. Celler and Z. Ma

Chapter 4
Materials and Characterization II

SOI and Other Semiconductor-On-Insulator Substrates Characterization *   79
A. Abbadie, Y. -M. Le Vaillant, C. Figuet and E. Guiot

Morphological and Electrical Comparison of GeOI Enriched Structures Obtained from SOI and sSOI Substrates   93
J. Damlencourt, B. Vincent, L. Clavelier, C. Le Royer, Y. Campidelli, S. Bernasconi, Y. Morand, T. Nguyen and S. Cristoloveanu

Chapter 5
Device Physics and Modeling I

Comparison of SOI FinFETs and Bulk FinFETs *   101
J. Lee, J. Lee, H. Jung and B. Choi

Mechanisms that Explain Short-Channel Effects of Sub-50-nm-Channel Ultra-Thin Symmetric Double-Gate SOI MOSFET   113
Y. Omura and Y. Tahara

Chapter 6
Device Physics and Modeling II

Advanced Design Methodology of High-Performance Sub-100-nm-Channel GAA MOSFET   121
S. Nakano, O. Hayashi, Y. Omura, S. Yamakawa and H. Wakabayashi

Comparison of the Electrostatics of Bulk and SOI Trigate MOSFETs   127
F. G. Ruiz, A. Godoy, I. Tienda-Luna and F. Gamiz

Gate Stack Influence on GIFBE in nFinFETs   133
J. A. Martino, M. Rodrigues, A. Mercha, E. Simoen, A. Veloso, N. Collaert, M. Jurczak and C. Claeys

Analytical Modeling of Double Gate Graded-Channel SOI Transistors for Analog Applications   139
F. A. Ferreira, A. Cerdeira, D. Flandre and M. Pavanello

Low-Temperature Measurements on Germanium-on-Insulator pMOSFETs: Evaluation of the Background Doping Level and Modeling of the Threshold Voltage Temperature Dependence   145
W. Van Den Daele, E. Augendre, K. Romanjek, C. Le Royer, L. Clavelier, J. Damlencourt, E. Guiot, B. Ghyselen and S. Cristoloveanu

The Wave SOI MOSFET: A New Accuracy Transistor Layout to Improve Drain Current and Reduce Die Area for Current Drivers Applications   153
S. P. Gimenez

Chapter 7
Device Technology II

SOI as Platform for Transition from Micro to Nano *   161
F. Balestra

Optimization of Blue/UV Sensors Using PIN Photodiodes in Thin-Film SOI Technology   175
O. Bulteel and D. Flandre

A Wide Range Temperature Sensor Using SOI Technology   181
R. Patterson, M. E. Elbuluk and A. Hammoud

vii

CHAPTER 1

PLENARY TALKS

SOI Technology Driving The 21st Century Ubiquitous Electronics

George K. Celler

Soitec USA, 2 Centennial Drive, Peabody, MA 01960, USA

Electronics is expanding into all spheres of our life, spreading far beyond computer systems, mobile telephony, and personal computers – into games, appliances, navigation systems, at-home health diagnostics, a variety of smart sensors at home, in cars and in the workplace. RFID chips are becoming integrated into paper documents, and various sensors are likely to be included in our clothing. Silicon-on-Insulator (SOI) based technology plays a growing role in many of these applications.

Introduction

SOI wafers consist of a layer of single crystalline silicon that is separated by a layer of amorphous SiO_2 from a silicon support or "handle" wafer. SOI is currently used in many mainstream electronic applications, such as high performance microprocessors, game and entertainment systems used in millions of homes, and smart power devices in automotive and lighting applications. SOI substrates significantly improve the performance of electronic circuits [1]. In the 1980s the main driving force for developing SOI material was to increase radiation hardness of integrated circuits, i.e., improved immunity to single event upsets caused by ionizing particles. Later, SOI applications entered the mainstream of digital electronics, where they provide increased switching speed of transistors and reduced power consumption – this is accomplished primarily because the source and drain capacitance is lowered by having an insulator layer below a MOSFET. As device gate length is scaled below 50 nm, thin SOI layers are becoming essential to maintain proper transistor action by suppressing the short-channel effects. Eventually, planar transistors may need to be replaced with vertical transistors or other multigate devices – SOI plays an important role in these novel device architectures as well [2,3].

The dominant commercial method of fabricating SOI structures is known as the Smart Cut™ technology, and it evolved from an invention by M. Bruel at LETI in Grenoble [4]. In the context of fabricating SOI structures, the Smart Cut™ technology utilizes ion implantation and wafer bonding to transfer thin silicon layers from a "donor" substrate to a new "handle" substrate.

In addition to electronic applications, SOI structures are also very important for MEMS devices, and for the emerging field of silicon photonics. Photonic integrated circuits can be built, in which light is channeled through waveguides that are etched in the SOI films, together with modulators, multiplexers, filters, and other key optical elements.

Another emerging field is flexible electronics; circuits that are built in thin silicon layers attached, for example, to polymer substrates. High quality Si films are obtained by transferring SOI from the handle wafers to new flexible supports.

Some History

The history of SOI developments can be best followed through the activities of topical conferences, such as the ECS SOI Science and Technology Symposium, held every two years over the last 30 years, and the annual IEEE International SOI Conference. The latter was started in 1975 as an SOS Technology Workshop, but after a few years its organizers realized that silicon-on-sapphire was only a subset of an emerging much broader field of silicon-on-insulator. For many years the close-knit community of SOI researchers presented at these two conferences and at other specialized meetings in Europe and Japan their improved methods of forming SOI structures and described design and fabrication of devices made in SOI. Much of this early work was supported by various government defense organizations in the US and Europe, as the main applications were for radiation–hardened electronics for military and space use [5]. SOS circuits were first to be qualified and used in various rad-hard systems in the 1980s and 1990s, and then they were slowly complemented or replaced by SOI based circuitry. Companies such as Honeywell started offering rad-hard SRAMs in SOI. At SOI conferences, performance advantages of devices on SOI were presented, but it took a major device and circuit design effort at IBM to break through the perception of SOI being suitable just for niche applications.

Currently, both the scientific landscape and the commercial activity are different. SOI has been in the commercial mainstream for several years now, something that the participants in SOI meetings in the 1980s and 1990s were hoping and preparing for, in spite of skeptics who believed that it would never happen. Because of the mainstreaming of SOI, technical papers on SOI technology, devices, and circuits are an integral part of major conferences in the respective areas, including such premier meetings like IEDM and VLSI Technology Symposium. SOI belongs now to the electrical engineer's toolbox, to be used in designing circuits for a variety of present and future applications.

Substrates

In the 1970s companies like Hughes, RCA, Rockwell, and Boeing had a need for rad-hard devices, and in order to satisfy this need silicon-on sapphire (SOS) substrates were developed [6]. Heteroepitaxy of Si on Al_2O_3 yielded very highly defective silicon films, and to mitigate this problem many creative solutions were developed such as partial amorphization of the epi films with Si ion implants, followed by solid phase epitaxial (SPE) regrowth from a less defective near-surface single crystalline layer [7].

During the same years laser annealing was first proposed and investigated, using either short pulses that melted the near surface silicon region, or with continuous (cw) beams that repaired implantation damage via SPE and/or defect diffusion and dissolution. Laser annealing activities led to the first non-epitaxial attempts to form thin single crystalline Si layers on insulator. Polycrystalline (or amorphous) Si films on oxidized Si wafers were the obvious choices for precursor wafers, and the success of the method depended on a clever control of temperature gradients and the design of seeding windows in the buried oxide (BOX) that linked the film to the handle wafer. These early attempts at creating SOI structures (as opposed to SOS) were soon supplemented by various other techniques, such as FIPOS, ELO, SPE, and others [1,8]. SIMOX technique was developed at NTT by Izumi et al. [9], and was later picked up by TI and others for further development and improvement. Wafer bonding and thinning (BESOI) [10,11] and also ELTRAN [12] emerged as other methods, in addition to SIMOX, that did not require any wafer patterning to form the SOI structure. Eventually the Smart Cut™ technology, based on Bruel's invention [4] proved itself as the dominant method for high volume manufacturing of SOI wafers.

The Smart Cut Technology

The Smart Cut™ technology is a high volume manufacturing method for thin layer transfer and SOI fabrication. It consists in defining a splitting region within the donor substrate by ion implantation (e.g. H^+ or He^+) that allows transferring a thin film to a handle wafer by means of bonding followed by splitting. Hydrogen doses $>5\times10^{16}$ cm^{-2} are typically used for splitting of silicon. For manufacturing SOI structures, the implanted surface is bonded to another wafer and splitting is achieved along the implanted zone [13]. The net result is that a thin layer of Si, defined precisely by the implant depth, is transferred from a donor wafer to a handle wafer. Since the thickness removed from the donor wafer is negligible compared to the total wafer thickness, the donor wafer can be reused many times. A range of thicknesses from 10 nm up to a few 10^3 nm for both top Si and BOX films is covered by this technology with standard industrial implanters and furnaces [14].

Since the Smart Cut technology is a layer transfer technique, it allows producing a variety of engineered wafers, with a layer of one material placed on a substrate of a different material. For example, biaxially strained Si layers or relaxed layers of SiGe alloys can be bonded to Si handle wafers with SiO_2 in between [15], germanium films on insulator (GeOI) can be formed [16], and compound semiconductor layers on oxidized Si handle wafers can be produced. Multilayer structures such as SOLES wafers that combine SOI and GeOI hold promise for monolithic integration of CMOS and III-V RF devices [17]. The dielectric film can be SiN or a composite film of SiO_2 and SiN, and in principle the handle wafer can be glass, fused silica, SiC or another flat and smooth substrate. This flexibility in combining dissimilar materials with specific properties facilitates fabrication of devices with improved performance or better matched to the unique need.

Current SOI Applications

High Performance microprocessors

The single largest application of SOI is in high performance microprocessors that are used in a broad spectrum of applications, from desktop PCs and game systems to servers, specialized scientific computers and the largest supercomputers. The use of SOI leads to higher performance, lower dynamic and static power and better immunity to soft errors.

Microprocessors in x86 architecture from AMD: These have been shipped since 2003, starting with single-core Opteron™ built in 130 nm technology, and moving to current dual, triple and quad cores, fabricated in many versions using 65 nm design rules. An example is AMD Phenom™ X4 that comes in 1.8 to 2.6 GHz clock versions.

Microprocessors in POWER architecture: Built by IBM and Freescale, and their foundry partners for use in some PCs, a variety of servers and blades, and customized mainframe computers. POWER6™ is the latest IBM version in production, with 65 nm design rules, and a clock frequency >4GHz [18]. An integrated communications processor MPC8569E PowerQUICC™ III from Freescale is built in 45 nm design rules, and it combines an e500 processor core built on Power Architecture™ technology with system logic required for networking, wireless infrastructure, and telecommunications applications.

Cell architecture microprocessors, Initially designed by IBM, Sony and Toshiba for game systems (Sony PlayStation 3), but also used or planned for HD TVs and a variety of other graphics intensive and multitasking applications. Cell is a heterogeneous chip multiprocessor that consists of an IBM 64-bit Power Architecture™ core, augmented with eight specialized co-processors based on a novel single-instruction multiple-data (SIMD) architecture called Synergistic Processor Unit (SPU), which is for data-intensive processing, like that found in cryptography, media and scientific applications. The system is integrated by a coherent on-chip bus [19].

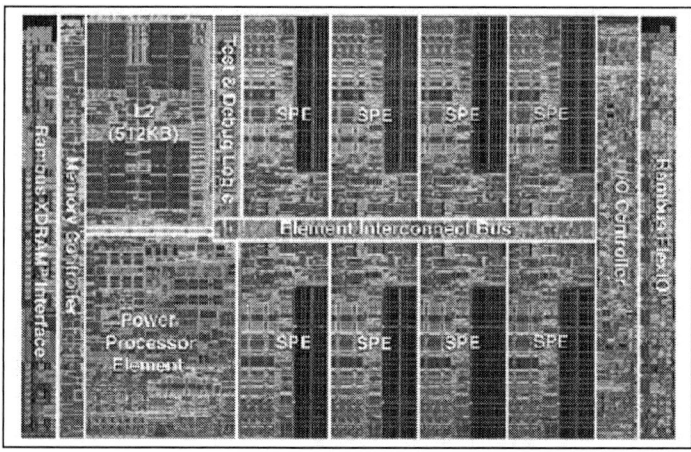

Fig. 1. Cell microprocessor with 9 CPU units [20].

Supercomputers: Four out of the top five most powerful supercomputers on the www.top500.org list are based on SOI microprocessor chips. First on the list is Roadrunner, which can operate at 1 petaflops (1E15 floating point operations/sec).. This system contains 6,948 dual-core Opteron™ processors and 12,960 PowerXCell™ 8i processors that are interconnected with about 57 miles of optical fiber. The Opteron processors handle standard processing such as file system I/O. The PowerXCell 8i processors accelerate mathematical and CPU-intensive processing. PowerXCell 8i is an HPC-specific modification of IBM's first-generation Cell Broadband Engine (Cell/B.E.) processor designed for the computer gaming market and the Sony PlayStation.

 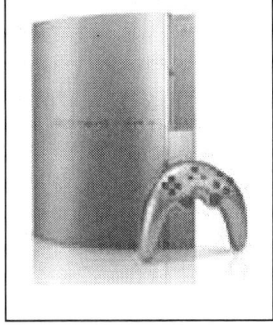 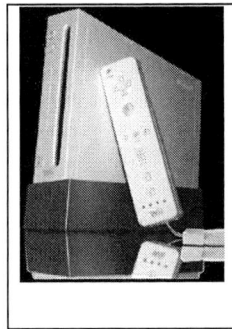

Fig. 2. Gaming computers: XBOX 360, Play Station 3, and Wii.

Low Power Applications

Performance and power are inversely related – operating a very high performance microprocessor at a reduced supply voltage Vdd cuts down on performance while greatly reducing power consumption during operation.

True low power circuits require that both the dynamic power Pdyn and the standby or static power Pstat are minimized:

$$P = P_{stat} + P_{dyn} = Ioff(Vdd)Vdd + afC_{load}Vdd^2$$

where f is the switching frequency and a is a proportionality coefficient.

The leakage current Ioff should be as low as 10 pA/μm in low standby power (LSTP) devices, and a few nA/μm in low operating power (LOP) devices. Fully depleted (FD) SOI transistors with undoped channels are ideally suited for low power applications [21]. Substrates for FD devices require very thin Si films, of the order of 20 nm with the uniformity range of <1 nm. After processing, transistor channel regions are as thin as 4-5 nm, which makes the S/D junction area and its contribution to Cload very small. FD SOI devices have very steep subthreshold slopes, with values in 60-65mV/decade range, close to the theoretical minimum. This allows setting Vt and Vdd very low, and assures that both Pstat and Pdyn are low.

Oki has specialized for many years in applications that are less demanding in performance but require ultra-low power. They have been building FD SOI system-on-a-chip (SOC) circuits for several popular lines of solar-powered radio-controlled Casio watches [22].

Memory Applications

Embedded and Stand-alone DRAM. Embedded memory consumes a dominant fraction of a microprocessor chip area. For example a memory cache occupies about 70% of the area in an Intel dual-core Itanium processor released in 2006. Although SRAM is faster than DRAM, it takes about 3-4 times more area per MB. Other benefits of DRAM include lower leakage power, lower soft error rates, and greater cell stability at low voltages. Best performance requires a combination of SRAM for lower level cache and DRAM for larger amount of storage. SOI microprocessors require SOI embedded SRAM and DRAM. SOI SRAM has been used very successfully in many generations of microprocessors [23]. More recently IBM has demonstrated that eDRAM with deep trench capacitors is simpler to fabricate (therefore less expensive) in SOI than in bulk silicon as BOX provides complete isolation of the capacitor plate [24]. Stand-alone SOI DRAM can also be implemented.

Floating body cell (FBC) memory. Further improvement in cell density can be achieved in embedded memories by taking advantage of the floating body effects in SOI transistors. In the early days of SOI, floating body effects that are intrinsically tied with PD SOI technology were a design challenge in building circuits. In FBC devices, each memory cell consists of one transistor, and the amount of charge in each transistor (floating body effect) determines the state of the memory cell. Elimination of storage capacitors greatly simplifies processing, including fewer lithography steps. FBC memory can be embedded, competing in speed with SRAM and with retention times of ~100 ms, or be designed as a stand-alone memory compatible in performance to conventional DRAMs. FBC memories often take advantage of back-gate biasing through the BOX. This requires that the BOX be very thin, as shown in the example presented in Fig. 3.

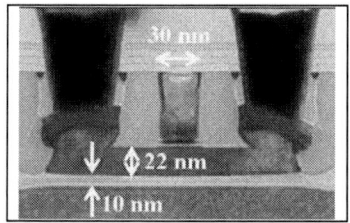

Fig. 3. Transistor for a floating body memory cell (after Intel [25]).

RF Applications

Cross talk between RF analog circuits and digital logic contained within the same "mixed signal" chips is reduced in SOI substrates. High impedance SOI, i.e. SOI in which the handle substrate has resistivity >1 kOhm-cm, enhances these advantages and improves performance of monolithically integrated passive components, such as inductors.

SOI technology provides complete oxide isolation, cutting off direct paths of substrate injection noise and a high resistivity substrate reduces the capacitive coupling, thus further reducing the substrate related RF loss. Because of the SOI-inherent isolation of the high impedance substrate, device latchup is not an issue.

Patterned ground shields (PGS) are used in Si bulk wafers to manufacture high quality factor Q inductors. In SOI these can be avoided and better Q factors are achieved even at higher frequencies. Typically 50% greater Q is achieved with high resistivity SOI (SOI HR) as compared to bulk: Coplanar transmission line measurements on SOI HR show that a loss better than 0.5dB/mm at 40 GHz and 1 dB/mm at 80 GHz can be achieved. At these values integrated passive components in SOI HR become comparable to what can be achieved on InP.

Recently, RF power application on thin SOI substrates were demonstrated, with good DC, small and large RF signal performance, thus showing that the integration of the antenna switch and the RF power amplifier is feasible with SOI technology [26].

Automotive applications

SOI-based electronics is attractive for automotive applications for several reasons [27]. Automotive electronics devices tend to be relatively high voltage (40 to 80V) or high power. Chips built on traditional bulk substrates, rather than SOI, require very large back-biased junctions to isolate one device from the next. Such p-n junctions isolate all devices in bulk silicon, but their thickness grows with operating voltage. SOI uses insulating BOX instead of junctions, leading to smaller, denser device structures. SOI devices can also operate at much higher temperatures than bulk devices. In bulk-based chips, current leakage through back-biased isolation junctions rapidly increases with temperature. As dielectric isolation (buried oxide) does not allow current flow regardless of temperature (to the first approximation), an SOI chip can be mounted in the high temperature environment under the hood.

From the automotive designer's perspective, SOI simplifies network integration. SOI enables the integration of different device types, such as power, analog, and digital, on a single die, bringing crucial protection to sensitive electronics, low power and superior EMC (electromagnetic compatibility) performance. Better EMC effectively eradicates potentially perilous miscommunication in the electrically noisy and hazardous automotive environment. Thanks to the enhanced EMC performance, automotive designers don 't need to protect wires and chips with complicated shielding systems, which in turn, brings down the cost of manufacturing. Another advantage of SOI-based chips is that they pull less power overall. With the rise in the shear amount of automotive electronics, respecting the power budget is an important consideration. Thus the key advantages are:
- *Reduced resistance*, Rds(on), for the transistor in the on –state. This means that much less waste heat is produced for a given amount of electrical output power.
- *Much greater packing densities*, especially for high voltages.
- *Latch-up-free behavior*, since there are no parasitic junctions between N-type and P-type devices.

- *Ability to handle voltage spikes* from the starter motor or alternator (up to 50V), for example, or caused by collapsing hysteresis field in the coils of motor drives.
- *Greatly improved heat tolerance.* In combination with the lower Rds(on), high -power handling ICs can be created without the need for heat sinks, reducing both size and costs.
- *Easy integration* of multiple power devices, bridge rectifiers, and flyback diodes on the same piece of silicon. In combination with a significant reduction in parasitic capacitance, this simplifies and speeds the chip design process.

High Temperature Applications. Modern automotive and avionics control systems often require operation at high temperatures. SOI technology allows chip temperatures of up to 200°C and ambient temperatures of up to 150°C. As a result, SOI devices can be placed in hot environments. In turbochargers or exhaust gas recirculation systems, many flaps have to be controlled by DC-motor driver ICs that are located very close to a hot engine. For example, Atmel's SMARTIS technology uses an SOI substrate instead of a standard BCDMOS bulk technology, and this greatly minimizes the junction leakage, a major issue in high-temperature designs. Furthermore, SOI devices provide superior latch-up immunity for increased reliability of circuits operating in hot environments.

Lighting

SOI-based HV/Smart Power chips have been long used to control fluorescent lamps made by Philips (now NXP). A new high voltage SOI circuit to drive a compact florescent light (CFL) bulb was recently reported [28] which increases lamp's lifetime and decreases operating temperature. The bulbs are rapidly replacing incandescent lamps, however they have lifetimes limited to less than 10,000 hours and generate as much as 150 degrees of heat. With new IC drivers not only these parameters are improved but also new functionality is added.

CMOS Imagers and 3D applications

CMOS imagers are in widespread use now in low cost cameras that have become ubiquitous in mobile phones and other portable electronic devices (dedicated higher quality cameras are more likely to utilize CCDs). Back-side illumination of CMOS imagers is the most efficient way to capture light unobstructed by transistors and metal interconnects. Device fabrication in SOI combined with subsequent layer transfer is the most practical way to obtain backside illumination. [29]. By stacking multiple SOI layers, 3D light sensing and processing circuits can be built, as shown by MIT LL [30].

Photonics

SOI substrates provide an excellent platform for optical waveguides and complete photonic circuits operating at the typical telecom wavelengths of 1.31 and 1.55 μm [31]. Mach-Zender and resonator type optical modulators, AWG mux/demux devices, delay lines, and other components can be fabricated directly in SOI, while photodetectors are typically formed in small regions of Ge that are grown by selective epitaxy on Si. Light sources (lasers) can be attached to SOI circuits or light can be provided remotely via optical fibers coupled to the Si waveguides.

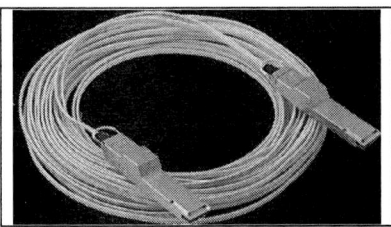

Fig. 4. Active optical cable. Each "connector" includes a transceiver chip with four channels for electrical to optical conversion at 10GB/s per channel. (After Luxtera [32]).

Flexible Electronics

Flexible thin-film transistors TFTs that can operate in the GHz regime are attractive for aerospace systems where large-area RF systems such as phased array antennas are employed for communication, remote sensing, and surveillance. Currently, these RF systems are made as stand-alone units from discrete components. They are typically bulky and are usually mounted as protrusions to the carrying vehicles. Flexible TFTs and passive components, if they can be monolithically integrated on flexible plastic substrates, will enable the large area RF systems to be directly and conformally mounted on the curved surfaces of mission vehicles. The realization of such flexible large-area systems will greatly reduce the weight and drag force during high-speed motion and increase the reliability of the mission vehicles. However, the stringent requirements for the active circuitry of the large-area RF systems such as capable of operating in the L-band (1–2 GHz) and even higher frequency regime have been posing a great challenge for existing flexible electronics technologies.

Using a single crystal Si membrane as an active layer that was preprocessed at high temperatures and transferred onto a low temperature PET substrate, flexible TFTs with f_T of 2.04 GHz and $f$max of 7.8 GHz have been demonstrated [33]. The influence of TFT layout on the high-frequency response has also been investigated. It was found that the source-to-gate and drain-to-gate distances play a crucial role in the high frequency response of the TFTs. By properly overlapping gate to source and drain, significant reduction of source/drain access resistance and output conductance has been achieved which results in significantly improved $f$max. Requirements of certain designated flexible RF applications may be met by suitably designing TFT layout.

RFID technology

There are indications that RFID chips will be the most ubiquitous of all electronic devices. The smaller such devices are, the more places they can be used in. Usami at Hitachi [34] has developed devices that are so small – 50×50 μm^2, and only 5 μm thick including top and bottom antenna electrode - that they are called "Powder LSI" [35]. SOI substrate provides benefits of dielectric isolation, but more importantly in this case it allows simple and precise removal of the handle wafer, with BOX serving as an etch stop. These super-thin devices can be easily embedded in paper sheets, banknotes, passports, tickets, or almost anything else. Each device stores one 128-bit long number in its ROM, and a database connected to a reader identifies the meaning of the information stored in RFID. Although some privacy issues are being raised, if used wisely RFID will have significant impact on commerce, security, and everyday life.

MEMS and MOEMS Applications

SOI is in many ways an ideal starting substrate for fabrication of MEMS and optical MEMS (MOEMS) devices [36]. Single crystalline low stress SOI layer that can be released by etching away the oxide below it is a very good mechanical material for a variety of applications, from accelerometers used in air-bags and a multitude of other systems to micromirrors that can be used as optical cross-connects in telecom systems. RF MEMS metal contacting switches and microresonators add new functionality to analog and mixed signal electronics [37]. Process sequences have been developed for fabrication of MEMS devices within conventional ICs.

Photovoltaic Applications

There have been attempts to utilize 5-20 μm thick freestanding SOI films that were lifted of a conventional SOI wafer as solar cells [37]. One motivation is to eventually reduce the cost of crystalline Si solar cells by minimizing silicon consumption. Another, possibly more significant motivation, is to produce ultra-light PV cells, for example for space applications.

Conclusions

The message from this brief and incomplete overview of SOI applications is that SOI substrates enable a broad spectrum of technologies that are used in many conventional areas of electronics, such as microprocessors, smart power, and high voltage because they offer better performance. In addition, SOI enables some unique applications that would be very difficult if not impossible in bulk Si, such as RF devices in high resistivity substrates, ultra-thin RFID chips, backside imagers, MEMS, photonic integrated circuits, and flexible electronics. With growing adoption of SOI, we are more and more likely to have SOI based devices in our homes and work places.

Acknowledgments

This review of applications would not be possible without helpful discussions with many colleagues at Soitec.

References

1. G. K. Celler and Sorin Cristoloveanu, *J. Appl. Phys.* **93**, 4955-4978 (2003).
2. Carlos Mazuré and André-Jacques Auberton-Hervé, in *Proc. of ESSDERC 2005*, edited by G. Ghibaudo, et al., pp. 29-38, (IEEE, Grenoble, Sept. 2005).
3. Carlos Mazuré and George K. Celler, *The Electrochemical Society Interface*, Vol. 15, No. 4, pp. 33-40 (January 2007).
4. M. Bruel, *Electron. Lett.* **31**, 1201 (1995).

5. M.L. Alles, D.R. Ball, R. D. Schrimpf, D. M. Fleetwood, R. A. Reed and B. Jun, in *Silicon-on-Insulator Technology and Devices XII,* ECS Proc Vol. 2005-03, edited by G. K. Celler et al. (The Electrochemical Society, Pennington, NJ, USA, 2005), pp.87-98.
6. Andreas G. Andreou, *Proc. IEEE-SOI Conf.*, paper 1.3 (2008).
7. A. Gupta and P. K. Vasudev, *Solid State Technol.* 26~2!, 104 (1983).
8. Jean-Pierre Colinge, *Silicon-on-Insulator Technology: Materials to VLSI,* 3rd edition (Springer, 2004).
9. K. Izumi, M. Doken, and H. Ariyoshi, *Electron. Lett.* **14**, 593 (1978).
10. J. B. Lasky, *Appl. Phys. Lett.* **48**, 78 (1986).
11. W. P. Maszara, G. Goetz, A. Caviglia, and J. B. McKitterick, *J. Appl. Phys.* **64**, 4943 (1988).
12. T. Yonehara, in *Silicon Wafer Bonding Technology for VLSI and MEMS Applications*, edited by S. S. Iyer and A. J. Auberton-Hervé, (INSPEC, London, UK, 2002), Chap. 4, p.53.
13. B. Aspar, H. Moriceau, E. Jalaguier, C. Lagahe, A. Soubie, B. Biasse, A. M. Papon, A. Claverie, J. Grisolia, G. Benassayag, F. Letertre, O. Rayssac, T. Barge, C. Maleville, and B. Ghyselen, *J. Electronic Materials*, 30, 834 (2001).
14. C. Maleville and C. Mazuré, *Solid-State Electron.* **48** (6), pp. 1055-1063 (2004).
15. S. Cristoloveanu and G. K. Celler, Chapter 4 of *Handbook of Semiconductor Manufacturing Technology, 2nd edition,* edited by R. Doering and Y. Nishi (CRC Press, Taylor and Francis Group, Boca Raton, Fl, 2007).
16. Takeshi Akatsu, Chrystel Deguet, Loic Sanchez, Frédéric Allibert, Denis Rouchon, Thomas Signamarcheix, Claire Richtarch, Alice Boussagol, Virginie Loup, Frédéric Mazen, Jean-Michel Hartmann, Yves Campidelli, Laurent Clavelier, Fabrice Letertre, Nelly Kernevez, and Carlos Mazuré, *Materials Science in Semiconductor Processing* 9 (2006) 444-448 (Proc. of E-MRS 2006 Symp T).
17. K.J. Herrick, T. E. Kazior, J. Laroche, A. W. K. Liu, D. Lubyshev, J. M. Fastenau, M. Urteaga, W. Ha, J. Bergman, B. Brar, M. T. Bulsara, E. A. Fitzgerald, D. Clark, D. Smith, R.F. Thompson, N. Daval, G. K. Celler, *ECS Transactions*, 16 (8), pp. 227-234 (2008).
18. R. Berridge, R. M. Averill III, A. E. Barish, M. A. Bowen, P. J. Camporese, J. DiLullo, P. E. Dudley, J. Keinert, D. W. Lewis, R. D. Morel, T. Rosser, N. S. Schwartz, P. Shephard, H. H. Smith, D. Thomas, P. J. Restle, J. R. Ripley, S. L. Runyon, P. M. Williams, *IBM J. Res. & Dev.* 51 (6), 685 (2007).
19. http://domino.research.ibm.com/comm/research.nsf/pages/r.arch.innovation.html
20. D. Pham, S.Asano,M. Bolliger, M. Day, H. Hofstee, C. Johns, J. Kahle, Kameyama, J. Keaty,Y. Masubuchi, M. Riley, D. Shippy, D. Stasiak, M.Wang1, J.Warnock, S.Weitzel, D.Wendel, T.Yamazaki, K.Yazawa, *ISSCC Tech. Digest*, paper 10.2 (2005).
21. O. Weber, O. Faynot, F. Andrieu, C. Buj-Dufournet, F. Allain, P. Scheiblin, J. Foucher, N. Daval, D. Lafond, L. Tosti, L. Brevard, O. Rozeau, C. Fenouillet-Beranger, M. Marin, F. Boeuf ,D. Delprat, K. Bourdelle, B.-Y. Nguyen and S. Deleonibus, *IEDM Tech. Digest.*, paper 10.4 (2008).
22. Masafumi Nagaya, *OKI Technical Review*, Issue 193, Vol.70, No.1, pp. 48-51 (2003).

23. R. V. Joshi, Y. Chan, D. Plass, T. Charest , R. Freese, R. Sautter, W. Huott, U. Srinivasan, D. Rodko, P. Patel , P. Shephard, T. Werner, *Proc. IEEE SOI Conf.* paper 1.2 (2006).
24. G. Wang, K. Cheng, H. Ho, J. Faltermeier, W. Kong, H. Kim, J. Cai, C. Tanner III, K. McStay, K. Balasubramanyam, C. Pei, L. Ninomiya, X. Li, K. Winstel, D. Dobuzinsky, M. Naeem, R.Zhang, R. Deschner, M.J. Brodsky, S. Allen, J. Yates III, Y. Feng, P. Marchetti, C. Norris, D. Casarotto, J. Benedict, A. Kniffin, D. Parise, B. Khan, J. Barth, P. Parries, T. Kirihata, J. Norum, and S.S. Iyer, *IEDM Tech. Digest*, paper 21.1, (2006)
25. U.E. Avci, I. Ban, D.L. Kencke, P.L.D. Chang, *Proc. IEEE-SOI Conf.* paper 3.1 (2008).
26. R.A. Bianchi, F. Monsieur , F. Blanchet , C. Raynaud, O. Noblanc, *IEDM Tech Digest,* paper 6.3 (2008).
27. George Celler, Soitec and Theo Lavrijsen, www.automotivedesignline.com March 2, 2006.
28. Rene Penning de Vries, *ISSCC Tech Digest*, paper 1.1 (2009).
29. Perceval Coudrain, Perrine Batude2, Xavier Gagnard, Cédric Leyris, Stéphane Ricq, Maud Vinet, Arnaud Pouydebasque, Norbert Moussy, Yvon Cazaux, Benoit Giffard, Pierre Magnan, Pascal Ancey, *IEDM Tech. Digest*, paper 11.3 (2008).
30. Vyshnavi Suntharalingam, Robert Berger, Stewart Clark, Jeffrey Knecht, Andrew Messier, Kevin Newcomb, Dennis Rathman, Richard Slattery, Antonio Soares, Charles Stevenson, Keith Warner, Douglas Young, Lin Ping Ang, Barmak Mansoorian, David Shaver, *ISSCC Tech. Digest*, paper 2.1 (2009).
31. Bahram Jalali and Sasan Fathpour, *J. Lightwave Techn.*, **24** (12), 4600 (2006).
32. www.luxtera.com
33. Hao-ChihYuan, Celler, Zhenqiang Ma, *J. Appl. Phys.* **102**, 034501 (2007).
34. Mitsuo Usami, Hisao Tanabe, Akira Sato, Isao Sakama, Yukio Maki, Toshiaki Iwamatsu, Takashi Ipposhi, Yasuo Inoue, *ISSCC Tech. Digest* paper 26.6 (2007).
35. T. Hornyak, *Scientific American*, February 2008, page 68.
36. G. Celler, *Solid State Technology*, Dec. 2003.
37. G. V. Herrera, Todd Bauer, M. G. Blain, P. E. Dodd, R. Dondero, E. J. Garcia, P. C. Galambos, D. L. Hetherington, J. J. Hudgens, F. B. McCormick, G. N. Nielson, C. D. Nordquist, M. Okandan, R. H. Olsson, K. Ortiz, M. R. Platzbecker, P.J. Resnick, R. J. Shul, M. J. Shaw, C. T. Sullivan, M. R. Watts, *Proc. IEEE-SOI Conf.* paper 1.2 (2008).

ECS Transactions, 19 (4) 15-26 (2009)
10.1149/1.3117388 ©The Electrochemical Society

Impact of Confinement of Semiconductor and Band Engineering on Future Device Performance

V. Sverdlov[a], O. Baumgartner[a], T. Windbacher[a], F. Schanovsky[b], S. Selberherr[a]

[a] Institute for Microelectronics
[b] C. Doppler Laboratory for TCAD at the Institute for Microelectronics
TU Wien, Gußhausstraße 27-29, 1040 Wien, Austria

A rigorous analysis of the subband structure in thin silicon films under stress is performed. Calculated subband effective masses are shown to depend on shear strain and thickness simultaneously. The effective masses and the subband splitting determine transport in silicon films. Decrease of the transport effective mass controlled by the shear strain component guarantees mobility enhancement even in ultra-thin silicon films. This increase of mobility and drive current combined with the improved channel control makes multi-gate MOSFETs based on thin films or silicon fins preeminent candidates for the 22nm technology node and beyond.

Introduction

The rapid increase in computational power and speed of integrated circuits is supported by the aggressive size reduction of semiconductor devices. Downscaling of MOSFETs as institutionalized by Moore's law is successfully continuing because of innovative changes in the technological processes and the introduction of new materials. The 32nm MOSFET process technology recently developed by Intel (1) involves new hafnium-based high-k dielectric/metal gates and represents a major change in the technological process since the invention of MOSFETs. Although alternative channel materials with a mobility higher than in Si were already investigated (2, 3), it is commonly believed that strained Si will be the main channel material even for MOSFETs beyond the 32nm technology node. With scaling apparently approaching its fundamental limits, the semiconductor industry is facing critical challenges. New engineering solutions and innovative techniques are required to improve CMOS device performance. Strain-induced mobility enhancement is the most attractive solution to increase the device speed and will certainly take a key position among other technological changes for the next technology generations. In addition, new device architectures based on multi-gate structures with better electrostatic channel control and reduced short channel effects will be developed. A multi-gate MOSFET architecture is expected to be introduced for the 22nm technology node. Combined with a high-k dielectric/metal gate technology and strain engineering, a multi-gate MOSFET appears to be the ultimate device for high-speed operation with excellent channel control, reduced leakage currents, and low power budget.
Confining carriers within thin Si films reduces the channel dimension in transversal direction, which further improves gate channel control. The quantization energy in ultra-thin Si films may reach a hundred meV. The parabolic band approximation usually employed for subband structure calculations of confined electrons in Si inversion layers becomes insufficient in ultra-thin Si films. A recent study of subband energies and transport in (001) and (110) oriented thin Si films reveals that even non-parabolic

15

isotropic dispersion is not sufficient to describe experimental data, and a direction-dependent anisotropic non-parabolicity must be introduced (4).

A comprehensive analysis of transport in multi-gate MOSFETs under general stress conditions is required for understanding the enhancement of device performance. Besides the biaxial stress obtained in silicon films grown epitaxially on a SiGe substrate, modern techniques allow the generation of large uniaxial stress along the [110] channel. Stress in this direction induces significant shear lattice distortion. The influence of the shear distortion on subband structure and low-field mobility has not yet been carefully analyzed. The two-band $\mathbf{k \cdot p}$ model (5-8) provides a general approach to compute the subband structure, in particular the dependence of the electron effective masses on shear strain. In case of a square potential well with infinite walls, which is a good approximation for the confining potential in ultra-thin Si films, the subband structure can be obtained analytically (9). This allows an analysis of subband energies, effective masses, non-parabolicity and the low-field mobility on film thickness for arbitrary stress conditions.

In the following we briefly review the main ideas behind the two-band $\mathbf{k \cdot p}$ model for a valley in the conduction band of Si. Then we shortly analyze the unprimed subband structure in (001) ultra-thin Si films, obtaining analytical expressions for the effective masses and non-parabolicity parameter. With these parameters the non-parabolic subband approximations for the subband dispersions are constructed. The non-parabolic subband dispersions are embedded into a subband Monte Carlo code in order to enable the computation of the low-field mobility. Results of the mobility enhancement calculations are finally analyzed.

Conduction Band in Silicon

Two-band Hamiltonian

The subband structure in a confined system must be based on accurate bulk bands including strain, where several options are available. The conduction band dispersions computed with several methods in [100] and [110] directions are compared in Fig.1. The method based on non-local empirical pseudo-potentials from (10) is the most accurate one as compared to DFT band structure results obtained with VASP (11). The $sp^3d^5s^*$ tight-binding model with parameters from (12) does not reproduce the anisotropy of the conduction band correctly. In addition, an accurate calibration of the parameters of the $sp^3d^5s^*$ model to describe the modification of the conduction band in strained Si is still lacking.

The $\mathbf{k \cdot p}$ theory is a well established method to describe the band structure analytically. As illustrated in Fig.1, the $\mathbf{k \cdot p}$ method reproduces the band structure accurately at energies below 0.5eV, which is enough to describe the subband structure and transport properties of advanced MOSFETs. From symmetry consideration the two-band $\mathbf{k \cdot p}$ Hamiltonian of a [001] valley in the vicinity of the X point of the Brillouin zone in Si must be in the form (6):

$$H = \left(\frac{\hbar^2 k_z^2}{2m_l} + \frac{\hbar^2 (k_x^2 + k_y^2)}{2m_t} \right) I + \left(D\varepsilon_{xy} - \frac{\hbar^2 k_x k_y}{M} \right) \sigma_z + \frac{\hbar^2 k_z k_0}{m_l} \sigma_y, \qquad [1]$$

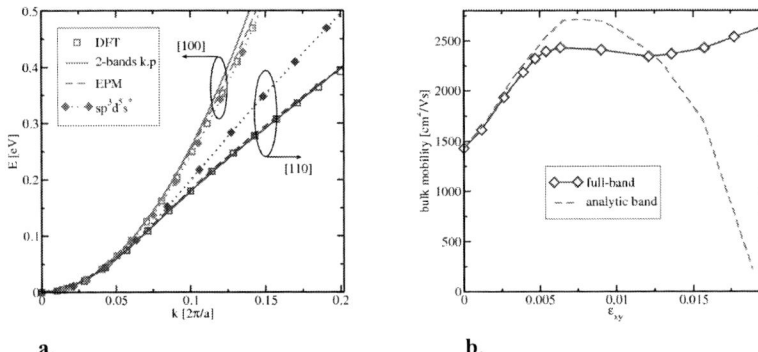

a.   b.

Figure 1 **a**. Comparison of bulk dispersions close at the minimum of the [001] valleys of the conduction band in [100] and [110] directions. DFT (11) and EPM (7,10) results are similar, while the $sp^3d^5s^*$ tight-binding model (12) underestimates anisotropy significantly;

b. Bulk mobility computed by accounting for the full-band structure (symbols) and approximated with non-parabolic dispersion of six valleys (dashed line).

where $\sigma_{y,z}$ are the Pauli matrices, I is the 2×2 unity matrix, m_t and m_l are the transversal and the longitudinal effective masses, $k_0 = 0.15\times2\pi/a$ is the position of the valley minimum relative to the X point in unstrained Si, ε_{xy} denotes the shear strain component, $M^{-1} \approx m_t^{-1} - m_0^{-1}$, and D=14eV is the shear strain deformation potential (5-8). The two-band Hamiltonian results in the following dispersion relations (6):

$$E = \frac{\hbar^2 k_z^2}{2m_l} + \frac{\hbar^2(k_x^2 + k_y^2)}{2m_t} \pm \sqrt{\left(\frac{\hbar^2 k_z k_0}{m_l}\right)^2 + \delta^2} , \qquad [2]$$

where the negative sign corresponds to the lowest conduction band,

$$\delta^2 = (D\varepsilon_{xy} - \hbar^2 k_x k_y / M)^2 . \qquad [3]$$

All moments as well as energies in [2] are counted from the X-point of the Brillouin zone. The classical parabolic approximation is obtained from [2], when coupling between the two conduction bands described by the parameter δ is neglected. Coupling between the bands is small, when the wave vectors $|k_x|$, $|k_y| \ll k_0 (M/m_l)^{1/2}$ and shear strain $\varepsilon_{xy} = 0$.

Due to band coupling the dispersion relations [2] become non-parabolic in strained Si, if the shear strain component is non-zero, and/or at higher energies. In order to check the accuracy of [2] we have carried out numerical band structure calculations with the empirical pseudo-potential method (EPM) with parameters from (7,10). Excellent agreement between the two-band **k·p** model [1] and the EPM results was found up to an energy of 0.5eV. Equation [2] is valid in a larger range of energies compared to a parabolic dispersion relation with isotropic non-parabolic correction and can be used to determine the subband structure in thin Si films.

Shear strain induced valley shift

It follows from [2] that the position of the conduction band minimum located at a distance k_0 from the X–point in unstrained silicon moves closer to the X–point for nonzero shear strain. Introducing dimensionless strain $\eta = m_t D \varepsilon_{xy} / \hbar^2 k_0^2$, one finds for the position of the minimum (7,8):

$$k_{min} / k_0 = \sqrt{1-\eta^2} \ . \qquad [4a]$$

At the same time, the minimum moves down in energy:

$$\Delta E_{min} = -\eta^2 \Delta / 4 \ , \qquad |\eta| \leq 1, \qquad [4b]$$

where $\Delta = 2m_t / \hbar^2 k_0^2$ is the gap between the two conduction bands at the minimum k_0. For $|\eta| \geq 1$ the conduction band minimum stays exactly at the X point, resulting in the following energy dependence:

$$\Delta E_{min} = -(2|\eta|-1)\Delta / 4 \ , \qquad |\eta| \geq 1. \qquad [4c]$$

Effective masses

Shear strain ε_{xy} modifies the effective masses of the [001] valleys. The transversal mass m_t acquires two different values along (+) and across (-) tensile stress direction (7,8):

$$m_t(\eta)/m_t = [1 \pm |\eta| m_t / M]^{-1}, \qquad |\eta| \leq 1; \qquad [5a]$$

$$m_t(\eta)/m_t = [1 \pm m_t / M]^{-1}, \qquad |\eta| \geq 1 \ . \qquad [5b]$$

The longitudinal mass m_l is expressed as

$$m_t(\eta)/m_l = [1 \pm \eta^2]^{-1}, \qquad |\eta| \leq 1; \qquad [6a]$$

$$m_t(\eta)/m_l = [1 \pm |\eta|^{-1}]^{-1}, \qquad |\eta| \geq 1 \ . \qquad [6b]$$

Non-parabolicity

From the two-band Hamiltonian the value of the non-parabolicity parameter is $\alpha_0 \approx 0.6 \, \mathrm{eV}^{-1}$ (13). It is close to the phenomenological value $\alpha_0 = 0.5 \, \mathrm{eV}^{-1}$ routinely used in calculations. Strain induced modification of the conduction band effective masses affects the nonparabolicity parameter α of the [001] valleys.
The expression for the density-of-states can be written in the form (14)

$$D(E) = \int \frac{dk_x dk_y}{(2\pi)^2} \delta(E - E_n(k_x, k_y)) = \frac{\sqrt{m_-(\eta)m_+(\eta)}}{(2\pi)^2} \int_{E=const} d\varphi \frac{1}{2} \frac{\partial \zeta^2(E,\varphi)}{\partial E}, \qquad [7]$$

where $\zeta^2(E,\varphi) = k_-^2 / m_t^-(\eta) + k_+^2 / m_t^+(\eta)$ is determined by the expression:

$$E = \frac{\zeta^2}{2} - \frac{\hbar^2 m_l}{8M^2 k_0^2 |1-q_n^2|} \left(m_- \cos^2 \varphi - m_+ \sin^2 \varphi\right)^2 \zeta^4. \qquad [8]$$

Substituting [8] into [7] and assuming the energy E close to the valley minimum so that $\alpha E \ll 1$, we obtain the following expression for the non-parabolicity parameter ratio:

$$\alpha = \alpha_0 \frac{1 + 2\left(m_t \eta / M\right)^2}{\left(1 - \left(m_t \eta / M\right)^2\right)^2}. \qquad [9]$$

The relative increase of α resulting in an increased density of states and scattering are responsible for a weaker mobility enhancement (Fig.1b) as compared to the mobility obtained with strain-independent α_0. However, for a stress larger than 3 GPa ($\eta \geq 0.5$) the energy difference from the minimum to the value at the X point becomes smaller than kT, and a full-band description is required for accurate mobility calculations (15).

18

Subbands in Ultra-Thin Silicon Films

Dispersion equations

For [001] silicon films the confinement potential gives an additional contribution $U(z)I$ to the Hamiltonian [1]. In the effective mass approximation described by [1] with the coefficient in front of σ_x set to zero, the confining potential $U(z)$ is known to quantize the six equivalent valleys of the conduction band of bulk silicon into the four-fold degenerate primed and the two-fold degenerate unprimed subband ladder (16). In ultra-thin films the unprimed ladder is predominantly occupied. In order to analyze the subbands, we approximate the confining potential of an ultra-thin silicon film by a square well potential with infinite potential walls. Generalization to include a self-consistent potential is straightforward though numerically involved (17).

Because of the two-band Hamiltonian, the wave function Ψ is a spinor with the two components $|0\rangle$ and $|1\rangle$. For a wave function with space dependence in the form $\exp(ik_z z)$ the coefficients A_0 and A_1 of the spinor components are related via the equation $H\Psi = E(k_z)\Psi$. For a particular energy E there exist four solutions k_i ($i=1,\ldots,4$) for k_z of the dispersion relation [2], so the spatial dependence of a spinor component is in the form $\sum_{i=1}^{4} A_\alpha^i \exp(ik_i z)$. The four coefficients are determined by the boundary conditions that both spinor components are zero at the two film interfaces. This leads to the following dispersion equations:

$$\tan\left(k_1 \frac{k_0 t}{2}\right) = \frac{k_2}{\sqrt{k_2^2 + \eta^2} \pm \eta} \frac{\sqrt{k_1^2 + \eta^2} \pm \eta}{k_1} \tan\left(k_2 \frac{k_0 t}{2}\right) \quad , \qquad [8]$$

where $\eta = m_l |\delta|/(\hbar k_0)^2$. If the value of

$$k_2 = \sqrt{k_1^2 + 4 - 4\sqrt{k_1^2 + \eta^2}} \qquad [9]$$

becomes imaginary at high η values, the trigonometric functions in [8] are replaced by the hyperbolic ones. Special care must be taken to choose a correct branch of $\sqrt{k_2^2 + \eta^2}$ in [9]: the sign of $\sqrt{k_2^2 + \eta^2}$ must be alternated after the argument becomes zero. Introducing $y_n = (k_1 - k_2)/2$, [8] can be written in the form (18):

$$\sin(y_n k_0 t) = \pm \frac{\eta y_n \sin\left(\dfrac{1 - \eta^2 - y_n^2}{1 - y_n^2} k_0 t\right)}{\sqrt{(1 - y_n^2)(1 - \eta^2 - y_n^2)}} \quad . \qquad [10]$$

Subband structure in thick films

We solve [10] by perturbation techniques. For small η and thick films the right-hand side in [10] can be ignored. The subband relations are found from the condition

$$y_n = \pi n /(k_0 t) . \qquad [11]$$

This results in the following approximate dispersion relation for unprimed subbands (16):

$$E_n(k_x, k_y) = E_n^0(k_x, k_y) - \Delta \eta^2 /[4\,|\,1 - q_n^2\,|] , \qquad [12]$$

where $q_n = (\pi n)/(t k_0)$ and E_n^0 is the subband dispersion relation for parabolic bands:

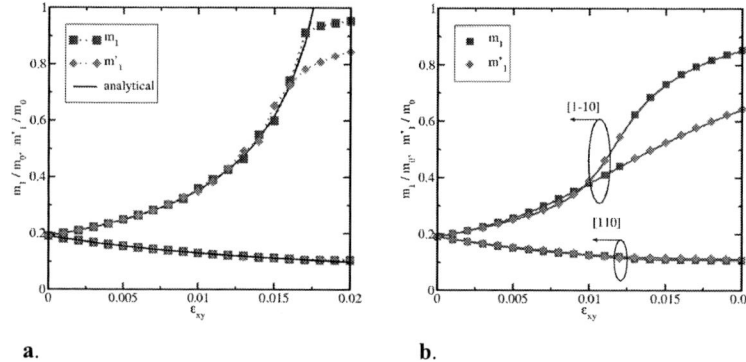

a.   b.

Figure 2 Effective masses of the two ground subbands. Symbols are obtained by numerical solution of the two-band Hamiltonian [1]:
a. t=10.86nm; solid lines are according to [16];
b. t=5.43nm; solid lines are obtained by numerically differentiating subband dispersions obtained from [8].

$$E_n^0(k_x,k_y) = \frac{\hbar^2\pi^2 n^2}{2m_l t^2} + \frac{\hbar^2(k_x^2+k_y^2)}{2m_t} - \frac{\hbar^2 k_0^2}{2m_l}.$$

[12] is valid when

$$\left(1-q_n^2\right)^2 >> \delta^2 m_t^2 / \hbar^4 k_0^4 . \qquad [13]$$

Equation [12] describes the subband quantization energy correction due to strain with respect to the valley minimum

$$\Delta E_n(\eta) = -\frac{\hbar^2\pi^2 n^2}{2m_l t^2}\frac{\eta^2}{|1-q_n^2|}, \qquad [14]$$

which is obtained after taking into account the strain-induced valley minimum energy shift ΔE_{min}. [14] can be absorbed into the quantization energy $\dfrac{\hbar^2\pi^2 n^2}{2m_l t^2}$ by introducing the longitudinal mass m_l depending on strain η and thickness t:

$$m_l(\eta,q_n) = \frac{m_l}{1-\eta^2/|1-q_n^2|}. \qquad [15]$$

[12] also describes dependencies of the transversal masses on strain η, the film thickness t, and subband number n:

$$m_t^{\mp}(\eta,q_n) = m_t\left(1\pm\eta\frac{m_t}{M}\frac{1}{|1-q_n^2|}\right)^{-1}. \qquad [16]$$

Here m_t^- is the effective mass along the direction [110] of tensile stress. In films the effective mass depends not only on strain but also on film thickness. We compare [16] (lines) with the results of the numerical solution of the two-band **k·p** model Hamiltonian (symbols) in a film with the thickness t=10nm for the ground (n=1) (shown in Fig.2a) and the second (n=2) subbands and find a good agreement. With the film thickness decreased and strain increased a substantial discrepancy between [16] and a numerical

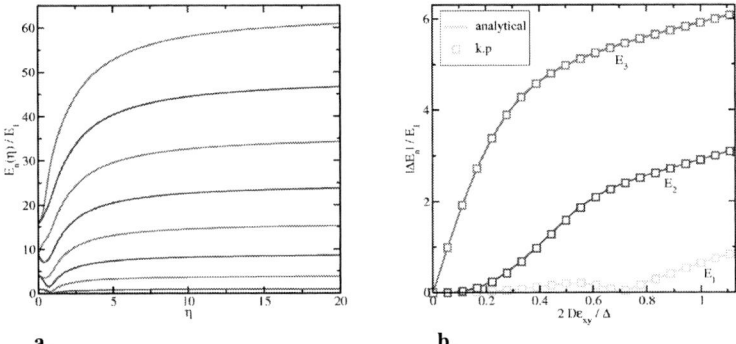

a.   b.

Figure 3 **a**. Normalized positions of the subband minima with respect to the strain-dependent conduction band minimum as function of dimensionless shear strain for a film of the thickness t=5.43nm.
b. Strain-dependent splitting between the minima of the unprimed subbands with the same n.

solution appears as shown in Fig.2b. This discrepancy is caused by the growing value of the right-hand side in [10] which cannot be ignored.

Splitting of unprimed subbands

Substituting [11] into the right-hand side of [10] and solving [10] for small strain η one obtains the following dispersion relation for the unprimed subbands n:

$$E_n^{\pm} = \frac{\hbar^2}{2m_l}\left(\frac{\pi n}{t}\right)^2 + \hbar^2\frac{k_x^2+k_y^2}{2m_t} \pm \left(\frac{\pi n}{k_0 t}\right)^2 \frac{\left|D\varepsilon_{xy} - \dfrac{\hbar^2 k_x k_y}{M}\right|}{k_0 t\left|1-\left(\pi n/k_0 t\right)^2\right|}\sin(k_o t) \; . \quad [17]$$

It follows that the subband degeneracy is preserved only, when shear strain is zero and either k_x=0 or k_y=0. [17] demonstrates that the unprimed subbands are *not equivalent*. We first analyze the splitting in energy between the two unprimed subbands with the same n, which is usually called the valley splitting (16). According to [17], shear strain induces a valley splitting linear in strain, for small shear strain values (18):

$$\Delta E_n = 2\left(\frac{\pi n}{k_0 t}\right)^2 \frac{D\varepsilon_{xy}}{k_0 t\left|1-\left(\pi n/k_0 t\right)^2\right|}\sin(k_o t) \; . \quad [18]$$

The valley splitting is inversely proportional to $(k_0 t)^3$ and oscillates with the film thickness, in agreement with earlier work (16,19).
To find the valley splitting at higher strain values, [10] must be solved numerically. Results shown in Fig.3 demonstrate that valley splitting can be effectively controlled by adjusting the shear strain and modifying the effective thickness t of the electron system. It is interesting to note that for extremely high strain values the dispersion of the lowest conduction band becomes parabolic again, and the quantization levels in a square well potential are therefore recovered in this limit. Although the value of strain in this limit is unrealistic, this result will be used to analyze dispersion relations for the primed subbands.

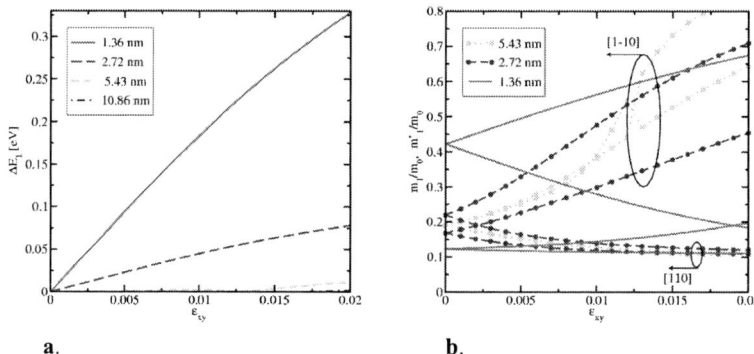

a. b.

Figure 4 **a**. Shear strain induced splitting of the ground subbands for several film thicknesses. In ultra-thin films splitting is larger than kT already for moderate stress. **b**. Effective masses of the two ground subbands. In ultra-thin films effective masses of the two ground subbands are different even without stress.

Uniaxial stress along [110] channel direction, which induces shear strain, is already used by industry to enhance the performance of modern MOSFETs. Therefore, its application to control valley splitting does not require expensive technological modifications. A possibility to introduce valley splitting larger than the Zeeman spin splitting makes Si promising for future spintronic applications (20).
As seen from Fig.4a, the valley splitting in ultra-thin Si films can be quite large already for reasonable stress values. In this case the higher subband becomes depopulated, prompting for mobility enhancement in (001) ultra-thin films strained along [110] direction.

Effective masses of unprimed subbands

Dispersion [17] predicts different effective masses in [110] direction for the unprimed subbands with the same n even without strain:

$$m_{(1,2)} = \left(\frac{1}{m_t} \pm \frac{1}{M} \left(\frac{\pi n}{k_o t} \right)^2 \frac{\sin(k_o t)}{k_o t \left| 1 - (\pi n / k_o t)^2 \right|} \right)^{-1}$$ [19]

Numerically found values of the masses for the two ground subbands are shown in Fig.4b. Contour plots of the subband dispersions for the two ground subbands are shown in Fig.5. It is to note, that the subband dispersions are not equivalent. This has a profound effect on the valley splitting. Without shear strain the Landau levels in the external [001] magnetic field B are determined using the Bohr-Sommerfeld quantization conditions:

$$E_m^{(1,2)} = \hbar \omega_c \left(m + \frac{1}{2} \right) \frac{\pi}{4 \arctan\left(\sqrt{m_{(1,2)} / m_{(2,1)}} \right)},$$ [20]

where $\omega_c = \dfrac{eB}{\sqrt{m_1 m_2} c}$ is the cyclotron frequency. Therefore, the magnetic field induces a valley splitting linear in the field strength B, in agreement with recent experimental results (20).

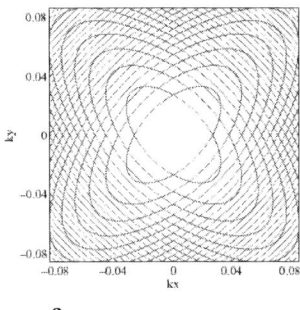 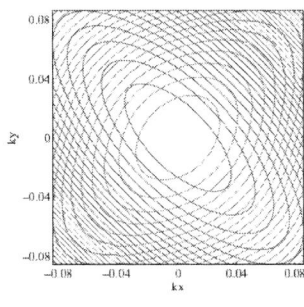

a.   **b.**

Figure 5 Dispersions of the two ground subbands for a film thickness of 1.36nm:
a. Without strain the subbands are degenerate at the minimum. The lower subband dispersion is described by the unification of the two ellipses with different masses [19], while the second subband is described by their intersection. The difference in quasi-classical orbits of the motion in the magnetic field is responsible for the subband splitting [20] in orthogonal magnetic field.
b. Shear strain of 1% removes the degeneracy between the minima of the ground subbands shown in Fig.4a. The subband dispersions are now characterized by the corresponding effective masses in [110] and [1-10] directions (Fig.4b).

A large value of the valley splitting observed by measuring conductance through a point contact can also be attributed to the difference in the subband dispersions, in particular, to the effective mass difference [19]. Indeed, confining the electron system laterally in [1-10] direction by the potential, the following dispersion relation of propagating modes within the point contact is obtained:

$$E_p^{(1,2)} = \frac{k^2 k_x'^2}{2m_{(2,1)}} + \hbar\omega_{(1,2)}\left(p+\frac{1}{2}\right)+V_b .$$ [21]

Here $\omega_{(1,2)}^2 = \kappa / m_{(1,2)}$ and V_b is a gate voltage dependent shift of the conduction band in the point contact (21). The energy minima of the two propagating modes with the same p are separated by $\Delta E_p = \hbar |\omega_1 - \omega_2|$ and they are resolved in the conductance of a point contact as two distinct steps. The difference in the effective masses [17] and, correspondingly, the valley splitting can be greatly enhanced by reducing the effective thickness t of the film.

Effective mass of primed subband

Shear strain in [110] direction does not affect the primed valleys along [100] and [010] direction, except for a small shift of the minimum (22). However, recent calculations of the primed subbands based on the density functional theory (23) and the "linear combination of bulk bands" method obtained with the empirical pseudo-potential calculations (24) reveal the dependence of the transport effective masses on silicon film thickness t. Here we briefly analyze the effective mass of the primed subbands based on the two-band Hamiltonian [1]. Let us assume the quantization direction along the X-axis. By formally replacing k_0/m_1 with k_y/M and $k_x k_y/M$ with $k_z k_0/m_1$ in [1] one finds the

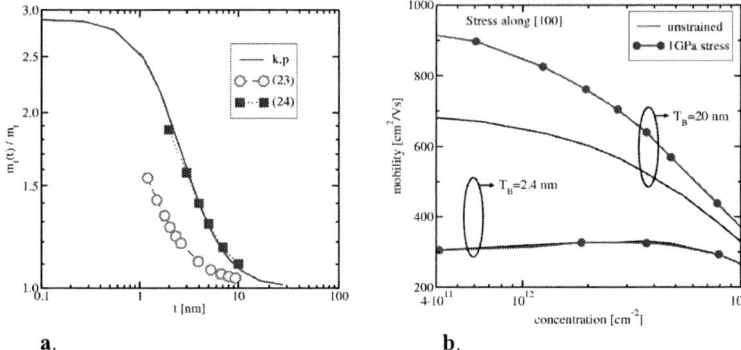

a.   b.

Figure 6 **a**. The thickness dependence of the effective mass of the lowest primed subbands computed with the two-band **k·p** model (solid line) is in excellent agreement with the full-band calculations (24) (filled symbols). Open symbols show calculations from (23).

b. The mobility enhancement in [100] direction due to [100] tensile stress of 1GPa is due to de-population of the primed subbands with an unfavourable effective mass in transport direction. In ultra-thin films, where the primed subbands are already de-populated ,mobility remains unchanged.

dispersion and the effective masses in the primed subbands, where results of calculations are shown in Fig.6a. The two-band **k·p** results are in excellent agreement with the "linear combination of the bulk bands" method with a potential barrier of 3eV at the film interface (24) and they are also consistent with the DFT calculations (23).

Mobility Enhancement due to Uniaxial Stress

A multi-subband Monte Carlo method designed for small signal analysis (25) was used to evaluate the mobility in MOSFETs with a thin Si film. The method is based on the solution of the linearized multi-subband Boltzmann equation, which is exact in the limit of vanishing driving fields. A particular advantage of the method is that it includes degeneracy effects due to the Pauli exclusion principle. Degeneracy effects are important for mobility calculations in ultra-thin films, especially at high carrier concentrations. The multi-subband method uses the subband wave functions and subband energies. They can be found by solving the Schrödinger equation and the Poisson equation self-consistently for each value of the gate voltage (17). The wave functions are then employed to evaluate scattering rates. Scattering with phonons and surface roughness is included. The surface roughness at the two thin film interfaces is assumed to be equal and uncorrelated. The parameters of the Gaussian surface roughness correlation function were calibrated to reproduce the universal mobility curve of Takagi (26) in the inversion layer. The same parameters are then used for mobility calculations in thin film MOSFETs.

Fig.6b displays the mobility in (001) silicon films for two film thicknesses under a tensile uniaxial stress of 1GPa in [100] direction. The stress shifts up and de-populates the two [100] primed valleys with unfavorable mass m_l in the transport direction providing the mobility enhancement in the stress direction for a 20nm film. In [010] transport direction the [100] primed valleys have a smaller transversal mass. Therefore, the de-population of the [100] primed subbands due to [100] tensile stress has a detrimental effect on the

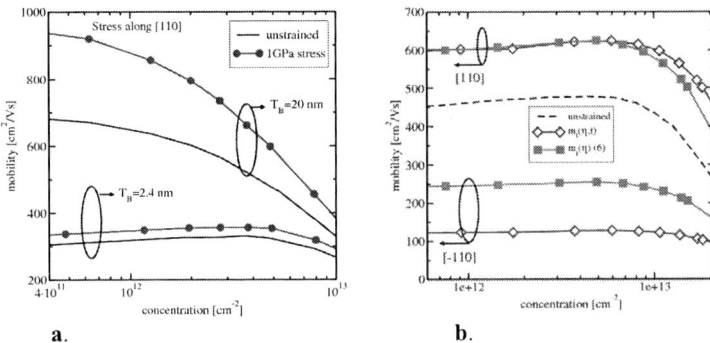

a.   b.

Figure 7 **a**. The mobility enhancement due to tensile stress of 1GPa in [110] transport direction due to transport mass decrease is preserved even in ultra-thin films.

b. Mobility in a 3 nm thick film as a function of electron concentration calculated for 0.5% tensile shear strain in [110] direction, with bulk effective masses [6] (filled symbols) and with strain- and thickness-dependent masses (Fig.4b) (open symbols). The increase of the density-of-states effective mass with strain enhances scattering which alters the mobility enhancement in [110] direction.

mobility in [010] direction. In ultra-thin films the primed subbands are nearly de-populated already without stress due to the large separation in energy between primed and unprimed subbands in a silicon film with a thickness of 2.4nm. Therefore, additional shifting up in energy of the [100] valleys does not have any effect on the mobility in ultra-thin films (Fig.6b).

Apart from shifting primed subbands with unfavorable transport masses in the (001) plane up in energy and de-populating them, tensile stress in [110] direction generates a shear component which modifies the transport effective masses of unprimed subbands (Fig.4b). The decrease of the effective masses in [110] direction induced by shear strain becomes more pronounced with the film thickness reduced guaranteeing the mobility enhancement even in ultra-thin films, as demonstrated in Fig.7a. However, the density-of-state effective mass $m^* = \sqrt{m_l^- m_t^-}$ in unprimed subbands increases with shear strain. This results in higher scattering rates which deteriorate the benefits of the thickness-enhanced transport mass decrease at higher stress values, as displayed in Fig.7b. However, the mobility enhancement remains substantial.

Conclusion

A rigorous analysis of the subband structure in thin silicon films is performed. The thickness dependence of the effective mass of primed subbands calculated within the two-band **k·p** model is in agreement with the earlier full-band calculations. It is demonstrated that within the two-band **k·p** model the unprimed subbands with the same quantum number n are not equivalent. A large splitting between the unprimed valleys of ultra-thin films can be introduced by a shear strain component. Calculated subband effective masses are shown to depend on shear strain and thickness simultaneously. Interestingly, the effective masses of the two unprimed valleys are different in ultra-thin silicon films even without strain. This results in a linear dependence of the subband splitting on the magnetic field strength and leads to large subband splitting in a laterally confined electron system in a point contact.

The mobility enhancement in strained MOSFETs with ultra-thin silicon films is investigated by a subband Monte Carlo method. The method based on the solution of the linearized Boltzmann equation includes the carrier degeneracy exactly. Transport in thin films is determined by subband structure, in particular by the effective masses. The decrease of the transport effective mass induced by the shear strain component is the reason for the mobility enhancement even in ultra-thin silicon films. This mobility and drive current increase combined with the improved channel control makes multi-gate MOSFETs based on thin films or silicon fins preeminent candidates for the 22nm technology node and beyond.

Acknowledgments

This work was supported in part by the Austrian Science Fund FWF, project P19997-N14, and by funds from FWF, project I79-N16, CNR, EPSRC and the EC Sixth Framework Programme, under Contract N. ERAS-CT-2003-980409 as part of the European Science Foundation EUROCORES Programme FoNE.

References

1. S.Natarjan, M.Armstrong, H.Bost, et al., *IEDM* 2008, pp.941-943.
2. M.K.Hudati, G.Dewey, S.Datta, et al., *IEDM* 2007, pp.625-628.
3. R.Chau, *ISDRS* 2007, p.3.
4. K.Uchida, A.Kinoshita, M.Saitoh, *IEDM* 2006, pp.1019-1021.
5. J.C.Hensel, H.Hasegawa, and M.Nakayama, *Phys.Rev.* **138**, A225-A238 (1965).
6. G.L.Bir, G.E.Pikus, Symmetry and Strain-Induced Effects in Semiconductors, J.Willey & Sons, NY, 1974.
7. E.Ungersboeck, S.Dhar, G.Karlowatz, et al., *IEEE T-ED* **54**, 2183-2190 (2007).
8. V.Sverdlov, E.Ungersboeck, H.Kosina, S.Selberherr, *ESSDERC* 2007, pp.386-9.
9. V.Sverdlov, G.Karlowatz, S.Dhar, et al., *Sol.State Electron.* **52**, 1563-1568 (2008).
10. M.Rieger and P.Vogl, *Phys.Rev. B* **48**, 14275-14287 (1993).
11. VASP (Vienna Ab-initio Simulation Program) G.Kresse and J. Hafner, Phys.Rev. B 47, 558 (1993); ibid. B 49, 14251 (1994); G.Kresse and J. Fertmueller, Phys.Rev. B 54, 11169 (1996); Computs.Mat.Sci. 6, 15 (1996).
12. T.B.Boykin, G.Klimeck, and F.Oyafuso, *Phys.Rev.B* **69**, 115201-1—10 (2004).
13. C.Jacoboni, L.Reggiani, *Rev.Mod.Phys.* **55**, 645-705 (1983).
14. M.V.Fischetti, S.E.Laux, *Journal of Applied Physics* **80**, 2234-2252 (1996).
15. V.Sverdlov, G.Karlowatz, E.Ungersboeck, H.Kosina, *SISPAD* 2007, pp.329-332.
16. T.Ando, A.B.Fowler, F.Stern, *Rev.Mod.Phys.* **54**, 437-672 (1982).
17. O.Baumgartner, M.Karner, V.Sverdlov, H.Kosina, *EUROSOI* 2009, pp.57-58.
18. V.Sverdlov and S.Selberherr, *Sol.State Electron.* **52**, 1861-1866 (2008).
19. M.Friessen, S.Chutia, C.Tahan, et al., *Phys.Rev.B* **75**, 115318-1—12 (2007).
20. S.Goswami, K.A.Slinker, M.Friesen, et al., *Nature Physics* **3**, 41-45 (2007).
21. B.J.van Wees, H.van Houten, C.Beenakker, *Phys.Rev.Lett.*, **60**, 848-850 (1988).
22. D.Rideau, M.Feraille, M.Michaillat, et al., *Sol.State Electron.* 2009, (online).
23. A.Martinez, K.Kalna, P.V.Sushko, et al., *IEEE T-Nano* 2009 (online).
24. J.van der Steer, D.Esseni, P.Palestri, et al., *IEEE T-ED* **54**, 1843-1851 (2007).
25. V.Sverdlov, E.Ungersboeck, H.Kosina, et al.,*Sol.StateElectron.***51**, 299-305 (2007).
26. S.-I.Takagi, A.Toriumi, M.Iwase, H.Tango, *IEEE T-ED* **41**, 2357-2362 (1994).

CHAPTER 2

MATERIALS AND CHARACTERIZATION I

28

Copper decoration combined with preferential etching for delineation of crystal defects in SOI and sSOI materials

H. Idrisi[a], M. Pejic, F. Uzelli, and B. O. Kolbesen[b]

Goethe University Frankfurt/Main, Institute of Inorganic and Analytical Chemistry,
Max-von-Laue-Str. 7, D-60438 Frankfurt, Germany
[a]idrisi@chemie.uni-frankfurt.de,
[b]kolbesen@chemie.uni-frankfurt.de

Preferential etching with a dilute Secco etch is the routinely applied procedure for the delineation of crystal defects in Silicon-On-Insulator (SOI)- and strained Silicon-On-Insulator (sSOI)-materials. Its capability can be improved with a preceding copper decoration step. By this combination even defects of some nanometer in size can be detected. However, a shortcoming of copper decoration may be the formation of artefacts resulting in a deceptively high defect density. Optimized copper decoration conditions are presented for crystal defects in SOI preventing the creation of artefacts. Furthermore, a combination of copper decoration with preferential etching was attempted to delineate all defect types in sSOI materials.

Introduction

Silicon-On-Insulator (SOI)- and strained Silicon-On-Insulator (sSOI)-wafers are used in the fabrication of advanced microelectronic devices (1, 2). Defects originating from the CZ silicon substrates, such as vacancy agglomerates (3-6), called COPs (Crystal Originated Particles), or caused by the SOI or sSOI fabrication process affect the quality of silicon wafers and may be harmful to devices. In SOI wafers crystal defects are delineated routinely by etching with a dilute Secco (7), a diluted version of the standard Secco etch, followed by a HF dip which undercuts the buried oxide (BOX) at the defect etch pit and improves the visibility of those defects due to its bright halo in optical microscopy (8). Different defect types can be revealed after etching sSOI-wafers: both threading dislocations (TD) and stacking faults (SF). For their delineation etching solutions such as dilute Secco or Organic Peracid Etching (OPE) solutions (9, 10) can be used.

Copper decoration is used to facilitate the delineation of crystal defects of some nanometer in size which are scarcely visualized by defect etching (11, 12). Copper precipitation occurs preferably at grown-in or process-induced crystal defects (13, 14) acting as nucleation sites for copper silicide (Cu_3Si) precipitates (15-17). These silicides cause an enhanced etch rate of the preferential etching solutions resulting in the magnification of the defects and improved delineation after etching. A shortcoming of this method is the possible formation of artefacts due to copper precipitation at the surface of the wafer (18). Hence, depositing too high copper concentrations on the fragments may give rise to a deceptively increased defect density (DD). (In the case of copper decoration of silicon substrates a certain layer is etched off to eliminate copper surface artefacts).

The aim of the present copper decoration study with SOI material was further optimization of the copper concentration used for decoration of all crystal defects and the implementation of precautions to minimize copper cross contamination due to the applied quartz tube. After accomplishment of the practical procedure described in the following defect characterization and determination of defect densities (DD) was carried out via light optical and scanning electron microscopy (SEM), resp..

In the present work, results of copper decoration experiments of crystal defects in SOI- and sSOI-wafers with subsequent preferential etching, using different etching solutions, are discussed.

Experimental

From SOI- and sSOI-wafers provided by SOITEC SA/Bernin, France, samples with a size of about 1 cm x 1 cm ("fragments") were prepared. The SOI-films used offered different thicknesses (62, 90, 118 and 149 nm), the thickness of the buried oxide (BOX) was 145 nm. The sSOI samples belonged to a period of early sSOI process development. The thickness of the sSOI-film was about 83.5 nm while the BOX thickness was approx. 145 nm. Copper decoration was accomplished by furnace annealing on both kinds of material. Since apart from copper-induced artefact formation copper cross contamination by the quartz and the furnace may disturb a reliable evaluation of defect densities precautions were taken: A second quartz tube inside the outer one was used. These inner quartz tubes were cleaned periodically to remove residual copper.

Copper decoration of crystal defects in SOI

In prior copper decoration studies (19) the following procedure was applied: $Cu(NO_3)_2$ solution was deposited on the backside of the SOI fragment and then dried by heating on a hot plate. Subsequently the fragments were annealed at temperatures in a range from 600-900 °C for 1 min in a furnace and afterward quenched in air to room temperature. In this previous work (19) the following set of parameters has been evaluated as most suitable: $Cu(NO_3)_2$ solution 0.5 μL, copper concentration $\leq$ 1 ppmw (1.57 x 10^{-5} mol Cu/L or $\leq$ 9.45 x 10^{18} copper atoms/L); furnace annealing temperature: 800 °C. These parameters were used also in the present work. In addition, in this work $Cu(NO_3)_2$ solutions in the concentration range from 0.0001-1 ppm (weight) corresponding to 1.57 x 10^{-9} to 1.57 x 10^{-5} mol/L of $Cu(NO_3)_2$ were used and for prevention of copper cross contamination the precautions outlined above were taken.

After copper decoration the SOI fragments were etched with dilute Secco (0.04 M Cr (VI)) from their initial layer thickness down to approx. 20-70 nm and treated with a dip in HF for approx. 90 s to delineate the defects. By the use of a HF dip the BOX at the etch pit site is underetched giving rise to a "halo" around the etch pit. Because of this halo the defects can be detected easily in a light optical microscope. Furthermore selected defects were analyzed in a SEM. The defect densities of copper decorated and non decorated ("references") SOI samples were compared.

Copper decoration studies of sSOI

A volume of 0.5-1.5 μL of $Cu(NO_3)_2$ solution in the concentration range from 0.0001-100 ppm (weight) corresponding to 1.57 x 10^{-9} to 1.57 x 10^{-3} mol/L of $Cu(NO_3)_2$ was applied on the backside of several sSOI fragments and dried by heating on a hot plate. Afterwards the fragments were annealed mainly at 800 °C (a few fragments at 860 °C) for 1 min in a furnace. Subsequently the fragments were quenched in air to room temperature. A part of the sSOI fragments was etched with dilute Secco (0.04 M Cr (VI)) from their initial layer thickness (83.5 nm) down to approx. 30 nm and treated with a dip in HF for approx. 23 s to reveal the defects primarily stacking faults (SF) and threading dislocations (TD). Moreover, sSOI fragments were also etched with different Organic Peracid Etching (OPE) solutions for defect delineation. The sSOI fragments were etched from their initial layer thickness (83.5 nm) down to approx. 37-65 nm. A subsequent HF dip is not necessary since "halos" are formed already during the etch by the HF in the etching mixture due to the long OPE etching time. OPE etching solutions without copper decoration delineate no SF but all TD. OPE B etching solution (containing: 114 mL acetic acid, 43 mL (50%) HF, 43 mL H_2O_2) provided best results. The defects can be readily detected in a light optical microscope. Selected fragments were analyzed in a SEM. The results of the sSOI fragments after copper decoration and etching with OPE B etching solution were compared a) to non-decorated OPE B etched fragments ("OPE B-reference") and b) to decorated and non-decorated dilute Secco etched fragments ("dil. Secco-reference").

Results and Discussion

Copper decoration of crystal defects in SOI

The defect densities of copper decorated and non-copper decorated SOI fragments were determined and compared after etching by a dilute Secco (0.04 M Cr (VI)). Two types of non-decorated "references" were used: a) "virgin" fragments (ref 1) and b) after furnace annealing without copper decoration (ref 2).

Table I shows the average of typical defect densities (DD) for SOI fragments with an initial SOI layer thickness of 62, 90 118 and 149 nm.

TABLE I. Comparison of average DD after Secco etching

Cu concentration [ppm]	$DD_{62 nm}$ [cm^{-2}]	$DD_{90 nm}$ [cm^{-2}]	$DD_{118 nm}$ [cm^{-2}]	$DD_{149 nm}$ [cm^{-2}]
0 (Ref1)	5651	8177	1696	1271
0 (Ref2)	13568	-	3012	2447
0.001	9607	18367	6638	3261
1	24571	22037	10945	7554

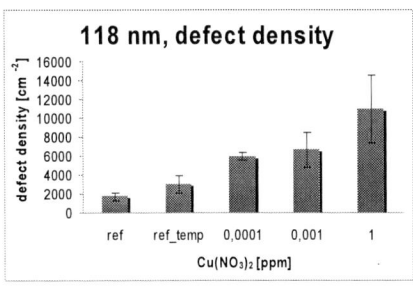
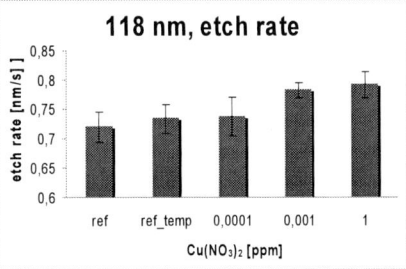

Figure 1. 118 nm-SOI; application of the obtained defect densities (1a, left) and etch rates (1b, right) against the $Cu(NO_3)_2$ concentration used.

In Figures 1a and 1b the defect densities and the etch rates of the dilute Secco etch, resp., determined in a 118 nm-SOI film for references and decorated samples are plotted against the $Cu(NO_3)_2$ concentrations in the solutions used.

From Tab. 1 and Figure 1 it can be concluded that the defect densities observed increase already after the furnace treatment without copper decoration (ref 2) and increase significantly after copper decoration with rising copper concentration, but decrease with increasing SOI film thickness. Moreover, no significant cross contamination effect could be observed for the low copper concentrations. According to Figure 1b furnace annealing and copper decoration produce also a distinct increase of the dilute Secco etch rate. An explanation for this etch rate enhancement may be a catalytic effect of copper in the etching process. Such an effect has been observed after intentional copper spiking of preferential etching solutions (19).

Optical micrographs of different SOI fragments taken after etching with dilute Secco and subsequent HF dip are shown in Figs. 2a and b: a non-decorated virgin fragment (ref 1) with a low defect density (DD: 1700 cm^{-2}) in Fig. 2a and a copper decorated SOI fragment (0.0001 ppmw of copper) in Fig. 2b. The higher defect density in the decorated sample (Fig. 2b) compared to the reference (Fig. 2a) is evident from the optical micrographs. In figure 1b a further defect type called "*red spots*" can be seen (19). These "*red spots*" appeared in some of the analyzed copper decorated SOI fragments. They may be attributed to copper silicide precipitates formed in the silicon at one of the interfaces SOI-film/BOX or BOX/Si-substrate. Occurrence of copper silicide precipitates at the Si/SiO_2 interface after a copper decoration step is well known (20, 21). However, it is not clear what kind of nuclei is acting in the formation of these precipitates in SOI material.

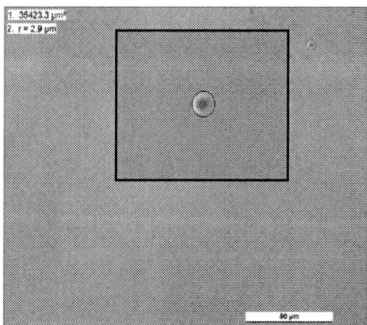 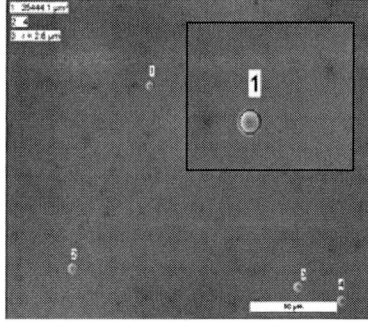

Figure 2: Samples with 118 nm-SOI film, etched with dilute Secco (0.04 M Cr (VI)); HF dip 90 s, light optical micrographs:
a) "virgin" sample (ref 1, left): defect density(DD): 1700 cm^{-2};.
b) Copper decorated sample (0.0001 ppm $Cu(NO_3)_2$,) (right): DD 6200 cm^{-2}; "*red spots*" visible, DD: 1.13 x 10^5 cm^{-2}.

Copper decoration studies of sSOI

Optical micrographs after Secco etching without or with copper decoration display a high density of stacking faults (SF) and threading dislocations (TD) in sSOI fragments of an early period of materials development (Figure 3). Using OPE solutions on non-copper decorated samples just some TDs are revealed (Figure 4). In the present work sSOI fragments were decorated with copper and subsequently etched with OPE B solution in order to reveal both TDs and SFs (Figure 5). (OPE B solution proved to be the best OPE etching solution for copper decorated fragments.) Furthermore the experimental parameters for copper decoration via furnace annealing were determined. The best results were achieved using a copper concentration of 10 ppmw (1.57×10^{-4} mol Cu/L) with a proper volume of about 0.5 or 0.75 µL of $Cu(NO_3)_2$ solution deposited on the sSOI fragment backside. A furnace annealing temperature of 800 °C seemed to be applicable because a higher annealing temperature (860 °C) caused degrading results. By etching down 30-40 % from the initial layer thickness (83.5 nm) the best defect delineation was obtained.

Fig. 3a shows a non-copper decorated and Fig. 3b a copper decorated and dilute Secco etched sSOI fragment where SFs and TDs are clearly visible as lines and dots (pits), respectively. A non-decorated sSOI fragment which was etched with an OPE B solution (ref 3) is displayed in figure 4. No SFs were revealed without prior copper decoration. Just a few TDs were delineated.

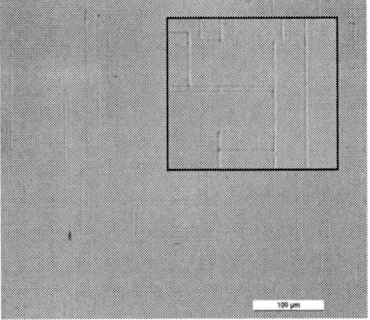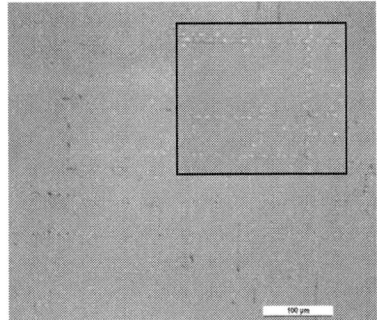

Figure 3. Samples with 83.5 nm sSOI film, etched with dilute Secco (0.04 M Cr (VI)), 54 s; HF dip: 8 s; light optical micrographs
a) "virgin" sample (ref1, left): residual layer thickness: 28 nm;
b) Copper decorated sample (right): 0.75 µL of 10 ppm $Cu(NO_3)_2$, 800 °C; dilute Secco (0.04 M Cr (VI)), 54 s; HF dip: 8s; residual layer thickness: 25 nm.

Figs. 5a and b display the results of defect delineation for sSOI fragments after copper decoration and subsequent etching with OPE B solution: Single bright (BOX undercut) and dark dots (sSi and BOX layer completely removed) as well as arrays of dot-like etch features are revealed which very likely correspond to TDs (dots) and SF-like defects (arrays of dots and occasionally even lines). Comparing non-copper decorated and copper decorated samples in Figs. 3 and Figs. 5 the etch features point to a disintegration of SF-like structures into arrays of dots in the decorated samples.

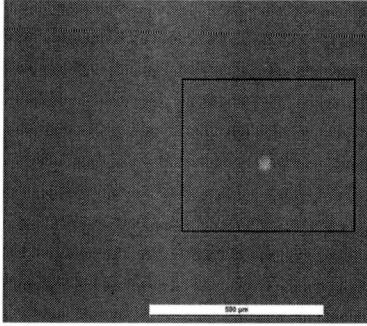

Figure 4. sSOI reference (ref 3): etched with OPE B, 14 min; no HF dip; residual layer thickness: 58 nm; light optical micrograph: bright dots correspond to halos of undercut BOX at TD etch pits.

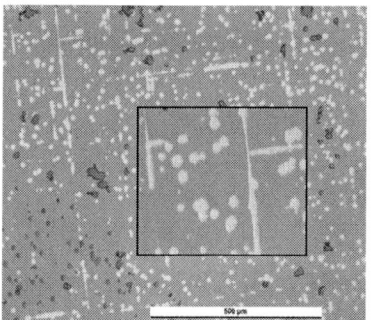 

Figure 5. Copper decorated sSOI fragment: 0.5 μL of 10 ppm $Cu(NO_3)_2$, 800 °C; etched with OPE B, 14 min; no HF dip; residual layer thickness: 48 nm; a) light optical micrograph (left): at bright features BOX is undercut, dark features reflect areas of completely removed sSi and BOX layer ; dots may correspond to TDs, arrays of dots and lines to SF-like defects. b) SEM shows similar features (right).

Summary

For the complete delineation of crystal defects in SOI materials a combination of copper decoration and preferential etching has been explored and the prior copper decoration procedure (19) was further improved. Copper cross contamination in the furnace used was minimized by appropriate precautions.

The defect densities observed in SOI fragments after a dilute Secco etch increase already after the furnace treatment without copper decoration (ref 2) and increase significantly after copper decoration with rising copper concentration. On the other hand, defect densities decrease with increasing SOI film thickness. Furnace annealing and copper decoration produce also a distinct increase of the dilute Secco etch rate. An explanation for this etch rate enhancement may be a catalytic effect of copper in the etching process.

A dilute Secco etch (0.04 M Cr (VI)) which is the standard procedure for sSOI material reveals stacking faults (SF) and threading dislocations (TD). Cr-less OPE solutions just delineate threading dislocations but not SF. In combination with copper decoration OPE B solution reveals stacking fault-like defects and threading dislocations.

Acknowledgments

The authors would like to thank SOITEC S.A., Parc technologique des Fontaines, Bernin for stimulating discussions and the supply of silicon substrates and Doris Ceglarek for providing the SEM images.

References

1. G. Celler and S. Cristoloveanu, *J. Appl. Phys.*, **93**, (9) (2003).
2. G. Taraschi, A. J. Pitera, L. M. McGill, Z. Cheng, M. L. Lee, T. A. Langdo and E. A. Fitzgerald, *J. Electrochem. Soc.*, **151**, G47-G56 (2004).
3. V. V. Voronkov and R. Faster, *J. Electrochem. Soc.*, **149**, (3), G167-G174 (2002).
4. P. Papakonstantinou, K. Somasun-dram, X. Cao and W. A. Nevin, *J. Electrochem. Soc.*, **148** , (2), G36-G42 (2001).
5. S. Sema, M. Porrini, F. Fogale and M. Servidori, *J. Electrochem. Soc.*, **148** , (9), G517-G523 (2001).
6. R. Falster and V. V. Voronkov, *Mater. Sci. Eng. B*, **73**, 87-94 (2000).
7. F. Secco d'Aragona, *J. Electrochem. Soc.*, **119**, (7), 948-951 (1972).
8. A. Abbadie, S. W. Bedell, J. M. Hartmann, O. Kononchuk, D. K. Sadana, F. Brunier, C. Figuet and I. Cayrefourq, *J. Electrochem. Soc.*, **154**, (8), (2007).
9. B. O. Kolbesen, D. Possner and J. Mähliß, *ECS Trans.*, **11**, 195-206 (2007).
10. D. Possner, B. O. Kolbesen, V. Klüppel, H. Cerva, *ECS Trans.*, **10** (1), 21 (2007).
11. F. Shimura, *Semiconductor Silicon Crystal Technology*, Academic Press (1989).
12. L. Mule'Stagno, *J. Electrochem. Soc. PV 2002-2*, 297 (2002).
13. A. A. Istratov and E. R. Weber, *J. Electrochem. Soc.*, **149** , (1), G21 (2002).
14. M. S. Kulkarni, J. Libbert, S. Keltner and L. Mule'Stagno, *J. Electrochem. Soc.*, **149** , (2), G153 (2002).
15. M. Seibt, M. Griess, A. A. Istratov, H. Hedemann, A. Sattler and W. Schröter, *Phys. Stat. Sol. (a)*, **166**, 171 (1998).
16. A. A. Istratov, H. Hedemann, M. Seibt, O. F. Vyvenko, W. Schröter, T. Heiser, C. Flink, H. Hieslmair and E. R. Weber, *J. Electrochem. Soc.*, **145**, (11), 3889 (1997).
17. W. Schröter,V. Kveder, M. Seibt, H. Ewe, H. Hedemann, F. Riedel and A. Sattler, *Mater. Sci. Eng. B*, **72**, 80-86 (2000).
18. K. Graf and P. Heim, *J. Electrochem. Soc.*, **141**, 2821 (1994).
19. H. Idrisi and B. O. Kolbesen, *Mater. Sci. Eng. B*, article in press (2009).
20. H. Wendt, H. Cerva, V. Lehmann and W. Palmer, *J. Appl. Phys.*, **65**, 2402 (1989).
21. H. Cerva and H. Wendt, *Mater. Res. Soc. Symp. Proc.*, **138**, 533 (1989).

36

ECS Transactions, 19 (4) 37-42 (2009)
10.1149/1.3117390 ©The Electrochemical Society

Simulation and Characterization of the Strain Induced by an Original "Embedded Buried Nitride" Technique

S. Baudot[a,b], F. Andrieu[b], M. Kostrzewa[b], J. Widiez[b], J.-C. Barbé[b], O. Faynot[b], Y. Lamrani[b], C. Vizioz[b], L. Brévard[b], H. Denis[b], V. Delaye[b], F. Rieutord[a] and J. Eymery[a]

[a] CEA-INAC, 17 rue des Martyrs, 38054 Grenoble Cedex 9, FRANCE
[b] CEA-LETI Minatec, 17 rue des Martyrs, 38054 Grenoble Cedex 9, FRANCE

An original technique is used to induce a compressive strain in the channel of Fully Depleted (FD)SOI MOSFETs thanks to an *embedded buried nitride* stressor. Strain measurements are performed by Grazing Incidence X-ray Diffraction (GIXRD) and compared to mechanical simulations. Both results are in agreement and prove that the strain level can achieve -0.6 % (-780 MPa) in the channel of a transistor, depending on the active region geometry and the nitride properties (intrinsic strain and thickness). This technique is thus very promising in order to boost pMOSFETs on thin films.

Introduction

Fully Depleted (FD)SOI is a promising device architecture for Low Power 22 nm MOSFETs technologies, mainly because of the low leakage current and the record matching performance already demonstrated (1,2). Some effective techniques were already experienced to boost the ON current through the strain in FDSOI, especially for nMOS (Contact Etch Stop Layer, sSOI substrates...) (3). However, for pMOS, the widely used embedded SiGe source/drain process is very challenging to integrate on FDSOI because of the small active body thickness. There is thus a huge interest for new process techniques to induce a compressive strain in the channel of the device.

In this study, an original *embedded buried nitride* stressor structure is proposed and characterized from the structural point of view. It is based on a stress transfer from a buried nitride stressor (embedded in the Burried OXide (BOX) of the substrate) to the transistor channel. The great interest of such stress transfer techniques is that the strain could be significant even for small channel dimensions, i.e. for nominal devices.

In order to check that this method can be applied to small dimensions, we accurately measured the strain in sub-100nm tests structures by Grazing Incidence X-Ray Diffraction (GIXRD). The strains in nano-structured and thin films can be obtained by several direct characterization techniques: Raman spectroscopy (4,5), High Resolution X-ray Diffraction (6,7), Grazing Incidence X-Ray Diffraction (8), or Transmission Electron Microscopy (TEM) based methods such as Convergent Beam Electron Diffraction or High Resolution (9). However, for this last technique, the lamella preparation induces a strain relaxation and fine mechanical simulations must be performed in order to interpret the data (10). For Raman spectroscopy, the complete lattice strain tensor is very difficult to obtain without any assumption (4). Among the different aforementioned techniques, GIXRD allows characterizing the complete lattice strain tensor with a high accuracy, without any assumption or special sample preparation. Here, we use GIXRD to check quantitatively the efficiency of the stress transfer from the nitride layer.

37

In this paper, we will first present the sample fabrication process, then the strain results obtained on patterned samples and finally the comparison between the GIXRD experimental data and finite element mechanical simulations.

Wafer fabrication

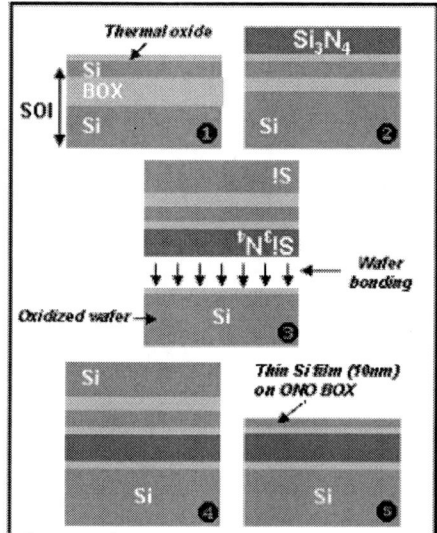

Figure 1a. (left) Schematic view of the process done to fabricate a SOI substrate with an ONO buried stack.
Figure 1b. (top) AFM scans ($1 \times 1 \mu m^2$, Z_{max}=5 nm) on the Si_3N_4 surface for the as-deposited layer (RMS=0.24 nm, PV=1.96 nm).

We fabricated SOI wafers with an Oxide/Nitride/Oxide (ONO) buried stack instead of the classical SiO_2 BOX. These specific substrates were made using the Bonded and Etched-back SOI (BESOI) technology (Figure 1a). The silicon nitride layers were obtained by Low Pressure Chemical Vapor Deposition (LPCVD) on 200 mm SOI donor wafers which were previously oxidized (3nm SiO_2 layer). Two Si_3N_4 thicknesses were processed: 30 nm (deposited at 750 °C) and 140 nm (deposited at 780 °C). Atomic Force Microscopy (AFM) was used to characterize the Si_3N_4 surface roughness (1×1 μm^2 scans): RMS values below 0.3 nm were measured on the as-deposited layers (Figure 1b). This very flat surface allows a direct chemical bonding without any polishing process. The base wafers were prepared by thermal growth of 5 nm thick silicon dioxide layers. We found that a standard RCA cleaning protocol on both donor and base wafers is sufficient for hydrophilic nitride bonding (11). A spontaneous bonding between thermal oxide and silicon nitride layers occurs at room temperature. The bonded pair was then annealed at high temperature (950 °C) in order to reinforce the bonding interface. To finish the process, the Si substrate and the SiO_2 BOX of the SOI donor wafer are removed by grinding and chemical etching. The final structure consists in a thin Si film (6nm after thinning steps) on an ONO buried stack with 4/26/6 nm and 4/140/6 nm thickness, measured by ellipsometry and confirmed by TEM (Figure 2).

Test structure patterning

The 6nm thin top Si layer and the ONO stack were etched by RIE to form a grid of 4 mm long and W=50,100,200 nm wide lines (see Figure 2). The objective of the etching is to relax the intrinsic tensile strain of the nitride layer and to transfer a compressive strain into the top Si layer (12). This concept is similar to the "Reverse Embedded SiGe" method (13,14). However, it is more compatible with thin film technologies because the stressor is directly integrated inside the BOX.

The efficiency of the stress transfer depends on the one hand on geometrical dimensions and on the other hand on the initial intrinsic tensile strain of the nitride layer. In this work, we have mainly investigated the influence of two geometrical dimensions: the nitride layer thickness and the active area width.

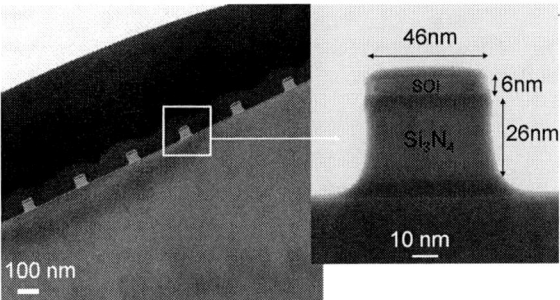

Figure 2. TEM cross-section of a sample with W=46nm narrow lines and 6/4/26/6nm Si/oxide/nitride/oxide thicknesses. The right figure is obtained in a Scanning TEM mode.

GIXRD Characterization

To check the efficiency of the stress transfer technique, strain measurements were performed by Grazing Incidence X-ray Diffraction (GIXRD) at the French CRG beamlines of the European Synchrotron Radiation Facility.

The X-ray energy is 11 keV. Grazing incidence (α_i=0.12°) and emergence angles (α_f=0.24°) close to the critical angles of total reflection allow measuring the diffraction of the (220) and (2-20) planes perpendicular to the surface (see inset of Figure 3a) of the top SOI layer. Larger grazing angles (α_i=0.3°, α_f=0.6°) are used to go through the amorphous nitride layer, in the substrate. The Si substrate gives an internal stress-free reference in the samples both in position and width allowing to check the setup alignment and the resolution function, and to decrease the error bar of the strain measurement. The ψ_{SOI} (ψ_{Sub}, respectively) Bragg peak angles of the SOI layer and substrate respectively are obtained by the optimization of intensity with radial and transverse scans at given grazing angles. Applying the Bragg law to the SOI layer and Si substrate, we get for the in-plane strain in the SOI layer:

$$\varepsilon = \sin(\psi_{Sub}/2)/\sin(\psi_{SOI}/2) - 1 \qquad [1]$$

This equation shows that the strain can be directly deduced from the detector angular position. The measurement of the (220) planes allow getting the strain in the so-called "longitudinal" direction, whereas the (2-20) planes give the strain in the "transverse" direction (see inset of Figure 3a).

The diffracted intensities corresponding to the planes along and perpendicular to the lines are shown in Figures 3a and 3b for 140 nm and 30 nm nitride thicknesses respectively and for different line widths. The Bragg peak relative to the Si substrate is also plotted as reference (lower sharp peaks). Along the lines, all the Bragg peaks of the top Si are at the same detector angular position ψ as the Si substrate. There is thus no significant strain in this direction, whatever the line width and the nitride thickness. In the other direction, the Bragg peaks of the Si on ONO are shifted from the Si substrate position toward larger ψ angles. This demonstrates that the *embedded buried nitride* technique is effective to induce a compressive strain in the narrowest direction. We extracted the average of the strain in the Si layer using [1] for each line width and nitride thickness. The experimental results have been plotted in Figure 5b (symbols). This figure shows a slightly higher compressive strain in the Si channel with 140 nm than with 30nm thick nitride. Moreover, this strain surprisingly decreases when the active width decreases, at least for W<200 nm. To explain these measurements and to predict the behavior above W=200 nm, we used finite element mechanical simulations.

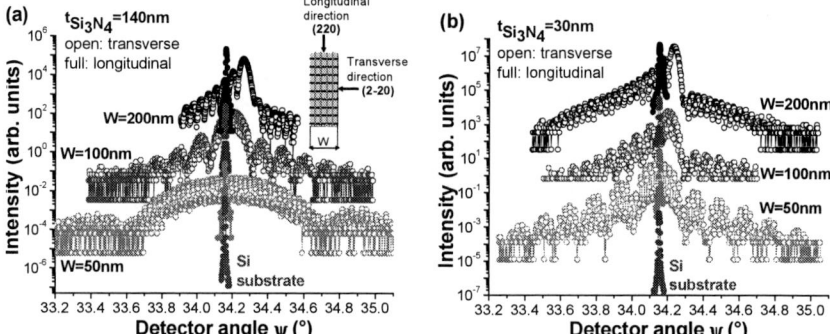

Figure 3. Measurements of (220) and (2-20) Bragg peaks of W=50,100,200nm SOI lines and Si substrate with a) 140 nm (left figure) and b) 30 nm (right figure) thick nitride layers. Only one peak corresponding to the Si substrate response is plotted at the bottom of each figure (always measured at the same position, whatever W). Inset: Strain measurement directions and top view of the corresponding crystallographic planes.

Mechanical Simulation

Finite element simulations have been performed with the ANSYS tool. The simulated structure was shown in Figure 2. Only the right half of the device was simulated because of the structure symmetry. The simulations were done for devices with 30 nm or 140 nm thick nitride and for different line widths W (see Figure 5b).

The simulation results confirm the aforementioned mechanism of the strain transfer. Indeed, the etching of the $SOI/SiO_2/Si_3N_4/SiO_2$ stack relaxes the strain of the Si_3N_4 layer from the edges (compare the red color before the etching in Figure 4a and the green/blue color after the etching in Figure 4b) transferring a compressive strain in the SOI layer (see the central blue region in Si in Figure 4b).

The main mechanical inputs of all these simulations were the intrinsic strain of the nitride layer (s=1100 MPa, measured on blanket wafers) and the elastic coefficients of the nitride (Young modulus E=160 MPa and Poisson ratio ν= 0.24). However, there is some discrepancy in the literature about the elastic coefficient values of the nitride. Moreover, the intrinsic strain value strongly depends on the deposition process and the layer thickness. In order to evaluate the uncertainty of the simulations, we studied the dependence of these coefficients on the final strain (Figure 5a). Finally, the simulation results have been compared with the GIXRD measurements taking into account this uncertainty on the mechanical inputs. Figure 5b show a very good overall agreement. Note that the absolute uncertainty of the GIXRD measurement (less than 0.02%) is lower than the simulation ones.

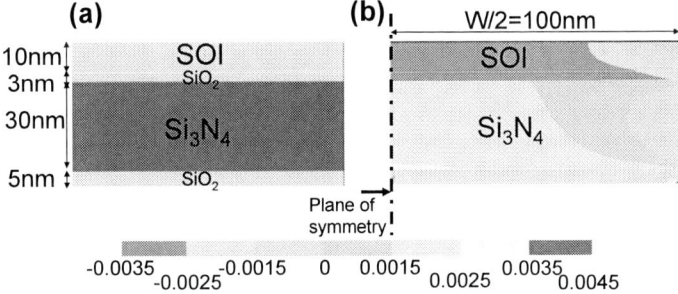

Figure 4. Simulation of the lateral strain in the *embedded buried nitride* structure (a) before and (b) after the W=200 nm large active area patterning. Before the etching, the strain in the nitride is uniform and tensile, and there is no strain in the SOI layer. After etching, the tensile strain of the nitride is reduced, and compressive strain is induced in the SOI layer.

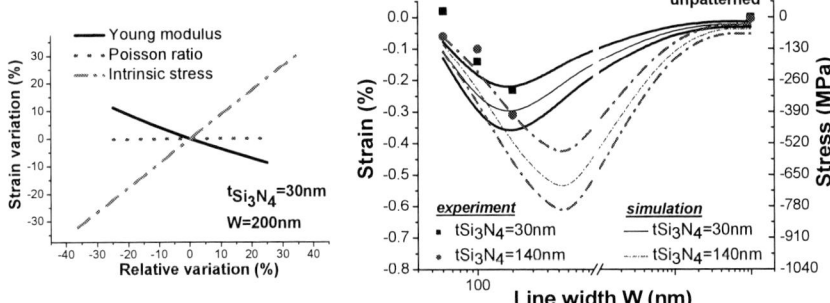

Figure 5a. Dependence of the mechanical coefficients on the simulated average strain in the SOI layer.

Figure 5b. Experimental and simulated average strain in the Si line vs. the line width for 30 nm and 140 nm nitride thicknesses (considering a 35% uncertainty on the intrinsic strain s and 25% on the Young modulus E).

The simulations highlight a non monotonous behavior of the strain vs. the active area width (W), with a strain optimum in the range 150 nm<W<800 nm, depending on the nitride thickness. This optimum is due to the balance between the aforementioned Si compression caused by the nitride relaxation (responsible for the growing part of the curve for wide lines) and the Si relaxation itself (responsible for the drop of the curve for narrow lines). This Si relaxation is also due to the etching and evidenced in Figure 4b, where we can see border effects in the SOI layer (green region). That is why, for sub-100nm patterns, this border effect leads to a rather small final strain in the SOI layer (see Figure 5b).

Conclusion

We propose an original technique to induce a compressive strain in the channel of FDSOI MOSFETs thanks to an *embedded buried nitride* stressor. The interest of this method is evidenced by the fabrication and strain measurement of a patterned substrate with a buried nitride layer. The in-plane deformations measured by GIXRD are in quantitative agreement with finite element mechanical simulations which show that the channel strain can reach -0.6 % (-780 MPa) with optimized geometrical features (i.e. the nitride and top Si thicknesses), and active area dimensions. A proper design adjustment must thus allow tuning the 2D strain in the channel of FDSOI MOSFETs in order to, in turn, boost CMOS performance.

Acknowledgments

This work was carried out in the frame of the OSEO Nanosmart project with SOITEC.

References

1. V. Barral, T. Poiroux, F. Andrieu, C. Buj-Dufournet , O. Faynot, T. Ernst, L. Brevard, C. Fenouillet-Beranger, D. Lafond, J.M. Hartmann, V. Vidal, F. Allain, N. Daval, I. Cayrefourcq, L. Tosti, D. Munteanu, J.L. Autran and S. Deleonibus, *IEDM Tech. Dig.*, p. 61, (2007).
2. O. Weber, O. Faynot, F. Andrieu, C. Buj-Dufournet, F. Allain, P. Scheiblin, J. Foucher, N. Daval, D. Lafond, L. Tosti, L. Brevard, O. Rozeau, C. Fenouillet-Beranger, M. Marin, F. Bœuf, D. Delprat, K. Bourdelle, B.-Y. Nguyen and S. Deleonibus, *IEDM Tech. Dig.*, p. 245, (2008).
3. F. Andrieu, O. Weber, T. Ernst, O. Faynot and S. Deleonibus, *Microelec. Eng.*, **84**, 2047, (2007).
4. I. De Wolf, H. E. Maes and S. K. Jones, *J. Appl. Phys.*, **79**, 7148, (1996).
5. K. Sawano, S. Koh, Y. Shiraki, N. Usami and K. Nakagawa, *Appl. Phys. Lett.*, **83**, 4339, (2003).
6. G. M. Cohen, P. M. Mooney, E. C. Jones, K. K. Chan, P. M. Solomon and H-S. P. Wong, *Appl. Phys. Lett.*, **75**, 787, (1999).
7. M. Gailhanou, A. Loubens, J.-S. Micha, B. Charlet, A.A. Minkevich, R. Fortunier and O. Thomas, *Appl. Phys. Lett.*, **90**, 111914, (2007).
8. S. Baudot, F. Andrieu, F. Rieutord and J. Eymery, *submitted to J. Appl. Phys.*
9. A. Armigliato, R. Balboni, G.P. Carnevale, G. Pavia, D. Piccolo, S. Frabboni, A. Benedetti and A.G. Cullis, *Appl. Phys. Lett.*, **82**, 2172, (2003).
10. L. Clément, R. Pantel, L. F. Tz. Kwakman and J. L. Rouvière, *Appl. Phys. Lett.*, **85**, 651, (2004).
11. O. Rayssac, H. Moriceau, M. Olivier, I. Stoemenos, A. M. Cartier and B. Aspar, Elctrochemical Society Proceedings Volume 2001-3
12. J-C. Barbe and T. Ernst, "Method for constraining a thin pattern", U.S. Patent 0091105, (2006).
13. R. A. Donaton, D. Chidambarrao, J. Johnson, P. Chang, Y. Liu, W. K. Henson, J. Holt, X. Li, J. Li, A. Domenicucci, A. Madan, K. Rim and C. Wann, *IEDM Tech. Dig.*, p. 465, (2006).
14. K-W. Ang, K-J. Chui, V. Bliznetsov, C-H. Tung, A. Du, N. Balasubramanian, G. Samudra, M. F. Li and Y-C. Yeo, *Appl. Phys. Lett.*, **86**, 093102, (2005).

CHAPTER 3

DEVICE TECHNOLOGY I

44

ECS Transactions, 19 (4) 45-54 (2009)
10.1149/1.3117391 ©The Electrochemical Society

**Advanced FinFET Devices for Sub-32nm Technology Nodes:
Characteristics and Integration Challenges**

A. Veloso[1], N. Collaert[1], A. De Keersgieter[1], L. Witters[1], R. Rooyackers[1],
M. J. H. van Dal[2], R. Duffy[2], B. J. Pawlak[2], R. J. P. Lander[2], T. Hoffmann[1],
S. Biesemans[1] and M. Jurczak[1]

[1]IMEC, Kapeldreef 75, 3001 Leuven, Belgium
[2]NXP-TSMC Research Center, Kapeldreef 75, 3001 Leuven, Belgium

We report a comprehensive evaluation and overview of the latest
developments and technology challenges of FinFET-based devices.
They offer improved electrostatics and steeper sub-threshold
slopes, attractive for enabling further CMOS scaling, but can also
suffer from higher parasitic resistance and parasitic capacitance for
narrow Fin devices. Critical solutions to minimize the impact of
the latter are here addressed, demonstrating their viability for
replacing planar CMOS devices. Multiple-V_T CMOS can be
achieved with capping technology, with aggressively scaled Ring
Oscillators (RO) and SRAM cells showing excellent performance
and matching behavior.

Introduction

FinFET-based multi-gate (MuGFET) devices are one of the most promising
candidates for enabling continued MOSFET scaling beyond the 32nm technology node,
thanks to their improved Short Channel Effects (SCE) behavior and the possibility to
control the channels potential without the use of heavy channel doping (1-8). The latter
allows minimizing V_T variability due to dopants fluctuation in scaled devices, crucial for
the cell stability of smaller SRAMs (4-8). In this paper, we review and report the latest
developments and technology challenges for these devices, namely by addressing: 1)
doping and contact strategies for limiting access resistance, and 2) work function (WF)
engineering options for multiple-V_T applications. Performance and V_{DD} scalability of
aggressively scaled FinFET-based circuits (RO, SRAM) are also discussed.

Device fabrication largely follows conventional bulk CMOS processing, with special
attention needed for specific process steps such as Fin & gate patterning and junction
formation, in addition to the co-integration with several advanced process modules such
as: high-k/metal gate (MG) stack, epitaxial raised source/drain (S/D), and strain
engineering. Fig.1 shows SEM images of a typical MuGFET and a dense, FinFET-based,
6T-SRAM.

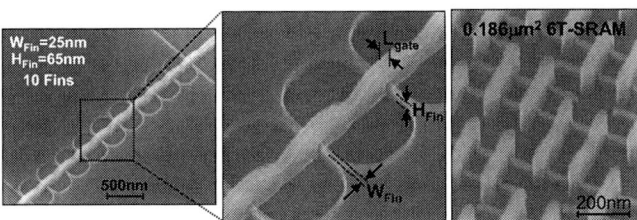

Figure 1. Tilted SEM images of a MuGFET device (left) and a dense, FinFET-based,
6T-SRAM cell (right) after gate patterning (SOI substrate).

45

The transition from extended bulk CMOS to non-classical device structures is currently not expected to occur at the same time for all applications and all chip manufacturers, because of the costs involved and compatibility with circuit design. In this context, FinFET technology appears as particularly attractive to help prolong SRAM scaling, since it can be implemented on both bulk-Si and SOI substrates (see Fig.2a), with functional FinFET SRAMs fabricated with planar FET peripheral circuits on the same bulk Si substrate recently demonstrated in Ref. (9). In both cases (bulk, SOI), Fin height (H_{Fin}) control can play a critical role in the gate etch process optimization to minimize MG undercut, as schematically illustrated in Fig.2b.

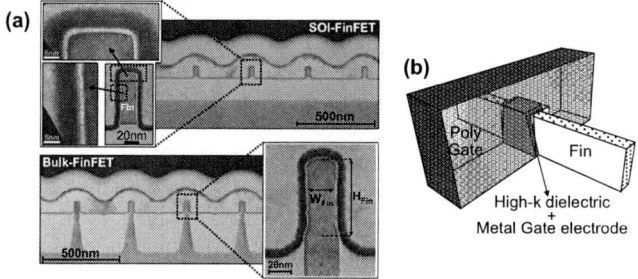

Figure 2. a) HRTEM images of SOI *vs.* bulk MuGFETs after full device fabrication, with b) schematically showing a tapered gate profile. The latter illustrates the 3D-topography impact on the gate etch process, with the extent of the MG undercut also dependent on H_{Fin} control.

Device characteristics & Technological challenges

Beyond the 32nm node, narrow Fins ($\leq$10nm) are required to fully benefit from the superior SCE control of multi-gate devices, as shown in Fig.3, where DIBL values of ~100mV/V are obtained for NMOS and PMOS MuGFET devices with $W_{Fin} \approx$ 6nm, $L_{gate} \approx$ 30nm, $H_{Fin} \approx$ 65nm and HfSiON/TiN gate stacks on SOI substrate.

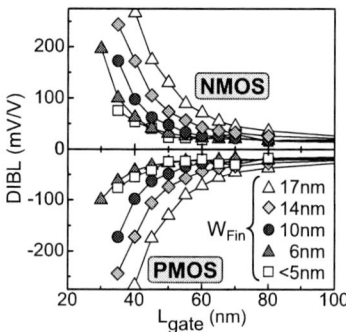

Figure 3. DIBL *vs.* L_{gate} dependence on W_{Fin} for NMOS and PMOS SOI-MuGFETs.

Spacer defined patterning is one of the processing techniques proposed to form such narrow, uniform and dense Fins with reduced line edge roughness and less stringent lithographic requirements (10-12), as shown in Fig.4.

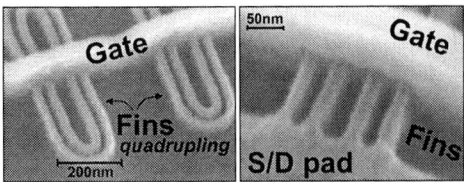

Figure 4. Tilted SEM images of Fins defined by spacer defined patterning technology with (right) and without (left) common S/D pads.

The superior SCE behavior of narrow Fin devices is however compromised by an increase of their access resistance R_{SD}, which leads to drive current degradation and compromises high-speed operation. As shown in Fig.5a, R_{SD} increase is more pronounced for NMOS than for PMOS devices, in agreement with the higher performance degradation seen for NMOS (13). This is currently attributed to a problematic re-crystallization after amorphizing ion implantation, resulting in defect formation and poor dopant activation, and with As-implanted narrow Fins (NMOS) more affected than BF_2 or B-implanted Fins (PMOS) (14-16). Confirmation is provided by the HRTEM images in Fig.5b, showing defects in the NMOS S/D regions after As implantation and activation anneal.

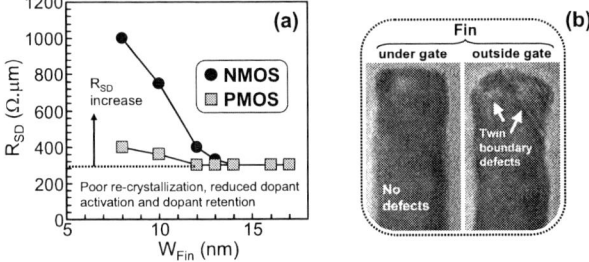

Figure 5. a) R_{SD} *vs.* W_{Fin} for NMOS and PMOS MuGFET devices, with R_{SD} extracted by extrapolating to $L_{gate}=0$ the total resistance R measured at $|V_{GS}|=3V$ and $|V_{DS}|=50mV$. b) HRTEM images of narrow Fin devices, showing twin-boundary defects in the S/D regions after As implantation and activation anneal.

Fig.6 shows R_{SD} results obtained for differently As-doped Fins and the corresponding HRTEM images of Fins after implantation and anneal: deep 0°tilt As implants result in the lowest resistance for wide Fins (planar-like) but the sharpest R_{SD} increase for scaling W_{Fin}; best R_{SD} scalability is obtained for shallower, less amorphizing, implants. In the HRTEM images, a larger amorphization of sub-20nm wide Fins is clearly seen for deeper As implants, with an expansion of the top portion of the Fins (encapsulated by spacers on the sidewalls) occurring after implantation. After 600°C, 60s anneal, the c-Si re-growth is incomplete and after 1050°C, spike anneal, the re-growth is complete but twin-boundary defects are present, and the top portion of the Fin is transformed into poly-Si for the most amorphizing implant condition. Dopants segregation at the Si grain boundaries can then occur, leading to higher resistance. This means that despite the presence of a lattice template under the gate and at the bottom of the Fin, the Fin sidewall surfaces suppress/retard crystal re-growth, promoting instead the formation of twin boundary defects in the implanted regions, with poly-crystalline silicon formation thought to be linked to untemplated re-crystallization known as Random Nucleation and Growth (15). Poor Fin re-crystallization during activation anneal therefore promotes the use of less-amorphizing, low energy extension implants to obtain improved Fin morphology and reduced R_{SD} resistance.

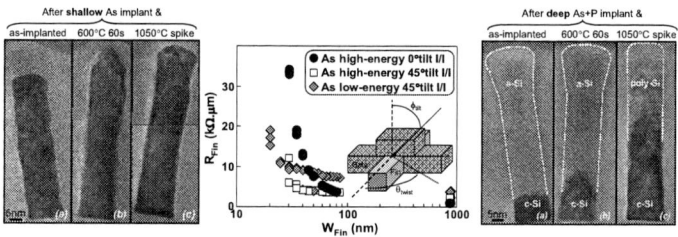

Figure 6. R_{Fin} dependence on W_{Fin} for different As implant conditions (centre plot), with HRTEM images of sub-20nm wide Fins after shallow and deep implants shown on the left and right, respectively.

Alternative doping techniques such as vapor phase doping or plasma doping (17,18) (see Fig.7) can bring some extra benefits in forming conformal junctions (with equal doping of top and side Fin surfaces) with limited amorphization of the Fins. But, even though they appear as possible solutions for that, they have yet to demonstrate improved performance-leakage trade-offs compared to conventional ion implantation. The latter remains a strong candidate for doping the Fins, despite the implants tilt angle restriction to avoid resist shadowing in dense-pitch structures. Most recently, a single-sided ion implantation strategy was also proposed to reduce V_T-variability in narrow, dense Fins (6).

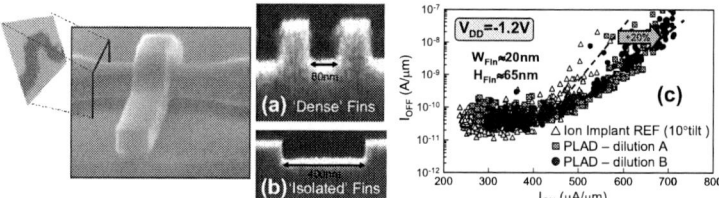

Figure 7. SSRM pictures of a pseudo-conformal pulsed plasma doping (PLAD) process measured on a) dense and b) isolated Fin structures. c) ITP behavior of PMOS MuGFET ($H_{Fin} \approx 65nm$) devices doped with optimized PLAD processes vs. a 10°tilt ion implantation process (small tilt angle used to avoid resist shadowing in SRAM cells).

It is widely acknowledged that raised S/D by Si-epitaxial growth (SEG) is also an effective way to reduce the parasitic resistance (14, 19). This can be seen in Fig.8a, where the drive current increase was primarily due to an almost 50% reduction in R_{SD} with SEG. Fig.8b also shows that NMOS devices are more sensitive to the SEG thickness, whereas for PMOS its impact on the R_{SD} reduction is minor. By providing more Si area, the contact area is increased, reducing at the same time the occurrence of over-silicidation in aggressively scaled Fins, and hence of excessive GIDL. This is shown by the high off-state leakage of the non-SEG devices in Fig.8a. The layout dependency of SEG can also be used advantageously to improve R_{SD} by merging neighboring Fins in logic devices as shown in Fig.9, while limiting the SEG thickness in SRAM cells to prevent them from merging there (respecting SRAM minimum Fin pitch).

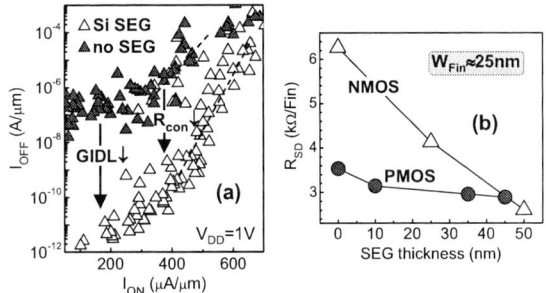

Figure 8. a) ITP of NMOS MuGFET devices with and without Si S/D SEG ($W_{Fin}\approx25nm$; SEG thickness=30nm). b) R_{SD} as a function of SEG thickness for NMOS and PMOS devices. R_{SD} extraction done at $|V_{GS}-V_T|=1V$ and $|V_{DS}|=50mV$.

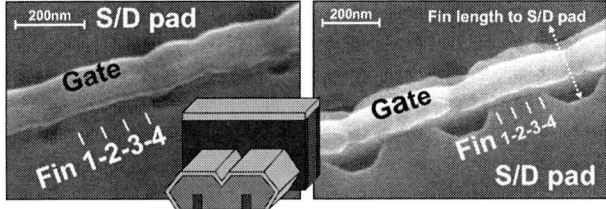

Figure 9. Si-epitaxial growth can be used advantageously to reduce the series resistance by merging neighboring Fins in logic devices, as shown in the SEM images taken for 2 slightly different layout configurations (different Fins length).

In MuGFETs, due to full depletion of the Fins, V_T tuning options for narrow Fins are limited to WF tuning with MG. Medium-V_T values are set by a mid-gap WF MG electrode such as TiN but, for flexible circuit operation, a wide V_T range (low-V_T - high-V_T) on the same wafer is required. Fig.10 shows that, compared with planar bulk devices, smaller WF shifts from mid-gap are needed to reach low-V_T targets in MuGFETs. Figs.11 and 12 show that can be successfully achieved with capping technology, where by introducing a thin ($\leq$1nm) capping layer in the gate stack, its WF can be successfully tuned towards band-edge. Data is shown here for the gate stack: HfSiO by MOCVD + TiN by PE-ALD. For low-V_T NMOS, a La_2O_3-cap by ALD or Dy_2O_3-cap by AVD is inserted in it; for low-V_T PMOS, an Al_2O_3-cap by ALD is used. Cap insertion is done immediately after the host dielectric (HfSiO) deposition or in-between 2 TiN metal layer

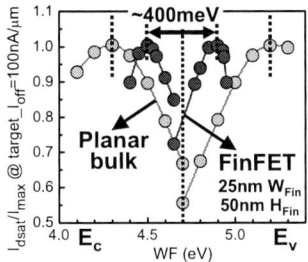

Figure 10. Compared to planar bulk devices, MuGFETs require smaller WF shifts for low-V_T applications, with optimum WF calculated to be ±200meV shifted from mid-gap.

depositions, such that the total TiN thickness is kept constant in all samples. For the [TiN/cap/TiN] sandwich stacks, several physical analysis techniques (e.g., TOF-SIMS in Fig.11b) have all confirmed diffusion of La, Dy and Al in TiN after final device fabrication (1050°C spike anneal used for junctions activation). Advantages of this sandwich configuration instead of HfSiO/cap/TiN are: comparable (or even slightly higher) WF shifts with reduced CET (see Fig.11a) and without impacting J_G, improved mobility, similar noise response and improved BTI behavior (20,21). Flexibility in the capping layers location (cap deposition after high-k or in-between TiN layers) also enables a simplified and more robust CMOS co-integration scheme to obtain low-, med- and high-V_T devices on the same wafer. By avoiding the underlying material (high-k or 1st TiN layer) from undergoing a cap removal process twice, impact/damage of the host dielectric with the selective cap removal process is more easily avoided. This multiple-V_T CMOS integration scheme (20) is schematically illustrated in Fig.12, with the capping layers (Al_2O_3 for low-V_T PMOS or high-V_T NMOS and La_2O_3 or Dy_2O_3 for low-V_T NMOS or high-V_T PMOS) selectively inserted in the flow at different locations in the gate stack.

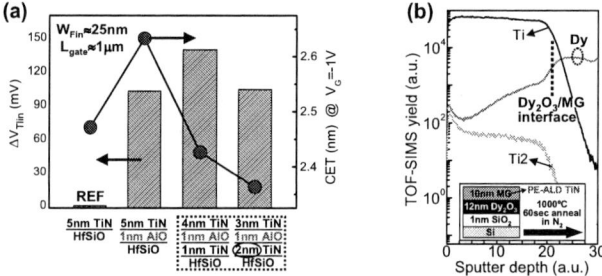

Figure 11. a) ΔV_{Tlin} and CET values of MuGFET devices with PE-ALD TiN gate electrode and an Al_2O_3-capping layer introduced in the gate stack using different deposition sequences. b) After applying a high thermal budget, diffusion of the cap in the metal layer for [TiN/cap/TiN] sandwich gate stacks is confirmed by several physical analysis techniques such as TOF-SIMS (data for Dy cap diffusion in PE-ALD TiN shown here).

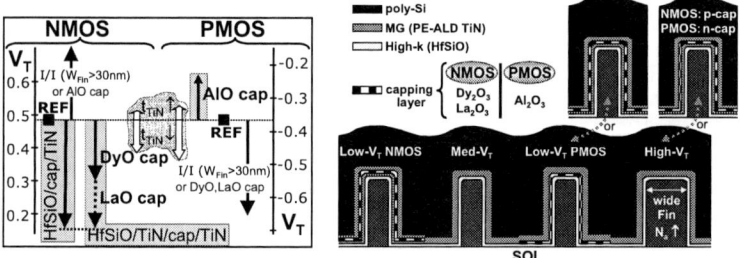

Figure 12. On the left: overview of WF engineering in MuGFETs using capping technology and PE-ALD TiN as gate electrode. On the right: proposed MuGFET CMOS integration scheme for multiple-V_T (Low-, Medium- and High-V_T) devices on the same wafer, exploring the flexible capping layers location in the gate stack (20).

MuGFETs are typically fabricated on substrates with (100)/<110> crystal orientation/current direction, with the Fins defined to have (110)/<110> vertical sidewalls and a $(100)/<110>$ top plane. The impact of the (110) sidewall surfaces on the carrier mobility increases for narrower Fin devices, affecting electrons and holes differently (22). For NMOS, as shown in Refs. (2,13), only a minimal degradation is observed in the high-field electron mobility for (110) dominated FinFETs *vs.* planar (100) devices. For PMOS, on the other hand, as hole mobility is substantially improved for the (110) transport plane, higher values have been consistently reported in narrower Fin devices as compared to (100) planar devices (2,13). Ultimately, to be able to obtain high-performance FinFETs that meet the stringent drive requirements of sub-32nm nodes, strain engineering will be needed. As for planar bulk, several strain or mobility enhancement techniques can be used for this purpose (2,23-26), including transverse and vertical stresses compatible with the FinFETs 3D-architecture. As an example, Ref. (23) showed that integration of a compressively strained SiGe layer in the S/D of a p-type MuGFET resulted in improved performance due to the combined effect of both improved hole mobility and series resistance reduction.

Circuit performance

As mentioned earlier, despite their excellent intrinsic behavior, there is some concern that the higher parasitics of FinFETs could be a potential show stopper for the introduction of these devices in future technology nodes. If the series resistance can be reduced by Si-epitaxial growth and layout optimization (27), the parasitic capacitance depends on the architecture. In FinFETs, Refs. (28,29) showed that the latter was dominated by the overlap and the fringing capacitance between the gate and Fin extensions (~60% of parasitics). However, as shown in Fig.13a and in Ref. (30), the additional capacitance penalty of FinFETs is limited when compared to planar transistors. Fig.13b shows that, despite the speed limiting FinFET parasitics, Ring Oscillators with an inverter stage delay comparable to that of planar devices can be obtained (29): 11ps at 100nW/stage and $V_{DD}=1V$ are achieved in scaled FinFET Ring Oscillators with an optimized gate stack, also used for SRAM fabrication (Fig.14).

Reports on the impact of FinFET device optimization (5,7) and design layout (4,5) on the overall SRAM cell stability support the choice of FinFETs as an attractive technology to help prolong SRAM scaling, with the added value that it can be implemented on both bulk-Si and SOI substrates. In Ref. (9), fabrication of functional FinFET SRAMs with planar FET peripheral circuits on the same bulk Si substrate showed that a higher SRAM beta ratio ($\beta>2.0$) can be achieved by tuning the effective channel width of each FinFET in the cell (through optimization of the initial Fin height and STI formation), without area

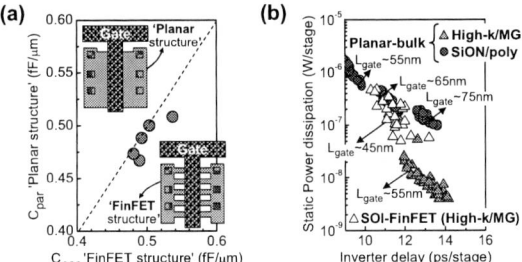

Figure 13. a) Parasitic capacitance (C_{par}) of FinFETs is slightly higher than that of planar devices (SOI substrate). b) Ring Oscillators Power/Speed trade-off ($V_{DD}=1V$) shows that FinFETs can achieve a comparable performance to planar devices.

penalty. Recently, we demonstrated 32nm-node ($0.186\mu m^2$), SOI-FinFET based, 6T-SRAM cells fabricated with advanced lithographic options (Immersion and Full-Field EUV) and excellent V_{DD} scalability down to 0.6V (8). Fig.15a shows the excellent V_T-mismatch results ($\sigma(\Delta V_T) \leq 30mV$) obtained for the correspondent Pull-Up, Pull-Down and Pass-Gate transistors ($W_{Fin} \approx 20nm$, $H_{Fin} \approx 40nm$, $L_{gate} \approx 45nm$) of these SRAM cells. FinFETs suitability for scaling SRAM applications is further confirmed in Fig.15b by the excellent V_T-mismatch behavior measured for scaled FinFETs ($AV_T < 2mV.\mu m$ for undoped SOI-FinFETs), when compared to planar devices.

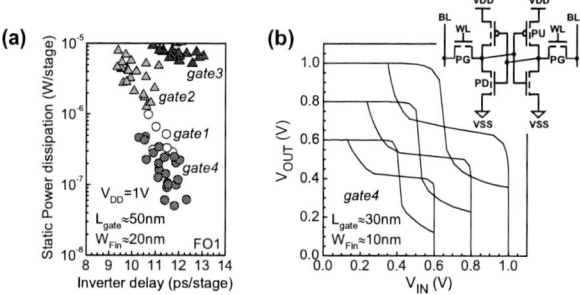

Figure 14. a) Inverter delay *vs.* Static Power dissipation for Ring Oscillators with different gate stacks, and b) SRAM cell butterfly curves for optimized gate stack.

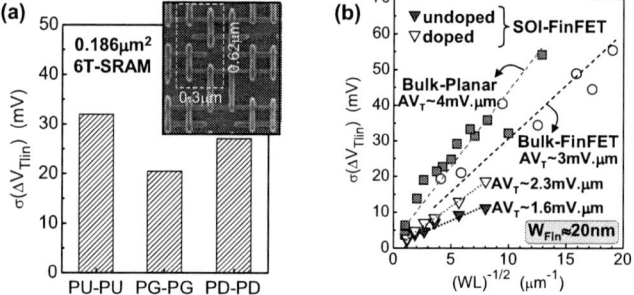

Figure 15. a) $\sigma(\Delta V_{Tlin})$ for Pull-Up (PU), Pass-Gate (PG) and Pull-Down (PD) transistors ($L_{gate} \approx 45nm$, $W_{Fin} \approx 20nm$) of $0.186\mu m^2$ SRAM cells. b) Pelgrom plot showing the excellent V_T-mismatch *vs.* area scaling performance for SOI-FinFETs ($AV_T < 2mV.\mu m$) when compared to bulk technology.

Conclusions

MuGFET devices are considered one of the most attractive candidates for enabling continued MOSFET scaling beyond the 32nm technology node, thanks to their improved electrostatics and steeper sub-threshold slopes. In this paper, we reviewed the latest developments and addressed some of the main technology challenges faced by these devices: 1) doping and contact strategies for limiting access resistance and improve performance of scaled devices, with SEG and careful implants optimization needed to avoid/minimize defects formation and dopants segregation; 2) use of capping technology for WF engineering to enable multiple-V_T applications on the same wafer; and 3) evaluation of the impact on circuit (RO, SRAM) performance, with excellent scalability and V_T-mismatch behavior reported.

References

1. L. Witters, N. Collaert, A. Nackaerts, M. Demand, S. Demuynck, C. Delvaux, A. Lauwers, C. Baerts, S. Beckx, W. Boullart, S. Brus, B. Degroote, J. F. de Marneffe, A. Dixit, K. De Meyer, M. Ercken, M. Goodwin, E. Hendrickx, N. Heylen, P. Jaenen, D. Laidler, P. Leray, S. Locorotondo, M. Maenhoudt, M. Moelants, I. Pollentier, K. Ronse, R. Rooyackers, J. Van Aelst, G. Vandenberghe, T. Vandeweyer, S. Vanhaelemeersch, M. Van Hove, J. Van Olmen, S. Verhaegen, J. Versluijs, C. Vrancken, V. Wiaux, P. Willems, J. Wouters, M. Jurczak and S. Biesemans, IEEE *VLSI Tech. Dig.*, p.106 (2005).
2. J. Kavalieros, B. Doyle, S. Datta, G. Dewey, M. Doczy, B. Jin, D. Lionberger, M. Metz, W. Rachmady, M. Radosavljevic, U. Shah, N. Zelick and R. Chau, IEEE *VLSI Tech. Dig.*, p.62 (2006).
3. W. Haensch, E. J. Nowak, R. H. Dennard, P. M. Solomon, A. Bryant, O. H. Dokumaci, A. Kumar, X. Wang, J. B. Johnson and M. V. Fischetti, *IBM J. Res. & Dev.* 50(4/5), p.339 (2006).
4. S. Inaba, H. Kawasaki, K. Okano, T. Izumida, A. Yagishita, A. Kaneko, K. Ishimaru, N. Aoki and Y. Toyoshima, IEEE *IEDM Tech. Dig.*, p.487 (2007).
5. N. Collaert, K. von Arnim, R. Rooyackers, T. Vandeweyer, A. Mercha, B. Parvais, L. Witters, A. Nackaerts, E. Altamirano Sanchez, M. Demand, A. Hikavyy, S. Demuynck, K. Devriendt, F. Bauer, I. Ferain, A. Veloso, K. De Meyer, S. Biesemans and M. Jurczak, IEEE *ICICDT Tech. Dig.*, p.59 (2008).
6. H. Kawasaki, M. Khater, M. Guillorn, N. Fuller, J. Chang, S. Kanakasabapathy, L. Chang, R. Muralidhar, K. Babich, Q. Yang, J. Ott, D. Klaus, E. Kratschmer, E. Sikorski, R. Miller, R. Viswanathan, Y. Zhang, J. Silverman, Q. Ouyang, A. Yagishita, M. Takayanagi, W. Haensch and K. Ishimaru, IEEE *IEDM Tech. Dig.*, p.237 (2008).
7. T. Mérelle, G. Curatola, A. Nackaerts, N. Collaert, M. J. H. van Dal, G. Doornbos, T. S. Doorn, P. Christie, G. Vellianitis, B. Duriez, R. Duffy, B. J. Pawlak, F. C. Voogt, R. Rooyackers, L. Witters, M. Jurczak and R. J. P. Lander, IEEE *VLSI Tech. Dig.*, p.241 (2008).
8. A. Veloso, S. Demuynck, M. Ercken, A. M. Goethals, M. Demand, J.-F. de Marneffe, E. Altamirano, A. De Keersgieter, C. Delvaux, J. De Backer, S. Brus, J. Hermans, B. Baudemprez, F. Van Roey, G. F. Lorusso, C. Baerts, D. Goossens, C. Vrancken, S. Mertens, J. J. Versluijs, V. Truffert, C. Huffman, D. Laidler, N. Heylen, P. Ong, B. Parvais, M. Rakowski, S. Verhaegen, A. Hikavyy, H. Meiling, B. Hultermans, L. Romijn, C. Pigneret, S. Lok, A. Van Dijk, K. Shah, A. Noori, J. Gelatos, R. Arghavani, R. Schreutelkamp, P. Boelen, O. Richard, H. Bender, L. Witters, N. Collaert, R. Rooyackers, P. Absil, A. Lauwers, M. Jurczak, T. Hoffmann, S. Vanhaelemeersch, R. Cartuyvels, K. Ronse and S. Biesemans, IEEE *IEDM Tech. Dig.*, p.861 (2008).
9. H. Kawasaki, K. Okano, A. Kaneko, A. Yagishita, T. Izumida, T. Kanemura, K. Kasai, T. Ishida, T. Sasaki, Y. Takeyama, N. Aoki, N. Ohtsuka, K. Suguro, K. Eguchi, Y. Tsunashima, S. Inaba, K. Ishimaru and H. Ishiuchi, IEEE *VLSI Tech. Dig.*, p.86 (2006).
10. Y.-K. Choi, T.-J. King and C. Hu, IEEE *Trans. Elect. Dev.* 49(3), p.436 (2002).
11. R. Rooyackers, E. Augendre, B. Degroote, N. Collaert, A. Nackaerts, A. Dixit, T. Vandeweyer, B. Pawlak, M. Ercken, E. Kunnen, G. Dilliway, F. Leys, R. Loo, M. Jurczak and S. Biesemans, IEEE *IEDM Tech. Dig.*, p.993 (2006).
12. A. Dixit, K. G. Anil, E. Baravelli, P. Roussel, A. Mercha, C. Gustin, M. Bamal, E. Grossar, R. Rooyackers, E. Augendre, M. Jurczak, S. Biesemans and K. De Meyer, IEEE *IEDM Tech. Dig.*, p.709 (2006).
13. M. J. H. van Dal, N. Collaert, G. Doornbos, G. Vellianitis, G. Curatola, B. J. Pawlak, R. Duffy, C. Jonville, B. Degroote, E. Altamirano, E. Kunnen, M. Demand, S. Beckx, T.

Vandeweyer, C. Delvaux, F. Leys, A. Hikavyy, R. Rooyackers, M. Kaiser, R. G. R. Weemaes, S. Biesemans, M. Jurczak, K. Anil, L. Witters and R. J. P. Lander, IEEE *VLSI Tech. Dig.*, p.110 (2007).

14. J. Kedzierski, M. Ieong, E. Nowak, T. S. Kanarsky, Y. Zhang, R. Roy, D. Boyd, D. Fried and H.-S. P. Wong, IEEE *Trans. Elect. Dev.* 50(4), p.952 (2003).

15. R. Duffy, M. J. H. van Dal, B. J. Pawlak, M. Kaiser, R. G. R. Weemaes, B. Degroote, E. Kunnen and E. Altamirano, *Appl. Phys. Lett.* 90, 241912 (2007).

16. M. J. H. van Dal, G. Vellianitis, R. Duffy, G. Doornbos, B. J. Pawlak, B. Duriez, L.-S. Lai, A. Hikavyy, T. Vandeweyer, M. Demand, E. Altamirano, R. Rooyackers, L. Witters, N. Collaert, M. Jurczak, M. Kaiser, R. G. R. Weemaes and R. J. P. Lander, *ECS Trans.* 13(1), p.223 (2008).

17. D. Lenoble, K. G. Anil, A. De Keersgieter, P. Eybens, N. Collaert, R. Rooyackers, S. Brus, P. Zimmerman, M. Goodwin, D. Vanhaeren, W. Vandervorst, S. Radovanov, L. Godet, C. Cardinaud, S. Biesemans, T. Skotnicki and M. Jurczak, IEEE *VLSI Tech. Dig.*, p.212 (2006).

18. Y. Sasaki, K. Okashita, K. Nakamoto, T. Kitaoka, B. Mizuno and M. Ogura, IEEE *IEDM Tech. Dig.*, p.917 (2008).

19. A. Dixit, K. G. Anil, R. Rooyackers, F. Leys, M. Kaiser, R. Weemaes, I. Ferain, A. De Keersgieter, N. Collaert, R. Surdeanu, M. Goodwin, P. Zimmerman, R. Loo, M. Caymax, M. Jurczak, S. Biesemans and K. De Meyer, IEEE *ESSDERC Conference Proceedings*, p.445 (2005).

20. A. Veloso, L. Witters, M. Demand, I. Ferain, N. J. Son, B. Kaczer, Ph. J. Roussel, E. Simoen, T. Kauerauf, C. Adelmann, S. Brus, O. Richard, H. Bender, T. Conard, R. Vos, R. Rooyackers, S. Van Elshocht, N. Collaert, K. De Meyer, S. Biesemans and M. Jurczak, IEEE *VLSI Tech. Dig.*, p.14 (2008).

21. A. Veloso, L. Witters, M. Demand, I. Ferain, N. J. Son, B. Kaczer, Ph. J. Roussel, C. Adelmann, S. Brus, O. Richard, H. Bender, T. Conard, R. Vos, R. Rooyackers, S. Van Elshocht, N. Collaert, K. De Meyer, S. Biesemans and M. Jurczak, IEEE *SOI Conference Proceedings*, p.119 (2008).

22. M. Yang, E. P. Gusev, M. Ieong, O. Gluschenkov, D. C. Boyd, K. K. Chan, P. M. Kozlowski, C. P. D'Emic, R. M. Sicina, P. C. Jamison and A. I. Chou, IEEE *Elect. Dev. Lett.* 24(5), p.339 (2003).

23. P. Verheyen, N. Collaert, R. Rooyackers, R. Loo, D. Shamiryan, A. De Keersgieter, G. Eneman, F. Leys, A. Dixit, M. Goodwin, Y. S. Yim, M. Caymax, K. De Meyer, P. Absil, M. Jurczak and S. Biesemans, IEEE *VLSI Tech. Dig.*, p.194 (2005).

24. N. Collaert, A. De Keersgieter, K. G. Anil, R. Rooyackers, G. Eneman, M. Goodwin, B. Eyckens, E. Sleeckx, J.-F. de Marneffe, K. De Meyer, P. Absil, M. Jurczak and S. Biesemans, IEEE *Elect. Dev. Lett.* 26(11), p.820 (2005).

25. N. Collaert, R. Rooyackers, A. De Keersgieter, F. E. Leys, I. Cayrefourcq, B. Ghyselen, R. Loo, M. Jurczak and S. Biesemans, IEEE *Elect. Dev. Lett.* 28(7), p.646 (2007).

26. M. Saitoh, A. Kaneko, K. Okano, T. Kinoshita, S. Inaba, Y. Toyoshima and K. Uchida, IEEE *VLSI Tech. Dig.*, p.18 (2008).

27. A. Dixit, K. G. Anil, N. Collaert, R. Rooyackers, F. Leys, I. Ferain, A. De Keersgieter, T. Y. Hoffmann, R. Loo, M. Goodwin, P. Zimmerman, M. Caymax, K. De Meyer, M. Jurczak and S. Biesemans, IEEE *SOI Conference Proceedings*, p.226 (2005).

28. W. Wu and M. Chan, IEEE *Trans. Elect. Dev.* 54(4), p.692 (2007).

29. B. Parvais, A. Mercha, N. Collaert, R. Rooyackers, I. Ferain, M. Jurczak, V. Subramanian, A. De Keersgieter, T. Chiarella, C. Kerner, L. Witters, S. Biesemans and T. Hoffmann, *Symposium on VLSI-TSA 2009* (accepted).

30. M. Guillorn, J. Chang, A. Bryant, N. Fuller, O. Dokumaci, X. Wang, J. Newbury, K. Babich, J. Ott, B. Haran, R. Yu, C. Lavoie, D. Klaus, Y. Zhang, E. Sikorski, W. Graham, B. To, M. Lofaro, J. Tornello, D. Koli, B. Yang, A. Pyzyna, D. Neumeyer, M. Khater, A. Yagishita, H. Kawasaki and W. Haensch, IEEE *VLSI Tech. Dig.*, p.12 (2008).

ECS Transactions, 19 (4) 55-64 (2009)
10.1149/1.3117392 ©The Electrochemical Society

ULTRA COMPACT FDSOI TRANSISTORS INCLUDING STRAIN AND
ORIENTATION: PROCESSING AND PERFORMANCE

C. Fenouillet-Beranger[b,a], L. Pham Nguyen[a,c], P. Perreau[b,a], S. Denorme[a], F. Andrieu[b],
O. Faynot[b], L. Tosti[b], L. Brevard[b], C. Buj[b], O.Weber[b], C. Gallon[a], V. Fiori[a], F. Boeuf[a],
S. Cristoloveanu[c], T. Skotnicki[a]

[a] STMicroelectronics, 850 rue Jean Monnet, 38926, Crolles, France
[b] CEA LETI, MINATEC, 17 rue des Martyrs 38054 Grenoble, France
[c] IMEP, MINATEC, 3 Parvis Louis Neel 38016 Grenoble, France
E-mail: Claire.fenouillet-beranger@st.com

The different ways to boost the devices performance, by using
local strain approach and crystalline orientations, are investigated.
We focus on the CESL (Contact Etch Stop Layer) induced
mechanical stress and on the specific technological features of
FDSOI (Fully-Depleted). The impact of device geometry, buried
oxide layer (BOX) and elevated source/drain is studied by using
finite element mechanical simulations. The significant amount of
stress transferred to the channel, observed by mechanical
simulation, has been validated by systematic measurements. An
important mobility improvement has been achieved for short gate
length which is a mandatory condition for advanced CMOS low-
power applications.

Introduction

With CMOS down scaling, thin film devices become more and more attractive
due to better control of short-channel effects. Using a high-k dielectric with a metal gate
is of great interest to improve FDSOI devices because of the gate leakage current
reduction, polysilicon depletion suppression and performance gain (1). In addition,
improving the carrier mobility is required by the ITRS roadmap for CMOS down scaling
(2). Many studies have focused on that topic, and mostly on the Process Induced Strain
(PIS) approach. In addition several papers proposed to combine strain CESL approach
with embedded SiGe source/drain (3), but this requires a selective epitaxy growth on a
recessed silicon film, which is not easily compatible with extremely thin SOI film (5-
10nm). Recently, publications on global stress approach have demonstrated that the use
of rotated or Strain SOI (SSOI) wafers becomes an additional option for mobility
improvement (4,5,6,7). In this paper we focus in a first part, on stress generated into the
channel by a low cost CESL process. The CESL is used as an etch stop layer for the
contact etch. A tensile CESL is more favourable for NMOS and a compressive CESL for
PMOS.
Several publications have presented electrical results on FDSOI. However, the
impact of the CESL on this type of devices has not been yet largely explored. The role of
the specific features of the FDSOI devices, as BOX layer, silicon film thickness and
elevated source/drain will be presented in the following by using mechanical simulations
and experimental data.

CESL: Mechanical simulations

Finite elements simulations using ANSYS (8) have been performed to evaluate the mechanical stress transmitted from the strained CESL into the channel and the impact of BOX (whose compliance is different from the Si one) in this transfer mechanism. The influence of the layout on the channel strain field is also studied in a FDSOI device with a raised Source/Drain architecture.

Modelling strategy and assumptions

In this 3D simulation, the assumptions are the following. First, the linear elastic analysis is focussed on the standalone effect of the nitride layer on the channel. More precisely, a strained CESL layer is deposited on a stress free structure. Then, due to equilibrium constraints, a relaxation phenomenon occurs and the strained layer partially transfers its energy to the channel. Under these assumptions, the contribution of the strained layer is isolated. Indeed, in order to get the absolute strain field in the silicon channel, a full process model, coupled to the use of suitable rheological models would be mandatory. On the other hand, the use of a 3D model enables to study actual complex architectures and layout effects. In particular, the specific effect along W and L_g directions can hence be decorrelated. Thanks to device symmetries and relevant boundary conditions, only a quarter of the full 3D transistor structure has been simulated as illustrated in Figure 1. Our study is performed on isolated NMOS transistors (as opposed to periodic patterns) with a 120 nm high poly gate associated with a 100 nm tensile (+1.2 GPa) SiN CESL with parameterized widths and lengths. Due to the linearity assumption, conclusions from the tensile strained layer can be extended to a compressive layer. Indeed, by replacing the tensile initial stress state by a compressive condition, induced strain effect in the structure would be strictly the opposite.

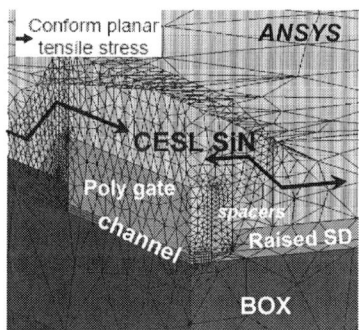

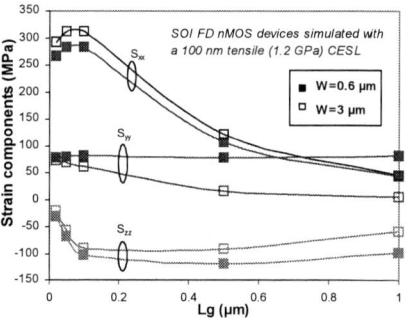

Figure 1. 3D schematic view of 1/4 simulated FDSOI transistor with raised Source/Drain. A conformal elastic strained layer is built to actually depict the CESL layer.

Figure 2. 3D simulations of silicon stress components versus L_g for $T_{Si} = 10$ nm, $T_{BOX} = 145$ nm and 25 nm Raised Source/Drain. Significant stress values are found, which confirm the transfer mechanism from the CESL to the underlying channel.

Simulations results and parametric study

Even though a raised (25 nm) S/D architecture was used, the stress was shown to be effectively transferred from the CESL to the underlying channel. Significant values of stress components in the channel are found (Figure 2), especially the S_{xx} component, which becomes more tensile as L_g decreases. These results also highlight that the stress level is strongly dependent on the layout (W and L_g). Generally two distinct strained regions are observed in the channel: a central area (central effects) with a compressive strain and an edge area (edge effects induced by "pocket stress effect") with a tensile strain. Hence, modelling results underline that edge effects are dominant for short devices, whereas the stress becomes less tensile for longer devices (even compressive in the centre of the channel). The balance between these two contributions is dependent on the geometrical parameters (W, L_g) of the devices. Consequently, the strain component can be either uni-axial, bi-axial or tri-axial, depending on device size and layout. On small devices with a tensile (compressive) CESL, the nMOS (pMOS) current performances should be improved by a tensile (compressive) strain transferred in the channel. The effect of the SiN thickness on the strain components, ranging from 50 to 150 nm with a tensile strained CESL of 1.2 GPa, is illustrated in Figure 3.

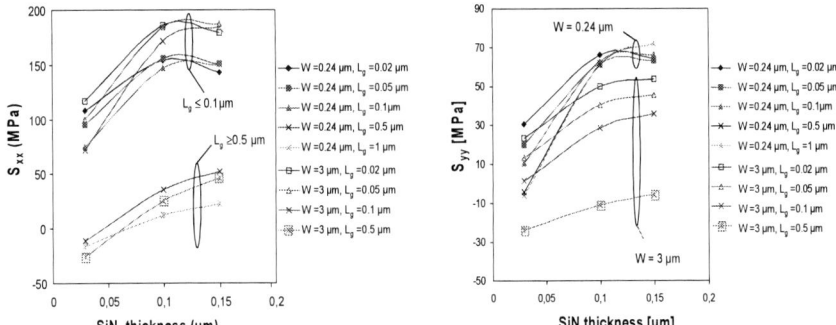

Figure 3. Influence of the CESL thickness on Sxx(left) and Syy(right) components (mean values) in the FDSOI channel. The results show an increase of the strain component value as the tensile CESL thickness increases.

Most of the results show an increase of the strain components with increasing SiN thickness. We have a confirmation that maximum tensile gain is obtained on smallest devices (Lg ≤ 0.1). However for these devices, we can notice a saturation of the gain on the strain components for SiN film thicker than 100 nm thickness. This could be explained by the fact that the tensile stress in short devices is governed by local edge effects. Therefore, thicker SiN layer are less effective.

In order to compare the strained CESL effect between bulk and SOI devices, the Buried OXide (BOX) under the channel was replaced by a full silicon layer in the model. The lower Young modulus of the BOX (compared to a silicon layer) induces an increase in the in-plane stress components of the channel. Indeed, since the channel lies on a smoother "substrate", the channel deformation is improved (by similarity with springs mounted in parallel). Hence, the average in-plane stress intensity in the channel of the SOI device is higher than for bulk devices (Figure 4). Note that the bulk devices simulated in this work are not fully representative of the state of the art architecture because of the presence of the raised SD. The aim of this BOX modification is essentially to show the effect of the stiffness of the material beneath the channel.

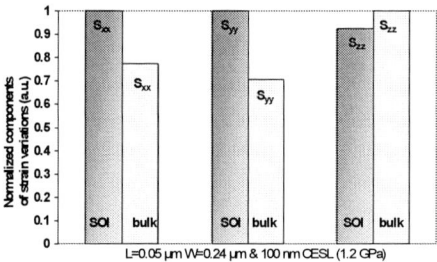

Figure 4. Strained CESL impact on strain components for FD SOI and bulk (the Buried OXide under the channel was replaced by a silicon layer).

In order to facilitate the salicidation of thin film and to reduce access resistance, elevated source/drain are grown by selective epitaxy. We observe on Figure 5, that the strain components are significantly dependent of the silicon film thickness and also of the source/drain height. Indeed, the thinner the silicon film and the source/drain, the higher the CESL-induced stress is.

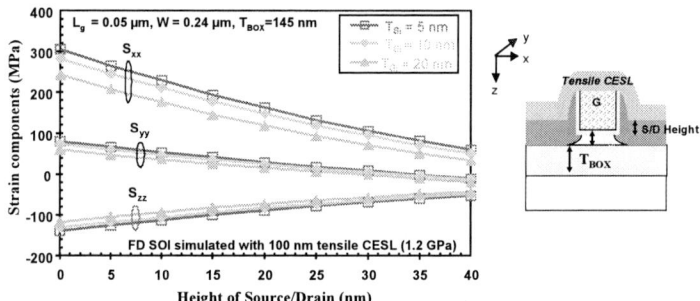

Figure 5. Mechanical simulations of strain components in the channel of a FD SOI with a tensile CESL. Strain transmission into the channel is impacted by both the height of S/D and the film thickness Tsi

The simulations reveal an important impact of the strained CESL in the device channel. We will now verify experimentally the accuracy of this mechanical simulation by a detailed analysis of FDSOI devices with a raised S/D architecture.

CESL: Experimental results

Polysilicon/SiON gate stack

We demonstrated in a previous paper (9-10) the impact of a strained CESL on FDSOI devices with polysilicon gate. Figure 6a highlights for nMOS devices a 10.5% I_{on} improvement at I_{off}=100 nA/µm using a 100 nm tensile liner of 1.2 GPa versus no stress liner and a loss of 10% using a compressive liner. For pMOS devices, the use of a 1.8 GPa compressive stress liner versus a low tensile stress liner (550 MPa) improves the performance by 17 % (Figure 6b).

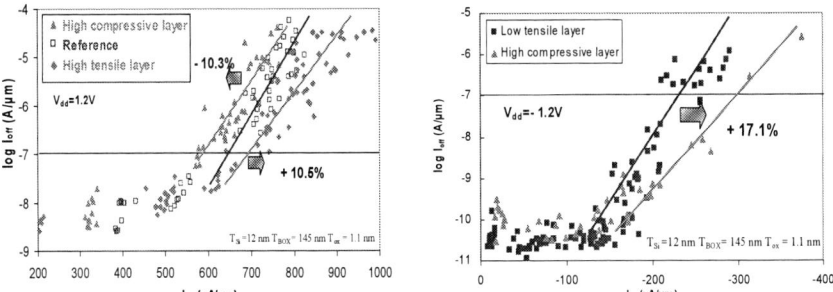

Figure 6. Impact of strained CESL on I_{on}/I_{off} characteristics at V_{dd}=1.2V for (a) nMOS and (b) pMOS devices.

High-k/Metal gate stack

Using a high-k dielectric with metal gate technologies is of great interest for FDSOI devices because of the gate leakage current reduction, polysilicon depletion suppression and performance improvement (1). While dual metal gates are required to adjust the threshold voltage of n and pMOSFETs of bulk technologies, the use of a single midgap metal gate such as TiN on FDSOI offers the advantage of adjusting the threshold voltage of the n-channel and p-channel devices simultaneously and keeping the channel undoped (1). To experimentally study the influence of CESL on the transistor performance, FDSOI devices with this gate stack and gate length around 30nm have been integrated. The starting materials are 300 mm <100> UNIBOND™ SOI wafers with 145nm BOX thickness.

• No channel doping
• **Hik deposition** HfO₂ or HfSiON
• **10nm TiN deposition**
• 100nm Polysilicon deposition
• Gate patterning
• Offset spacer formation
• **Optional Selective Si extension epitaxy**
• LDD implantation
• Dshape spacer formation
• **Optional Selective Si S/D epitaxy**
• S/D implantation
• Spike anneal
• NiPt silicidation
• **Strain CESL deposition**
• Std backend

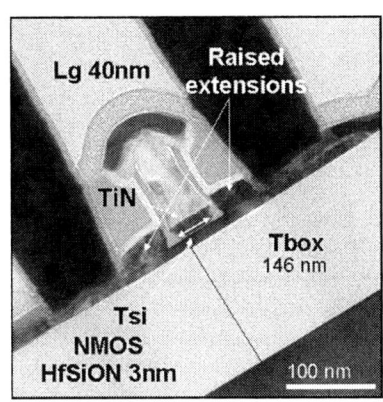

Figure 7. FDSOI process integration with Hight-K / Midgap Metal Gate stack.

Figure 8. TEM cross section of a nMOS transistor with High-K/ Midgap Metal Gate.

SOI films were thinned down by thermal oxidation and wet etching to achieve a final thickness of around 8-10 nm. After STI isolation, a high-k dielectric (HfSiON) of approximately 2.5 nm was deposited. A metal gate (ALD TiN 10nm) and a poly-Si layer of 100 nm were deposited for gate fabrication. A 193 nm lithography combined with trimming was performed to achieve the desired gate dimensions. The minimum gate length dimension measured on the wafers is around 30 nm. After a 10 nm thin offset spacer realization, a selective epitaxy of 10nm is performed in extension regions in order to reduce access resistance. Raised extensions are implanted. To finish a D-shape spacer, S/D implantation (activated by a 1080°C RTP spikes anneal) and silicidation (NiPtSi) are realized. Nitride layers (Compressive and Tensile) are added to boost the performance. The process flow scheme is summarized in Figure 7. A cross section of the device is seen on Figure 8.

Ion/Ioff performance

The integration of highly tensile (compressive) liners allows 12% and 25% drive current improvement on nMOS and pMOS respectively (Figures 9 & 10). Good LSTP (Low Standby Power) performance with respect to previous published results (11)-(13) is obtained at V_{dd}=1.1V for an I_{off} = 6.6pA/µm for both nMOS (I_{on}=400µA/µm) and pMOS (I_{on}=256µA/µm). Moreover, this technology was used to fabricate the highest density (0.179µm²) FDSOI high-k/metal gate 6T-SRAM bit-cells ever reported in literature with 32 nm design rules (1).

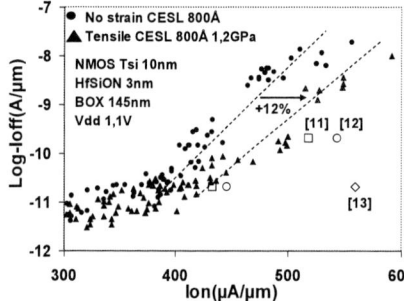
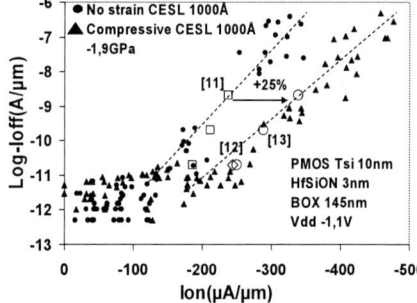

Figure 9. Impact of a high tensile CESL on the I_{on} vs I_{off} trade-off for the nMOS channel.

Figure 10. Impact of a high compressive CESL on the I_{on} vs I_{off} trade-off for the pMOS channel.

Mobility extraction

To better understand the impact of tensile and compressive CESL on the mobility behaviour, front and back channel mobilities were extracted in these advanced FDSOI MOSFETs by the $I_d/(g_m)^{1/2}$ method (14). Measurements were performed on NMOS and PMOS transistors, with different gate lengths (from 10 µm down to 40 nm). For NMOS, we observe a mobility degradation as the gate length is reduced. This effect has been attributed to neutral defects induced by source/drain implantation (15). However a clear mobility improvement is observed for short channel length with tensile CESL compared to compressive CESL (Figure 11a). In addition, a lower front-channel electron mobility compared to back-channel mobility is observed which suggests an additional scattering mechanism due to high-k dielectric (Figure 11b).

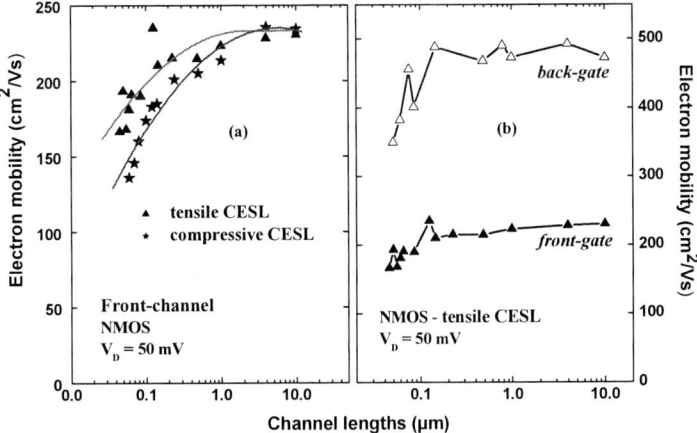

Figure 11. Low-field electron mobility variation with channel length (a) for tensile/compressive strain and (b) for front/back channels (15).

For pMOS, the use of compressive CESL offers a clear mobility enhancement compared to tensile CESL (Figure 12a). It reveals that the "pocket stress effect" (as described previously) localized in the channel near the gate edges becomes very effective for short gate lengths. In addition, the impact of compressive CESL is also visible in back channels which exhibit similar behavior (Figure 12b). This can be explained by the total transmission of the stress across ultra-thin silicon films (15).

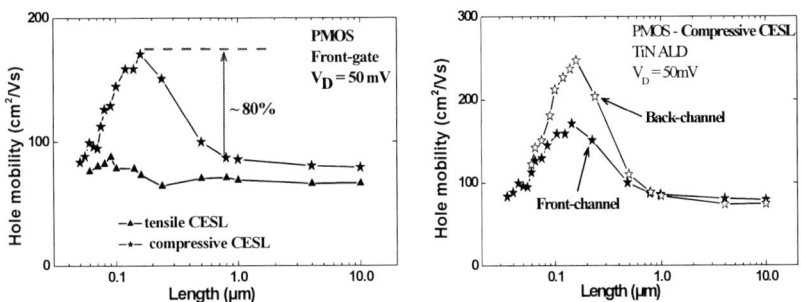

Figure 12. (a) Hole mobility versus channel length for tensile and compressive strain. (b) Front/back hole mobility versus channel length for compressive CESL.

SOI rotated substrates

Changing the substrate orientation is an alternative way to modify the carrier mobility. It is also important to decorrelate the substrate orientation and the channel direction because they influence together the carrier transport.

Figure 13 illustrates a (100) oriented substrate (non rotated wafer) where the MOSFET channel is in the <110> direction.

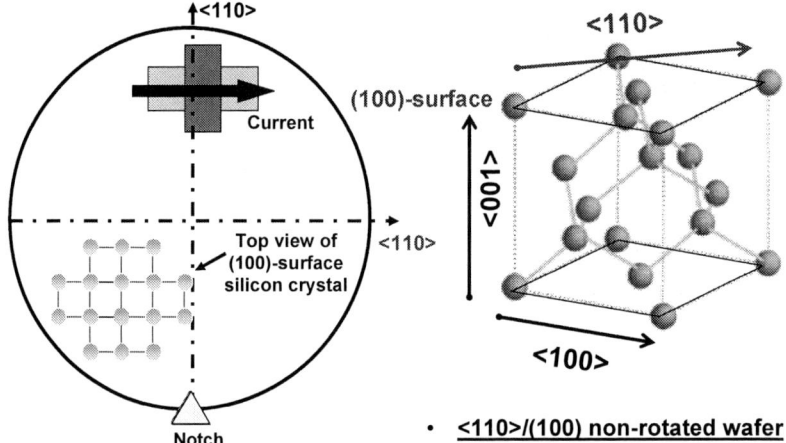

Figure 13. (100) non rotated wafer with the channel in the <110> direction.

Figure 14 shows the different case of a (100) 45° rotated substrate with the MOSFET channel in the <100> direction.

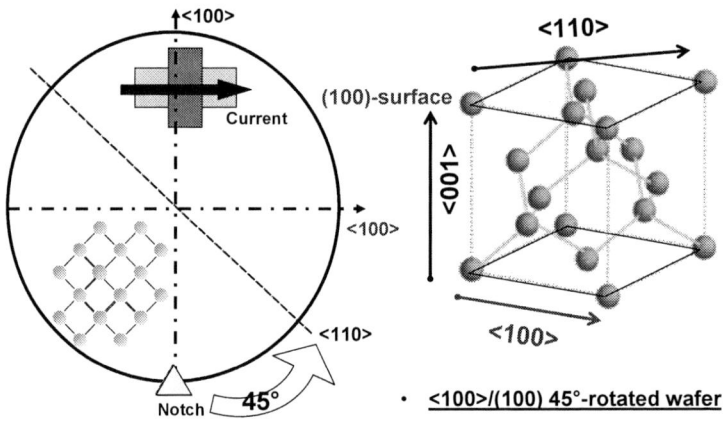

Figure 14. (100) 45° rotated wafer with the channel in the <100> direction.

The hole mass in the <100> direction is lower than in the <110> direction and enables to improve the PMOS device performances (16). An Ion gain of ~20% has been demonstrated on PMOS transistors oriented in the <100> direction for a 45nm gate length (17). It is very interesting to use this type of substrate because hole mobility is improved in the <100> direction and the electron mobility is not impacted by modifying the channel orientation from <110> to <100> direction.

We have integrated NMOS and PMOS FDSOI featuring high-k/metal gate stack on rotated SOI substrate. An Ion improvement of 11% is obtained on PMOS in the <100> channel direction compared to standard <110> (Figure 15a), without degradation of NMOS performance (Figure 15b) (18).

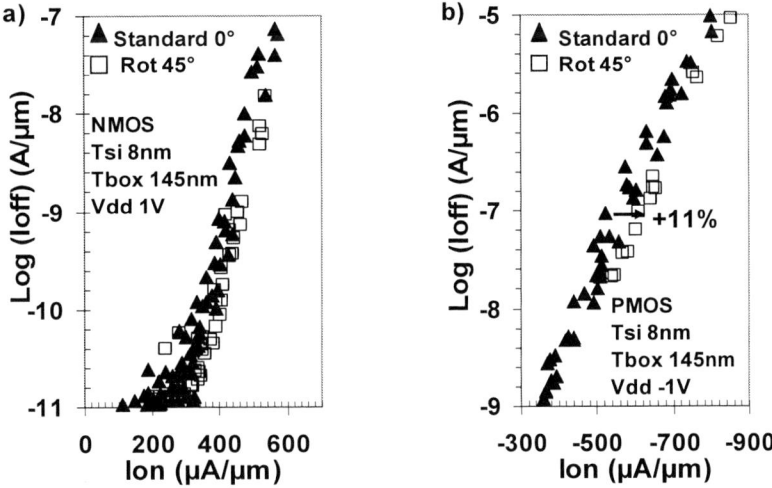

Figure 15. Impact of the substrate orientation on the Ion/Ioff performance on (a) NMOS and (b) PMOS FDSOI devices.

In order to increase further the carrier transport properties, it has been proposed to combine the effects of the mechanical strain with the oriented substrates (3). However, the impact of tensile/compressive effect on this type of substrate should be reconsidered.

Conclusions

We investigated in this paper realistic solutions to improve the channel mobility which is a necessary step for increasing the device performance. One of the most used knobs is the mechanical strain introduction. If the strain effect is well known on bulk devices few studies have produced results on ultrathin FDSOI transistors. This is why we showed, by mechanical simulations, the important role of the device geometry and the transistor architecture (thin film and optimized thickness of raised source/drain) on the CESL impact into the channel. The main tendencies observed by simulation have been validated experimentally, revealing a nMOS Ion improvement by 12% by using tensile CESL and pMOS 25% gain by using compressive CESL; these results are associated with a mobility improvement. Finally the PMOS performance improvement by 11% has been demonstrated on rotated substrate without adding strain CESL. The combination of rotated substrate with strain CESL seems very promising but it requires a reconsideration of the CESL strain effects versus channel directions on NMOS and PMOS respectively.

Acknowledgements

This work was partially supported by European projects IST-Pullnano and Nanosil.

References

1. C. Fenouillet-Beranger, S. Denorme, B. Icard, F. Boeuf, J. Coignus, O. Faynot, L. Brevard, C. Buj, C. Soonekindt, J. Todeschini, J.C. Le-Denmat, N. Loubet, C. Gallon, P. Perreau, S. Manakli, B. Minghetti, L. Pain, V. Arnal, A. Vandooren, D. Aime, L. Tosti, S. Deleonibus, T. Skotnicki, H. Mingam et al, *IEDM Tech Dig*, 267 (2007).
2. http://www.public.itrs.net
3. S. Pidin, T. Mori, R. Nakamura, T. Saiki, R. Tanabe, S. Satoh, M. Kase, K. Hashimoto, T. Sugii, *Proc of VLSI Tech Dig,* 54 (2004).
4. K. Rim, K. Chan, L. Shi, D. Boyd, J. Ott, N. Klymko, F. Cardone, L. Tai, S. Koester, M. Cobb, D. Canaperi, B. To, E. Duch, I. Babich, R. Carruthers, P. Saunders, G. Walzer, Y. Zhang, M. Steen, M. Ieong, *IEDM Tech Dig*, 49 (2003).
5. E. Augendre, G. Eneman, A. De Keersgieter, V. Simons, I. De Wolf, J. Ramos, S. Brus, B. Pawlak, S. Severi, F. Leys, E. Sleeckx, S. Locorotondo, M. Ercken, J-F de Marneffe, L. Fei, M. Seacrist, B. Kellerman, M. Goodwin, K. De Meyer, M. Jurczak, S. Biesemans, *Proc of ESSDERC*, 301 (2005).
6. F. Andrieu, C. Dupré, F. Rochette, O. Faynot, L. Tosti, C. Buj, E. Rouchouze, M. Cassé, B. Ghyselen, I. Cayrefourcq, L. Brevard, F. Allain, J.C. Barbé, J. Cluzel, A. Vandooren, S. Denorme, T. Ernst, C. Fenouillet-Beranger, C. Jahan, D. Lafond, H. Dansas, B. Prévitali, J.P. Colonna, H. Grampeix, P. Gaud, C. Mazuré, S. Deleonibus, *Proc of VLSI Tech Dig,* 168 (2006).
7. F. Andrieu, O. Faynot, F. Rochette, J-C. Barbé, C. Buj, Y. Bogumilowicz, F. Allain, V. Delaye, D. Lafond, F. Aussenac, S. Feruglio, J. Eymery, T. Akatsu, P. Maury, L. Brevard, L. Tosti, H. Dansas, E. Rouchouze, J-M. Hartmann, L. Vandroux, M. Cassé, F. Bœuf, C. Fenouillet-Beranger, F. Brunier, I. Cayrefourcq, C. Mazuré, G. Ghibaudo, S. Deleonibus, *Proc of VLSI Tech Dig,* 50 (2007).
8. http://www.ansys.com
9. C. Gallon, C. Fenouillet-Beranger, S. Denorme, F. Bœuf, V. Fiori, N. Loubet, A. Vandooren, T. Kormann, M. Broeckaart, P. Gouraud, F. Leverd, G. Imbert, C. Chaton, C. Laviron, L. Gabette, F. Vigilant, P. Garnier, H. Bernard, A. Tarnowka, R. Pantel, F. Pionnier, S. Jullian, S. Cristoloveanu, T. Skotnicki, *Jap Journ of Appl Phys*, vol **45**, n°4B, 3058 (2006).
10. C. Gallon, C. Fenouillet-Beranger, A. Vandooren, F. Bœuf, S. Monfray, F. Payet, S. Orain, V. Fiori, F. Salvetti, N. Loubet, C. Charbuillet, A. Toffoli, F. Allain, D. Delille, F. Judong, C. Perrot, M. Hopstaken, P. Scheiblin, P. Rivallin, L. Brevard, O. Faynot, S. Cristoloveanu, T. Skotnicki, *IEEE Int SOI Conference*, 17 (2006).
11. Y. Yasuda, N. Kimizuka, Y. Akiyama, Y. Yamagata, Y. Goto, K. Imai, *IEDM Tech Dig,* 73 (2005).
12. T. Hoffmann, A. Veloso, A. Lauwers, H. Yu, M. Van Dal, H. Tigelaar, T. Chiarella, C. Kerner, R. Mitsuhashi, I. Satoru, M. Niwa, A. Rothschild, B. Froment, J. Ramos, A. Nackaerts, S. Brus, C. Vranken, P.P. Asbil, M. Jurczak, J. A. Kittl, S. Biesemans, *Proc of VLSI Tech Dig,* 194 (2006).
13. H. Nakamura, Y. Nakahara, N. Kimizuka, T. Abe, I. Yamamoto, T. Fukase et al, *Proc of VLSI Tech Dig*, 198 (2006).
14. C. Mourrain, B. Cretu, G. Ghibaudo, P. Cottin, *Proc of ICMTS*, p. 181, (2000).
15. L.Pham-Nguyen, C. Fenouillet-Beranger, G. Ghibaudo, T. Skotnicki, S. Cristoloveanu, *EUROSOI,* 43 (2009).
16. M. V. Fischetti, Z. Ren, P. M. Solomon, M. Yang, K. Rim, *Journal of Applied Physics*, vol 94, 1079 (2003)
17. T. Komoda, A. Oishi, T. Sanuki et al, *Proc of ESSDERC*, 217 (2004).
18. F. Andrieu, F. Allain, C. Buj-Dufournet, O. Faynot, F. Rochette, M. Cassé, V. Delaye, F. Aussenac, L. Tosti, P. Maury, L. Vandroux, N. Daval, I. Cayrefourcq, S. Deleonibus, *Proc of SSDM*, 888 (2007).

ECS Transactions, 19 (4) 65-70 (2009)
10.1149/1.3117393 ©The Electrochemical Society

Impact of New Approach to Improve MOSFETs Performance with Ultrathin Gate Insulator

Weitao Cheng[1], Akinobu Teramoto[1] and Tadahiro Ohmi[1,2]

[1]New Industry Creation Hatchery Center, Tohoku University
[2]WPI Research Center, Tohoku University
6-6-10 Aza-Aoba, Aramaki, Aoba-ku, Sendai, Japan

In this study, we focus on the difference behavior of the gate-channel capacitance (C_{gc}) characteristics in the inversion-mode (IM) and the accumulation-mode (AM) SOI MOSFETs. We experimentally demonstrate that the gate-channel capacitance in the AM MOS device is obviously larger than that in IM MOS device using the same gate insulator on the same wafer. The increase of the C_{gc} in AM MOSFETs is much more remarkable as downscaling the gate insulator due to the different inversion and accumulation layer quantization effect and poly-Si gate depletion effect. As the result, the current drivability has been obviously improved in the AM MOSFETs resulted from its enhanced gate-channel capacitance compared with conventional IM MOSFETs.

I. Introduction

The MOSFETs current drivability strongly depends on the gate-channel capacitance. After several decades of continuous device size downscaling, the gate insulator thickness is continuously reduced to suppress the short-channel effect and keep a good subthreshold slope. At the conventional dual gate CMOS process, implanted n^+- and p^+-type polysilicon are introduced as gate electrode. However, it has been reported that the polysilicon gate electrode slightly depletes even its doping concentration is high enough over 10^{20} cm^{-3}. On the other hand, it has been reported that the inversion layer capacitance degrades induced by the inversion layer quantization effect. The loss of C_{gc} becomes much more significant while the gate insulator thickness is reduced and results in the increase of degradation of the current drivability [1]. It is one of the most important limitations have to be solved, what limits the device performance.

In this paper, we demonstrate that the gate-channel capacitance in the AM SOI device structure is much larger than that in the IM SOI devices and results in the improvement of the drain current.

II. Experimental

65

p- and n-MOSFETs on SOI-Si(100) is employed for this experiment. The thickness of SOI layer is adjusted to 40 nm by repeated sacrifice radical oxidation. The same 2.5 nm gate oxide (optical thickness measured by ellipsometer) has been formed by microwave-excited high-density plasma oxidation (radical oxidation) at 400 °C after wet cleaning to realize high-quality thin gate oxide [2] for both the inversion-mode (IM) and accumulation-mode (AM) SOI MOSFETs. For AM MOSFETs, P^+ and B^+ ions (5.0×10^{15} cm^{-2}) are implanted to gate Poly-Si layer (150 nm) and As^+ and BF_2^+ (1×10^{15} cm^{-2}) ions are implanted to Source/Drain region for n- and p-MOSFET, respectively. Based on the calculation, the gate Poly-Si layer impurity concentration is over 10^{20} cm^{-3} in this experiment. Figure 1 shows the schematic of (a) AM n-MOSFETs and (b) AM p-MOSFET device structures.

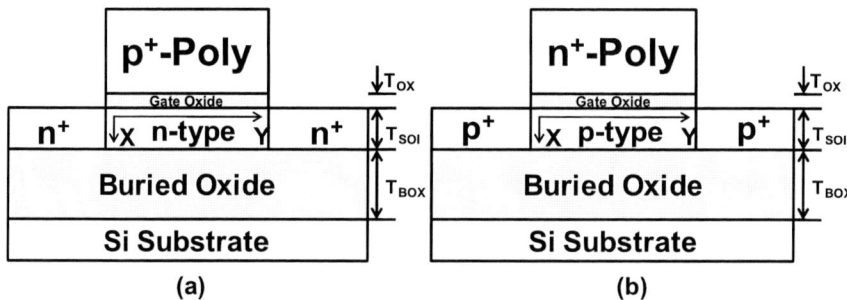

Fig. 1. Schematic Accumulation-mode FD-SOI device structures, (a) n-MOSFET and (b) p-MOSFET.

III. Results and Discussions

Gate-channel capacitance and mobility characteristics are the most important parameters for MOSFETs. It has been reported that the MOSFET device continues to scale down toward the sub-10 nm dimensions, the device performance is still strongly affected by gate-channel capacitance and mobility. [3-5]. Commonly, C_{gc} increases by scaling down the gate insulator thickness and introducing metal gate electrode to suppress the gate electrode depletion. However, as it had been reported, gate-channel capacitance of the IM MOSFETs reduces even with the metal gate electrode without depletion because of the inversion layer quantization effect [1]. Considering the gate insulator must have to be scaled down to suppress the short-channel effect, it is a very urgent problem to improve the gate-channel characteristics in order to avoid the device performance degradation. It has been reported that the quantization effect is different in inversion and accumulation layer at MOS capacitors [6-10]. At the theoretical calculation with QM model, the gate insulator thickness calculated from inversion region is about 0.3-0.4 nm thicker due to degradation resulted from the inversion carrier peaks at approximately 1 nm below the silicon and oxide interface [1]. The similar effect appears at the polysilicon gate electrode else. These effects obviously reduce the gate capacitance in MOS device, especially with ultra-thin oxides. However, almost no experimental data has been reported for the gate-channel capacitance characteristics in the AM SOI MOSFETs. Hence, in this study, we focus on the gate-capacitance characteristics in both the IM and AM SOI MOSFETs and the relation with current drivability.

At first, we show the energy-band diagrams for (a) IM and (b) AM p-MOSFETs at fig. 2 when the negative gate bias is applied to gate electrode. At conventional CMOS fabrication process, the impurity concentration of p^+-type poly-Si gate electrode in the p-MOSFET is slightly lower compared with n-MOSFET with n^+-type poly-Si gate electrode. Hence, the poly-Si gate depletion in p-MOSFET is considered severer than that in n-MOSFET. At IM p-MOSFET, the gate electrode changes from a slight depletion state to a inverted state and SOI layer is in strong inverted state when increase the gate bias . On the other hand, both the gate electrode and SOI layer are in accumulated state in AM p-MOSFET. As the result, accumulation-mode device structure can essentially solve the problem of gate depletion and are expected to reduce the quantization effect as gate insulator scaled down.

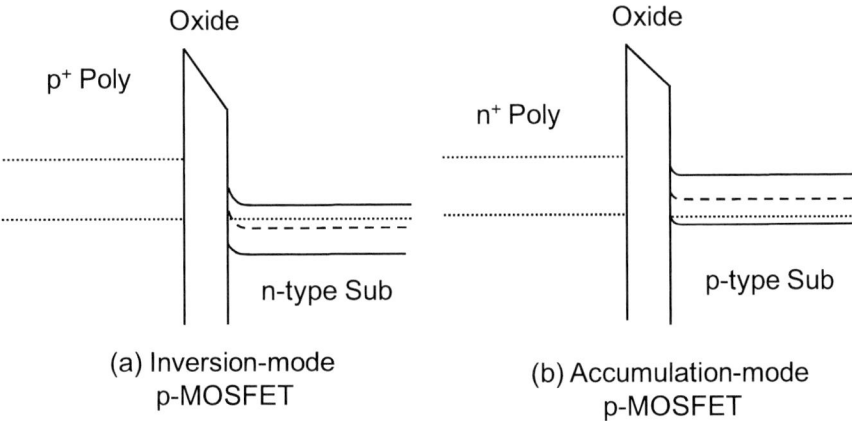

Fig. 2. The energy-band diagrams for (a) IM and (b) AM p-MOSFETs when the negative gate bias is applied to gate electrode. Both the gate electrode and SOI layer is in accumulated state in AM p-MOSFET.

Figure 3 shows the measured experimental data of the gate-channel capacitance characteristics in IM and AM SOI p-MOSFETs. The capacitance-voltage measurement of the gate-channel capacitance was carried out with the measurement frequency at 1 MHz and 20 mV voltage step at 297 K. Both the IM and AM p-MOSFETs are fabricated on the same wafer with the same gate insulator. The impurity doping concentration of the poly-Si gate electrode in both the IM and AM p-MOSFETs is over 10^{20} cm^{-3} so that the poly-Si gate electrode depletion effect is very limited in this experiment. The optical thickness of the gate insulator was measured by ellipsometer (with the fixed index of refraction of 1.46). Total gate capacitance can be simply written as,

$$\frac{1}{C_{total}} = \frac{1}{C_{gate}} + \frac{1}{C_{SiO2}} + \frac{1}{C_{inv}}$$ (1)

where C_{total} is the total gate capacitance, C_{gate} is the capacitance of the gate electrode depletion layer, C_{SiO2} is the capacitance of gate insulator, and C_{inv} is the capacitance of inversion layer. In IM MOSFETs, the poly-Si gate electrode depletion and inversion layer quantization effects result in an increase of the effective oxide thickness. The gate-channel capacitance does not increase with increasing the gate bias. However, the gate oxide thickness measured in AM p-MOSFETs is as the same as its optical thickness (2.5nm) measured by ellipsometer and that of IM p-MOSFETs is 3nm, which is about 0.5nm thicker than the optical thickness. Poly-Si gate electrode turns to accumulated state while increases the gate bias. The accumulation layer quantization effects are much smaller compared with the inversion layer [7]. As the result, the gate-channel capacitance increases with increasing the gate bias. The gate-channel capacitance characteristics measured in IM and AM p-MOSFETs agree with the gate capacitance measured at inverted and accumulated states of MOS capacitor very well what had been reported [7]. As a result, the gate-channel capacitance is about 1.2 times larger at the gate bias of -3V in AM p-MOSFET compared with the IM p-MOSFET with the same thin gate oxide. The same result has been observed at the IM and AM n-MOSFETs else.

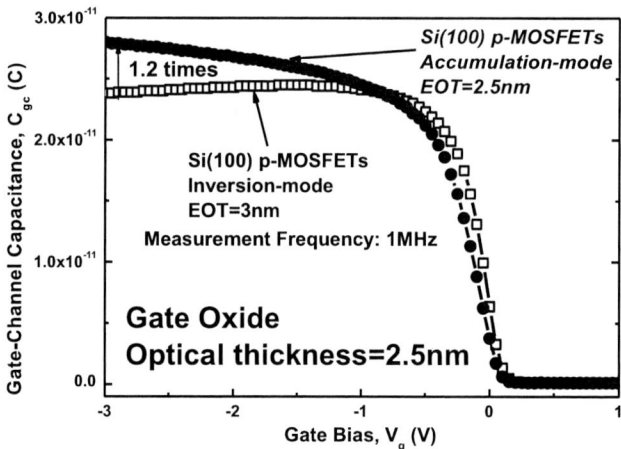

Fig. 3. The gate-channel capacitance characteristics in IM and AM p-MOSFETs. The gate oxide thickness measured in AM pMOSFETs is as the same as its optical thickness (2.5nm). The oxide in IM pMOSFET is 0.5nm thicker.

Fig. 4 (a) shows the experimental data of measured I_d-V_d characteristics of IM and AM p-MOSFETs on Si(100) surfaces at 297 K. The current drivability is improved about 1.35 times in AM p-MOSFET compared with IM p-MOSFET with the same thin gate oxide. We consider that both the enhanced mobility [11] and gate-channel capacitance result in the improvement of the drain current in AM MOSFETs. (b) shows the transconductance characteristics as a function of gate bias. It is observed that the values of transconductance in AM p-MOSFETs are about 30 % larger than those in IM p-MOSFET.

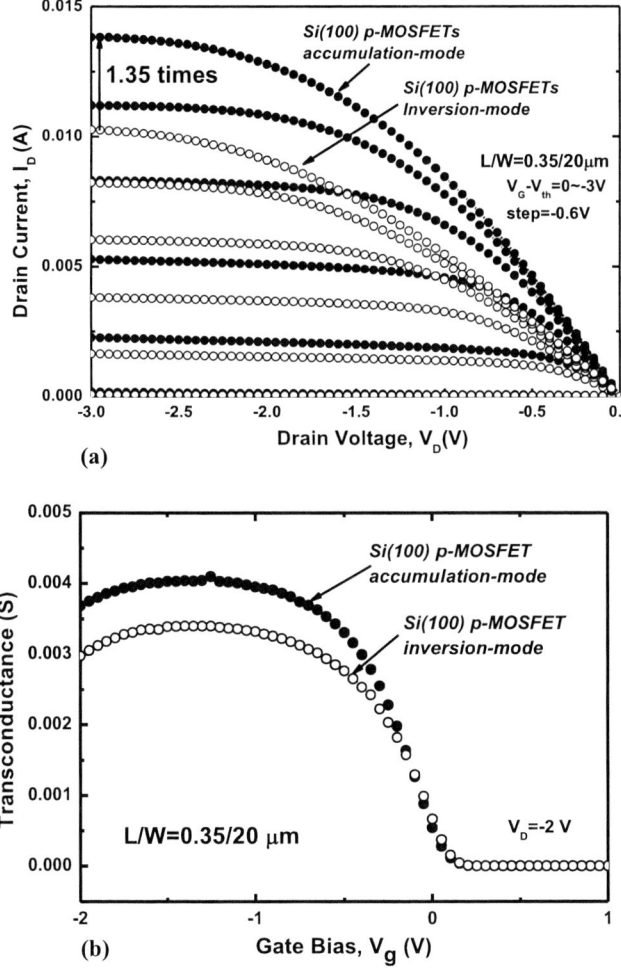

Fig. 4. (a) I_d-V_d characteristics of pMOSFETs on Si(100) surfaces. The drain current drivability is improved about 1.35 times in AM p-MOSFET compared with IM p-MOSFET with the same thin gate oxide. (b) Transconductance characteristics in AM p-MOSFET is improved compared with that in IM p-MOSFET.

IV. Conclusion

In this paper, we focus on the difference behavior of the gate-channel capacitance characteristics in the IM and AM SOI MOSFETs. We experimentally demonstrate that the gate-channel capacitance in the AM MOS device is obviously increased than that in IM MOS device resulted from the different inversion and accumulation layer quantization effect and poly-Si gate depletion effect. And the drain current drivability has been obviously improved in the AM MOSFETs resulted from its enhanced mobility and vastly increased gate-channel capacitance, especially in the MOSFETs with the ultrathin gate insulator.

Acknowledgments

This work was conducted as a part of the project under Grant-in-Aid for Specially Promoted Research (project No. 18002004) and the project under Grant-in-Aid for Young Scientists (A) (project No. 19686019), supported by Japanese Ministry of Education, Culture, Sports, Science and Technology.

REFERENCES

[1] Y. Taur, D. A. Buchanan, W. Chen, D. J. Frank, K. E. Ismail, S. Lo, G. A. Halasz, R. G. Viswanathan, H. C. Wann, S. J. Wind, and H. Wong: *Proceeding of the IEEE.*, pp. 486 - 504, (1997).

[2] T. Ohmi, M. Hirayama, and A. Teramoto: *J. Physics D: Applied Physics*, R1-R17, (2006).

[3] J. Guo, S. Datta, and M. Lundstrom: *IEEE Trans. Electron Devices*, Vol. 51, pp. 172 - 177, (2004).

[4] M. Lundstrom: *IEEE Electron Devices Letters*, Vol. 18, pp. 361 - 363, (1997).

[5] S. Tanaka, and M. Lundstrom: *IEEE Trans. Electron Devices*, Vol. 42, pp. 1806 - 1815, (1995).

[6] T. Ando: *Surface Science*, Vol. 58, pp. 128 - 134, (1976).

[7] F. Pregaldina, C. Lallement, and D. Mathiot: *Solid-State Electronics.*, Vol. 48. pp. 781 – 787, (2004).

[8] K. Ahmed, E. Ibok, G. Bains, D. Chi, B. Ogle, J. J. Wortman, and J. R. Hauser: *IEEE Trans. Electron Devices*, Vol. 47, pp. 1349 - 1354, (2000).

[9] N. D. Arora, R. Rios, and C. Huang: *IEEE Trans. Electron Devices*, Vol. 42, pp. 935 - 943, (1995).

[10] F. Li, S. Mudanai, L. F. Register and S. K. Banerjee: *IEEE Trans. Electron Devices.*, Vol. 52. pp. 1148 – 1158, (2004).

[11] W. Cheng, A. Teramoto, M. Hirayama, S. Sugawa and T. Ohmi: *Jpn. J. Appl. Phys.* Vol. 45, pp. 3110-3116, (2006).

ECS Transactions, 19 (4) 71-76 (2009)
10.1149/1.3117394 ©The Electrochemical Society

Process and Characterization of Hybrid-Oriented Complementary Single-Crystal Si Thin-Film Transistors on A Plastic Substrate

H. Pang[a], G. K. Celler[b] and Z. Ma[a,*]

[a] Department of Electrical and Computer Engineering, 1415 Engineering Dr., Madison, WI 53705, USA
[b] Soitec USA, 2 Centennial Dr., Peabody, MA 01960, USA
* mazq@engr.wisc.edu

We report the first flexible single-crystal Si complementary thin-film transistors (TFTs) based on hybrid-orientation technology (HOT) fabricated on plastic substrate. Silicon nanomembranes (SiNMs) from Si (001) and (110) source SOI wafers were integrated onto one plastic substrate. The complementary inverters consist of n-TFTs with a channel direction along Si (110)/<110> aligned side-by-side with p-TFTs along (001)/<110>, The gate lengths of both devices are 5 μm and gate widths are 15 μm and 30 μm for n- and p-TFTs respectively. Typical field-effect transistor performances were achieved from individual n- and p-TFTs. This work shows the potential hybrid inverters composed of n-TFTs along Si (001)/<110> and p-TFTs along (110)/<110>, which is believed to have an enhanced switching speed.

Introduction

Flexible, large-area electronics, also defined as macroelectronics, is beginning to show tremendous promise for a variety of commercial and military applications [1]-[3]. Owing to the recent transfer technique, single-crystal Si nanomembranes (SiNMs), which has the highest intrinsic mobility, can be peeled off from silicon-on-insulator (SOI) [4] or oriented bulk Si [5] wafers and transferred onto any new hosts. Hence, high performance devices and circuits can be realized on flexible substrates. So far, the fastest thin-film-transistors (TFTs) made on plastic have reached RF level cut-off/maximum oscillation frequency (f_T/f_{max}) values of 2.0/7.8 GHz [6]. In the circuit level, stretchable and foldable integrated circuits have been demonstrated, including logic gates, ring oscillators and differential amplifiers [7]. Nevertheless, the switching speed of the complementary inverter is limited by the low hole mobility of silicon (001) which is the most commonly used orientation for flexible circuits. Therefore, hybrid-orientation technology (HOT) [8] has been employed to flexible applications, and complementary TFTs using Si (110) nanomembranes have been fabricated on plastic, showing high gains and large noise margins at low supply voltages [9]. In order to achieve the highest speed of the complementary logic circuit, it is necessary to transfer SiNMs with different orientations together, where the highest mobility of both electrons and holes can be integrated in one inverter.

In this work, we introduce the fabrication process with two-time transfer of flexible hybrid-orientation complementary inverters employing both SiNMs (001) and (110) on one plastic substrate. Basic DC performances of individual n- and p-TFTs were measured. The process described in this work makes it possible to further enhance the logic speed by integrating SiNMs with different orientations.

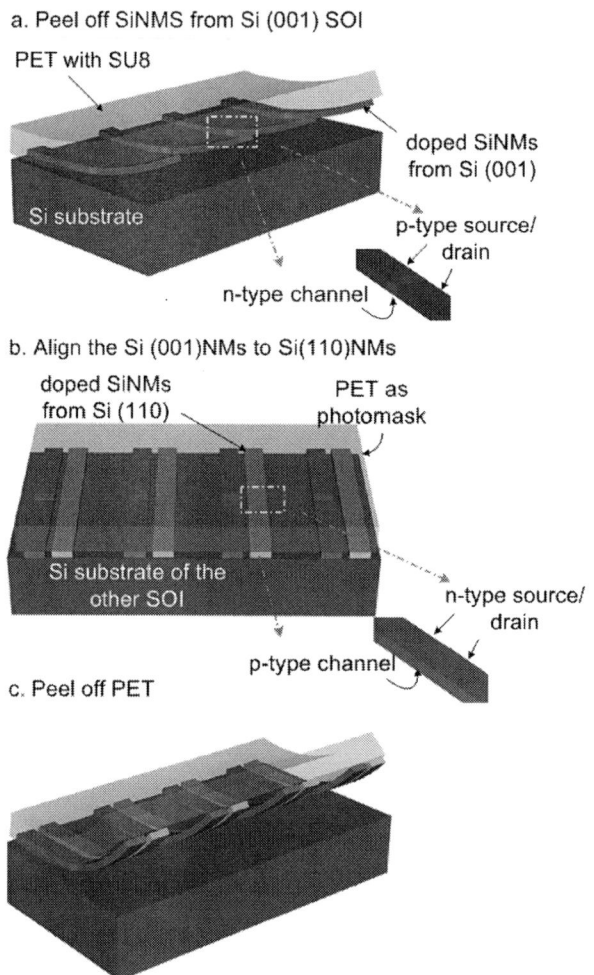

Figure. 1 Illustration of the transfer process: a, transfer the p-type doped SiNMs from SOI (001) wafer to PET substrate with SU8 as glue layer; b, using the PET as a photo mask, align SiNMs (001) to (110) on the n-type doped SOI (110) wafer; c, peel off the PET from the substrate.

Device Fabrication

The fabrication of the TFTs in this study starts from two SOI wafers with lightly p-type doped 270 nm Si (001) and 190 nm Si (110) template layers on top of 200 nm buried oxide (BOX) layers. According to the previous doping study [9], the source/drain regions were ion implanted with phosphorus with a dosage of $1 \times 10^{16}/cm^2$ at 12 KeV and boron with a dosage of $5 \times 10^{15}/cm^2$ at 5 KeV for n- and p-TFTs respectively. Low-energy doping condition is chosen to suppress the ion damage, leaving more single-crystal material for a better recrystallization process, hence lower contact resistances and sheet resistances [10]. In order to form n-well for p-TFTs, the p-type template layers were further lightly doped with $3 \times 10^{11}/cm^2$ phosphorus at 30KeV at designated regions to convert the channel into n-type. The samples were furnace annealed at 950°C for 45 min after the n-type implantation, and another 30 min after p-doping. Since the backside of the SiNMs will be exposed after the transfer, a shorter annealing time for p-type doping is preferred to prevent boron dopants from diffusing into the BOX, thus to ensure a high doping concentration at the Si/SiO_2 interface [11]. Both samples were then patterned into strips, and the BOX layers were selectively removed by HF.

The following transfer method was the most critical step, as illustrated in Figure. 1. First, released SiNMs on one of the samples (Si (001) in Figure. 1a) was transferred onto PET substrate with soft SU8 on top as a glue layer. Next, the transparent plastic host with transferred Si strips was served as a photo mask, and the SiNMs of the other sample (Si (110) in Figure. 1b) were aligned to it with two channel regions side-by-side. Then, the second sample was attached to the plastic "mask" with a firm bonding, hence another arrays of SiNMs were transferred onto the same PET substrate. The plastic was then peeled off from the Si handling substrate again, with two types of SiNMs aligned next to each other, forming pairs of hybrid-oriented inverters (see Figure 1c). After the two-time transfer, the SU8 layer was cross-linked under UV exposure so that the SiNMs can stay on the plastic unsolvable. The gate stacks were e-beam evaporated in room temperature followed by standard lift-off procedure, consisting of 120 nm SiO, 30 nm Ti and 120 nm Au. Finally, source/drain metal pads were deposited with Ti/Au of 30 nm/150 nm. Figure. 2 shows the optical microscope image of the finished complementary inverter. The top pink strip is from p-doped Si (001) while the bottom green one from n-doped Si (110). Both types of devices have gate lengths of 5 μm, while the gate widths were designed non-symmetric with 15 μm for n-TFT and 30 μm for p-TFT to balance the switching speed.

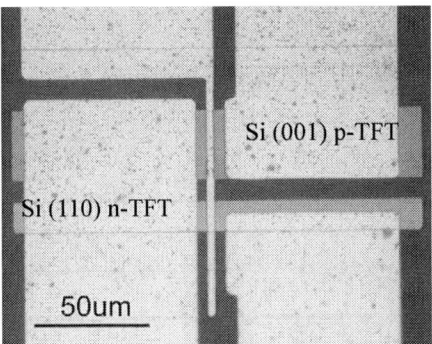

Figure. 2 Optical microscopic image of a finished hybrid complementary inverter on plastic substrate. The dark regions are probing stage seen through the clear PET substrate. The top pink strip is p-TFT from SOI (001) source wafer, and the bottom green one is n-TFT from SOI (110).

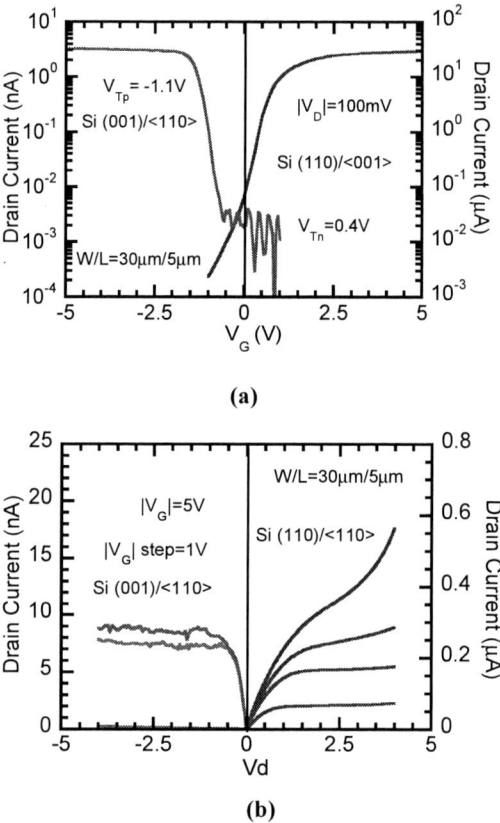

(a)

(b)

Figure. 3 (a) Transfer and (b) I-V characteristics of individual flexible n- and p-TFTs along Si (110)/<110> and Si (001)/<110> respectively. The gate lengths and widths are 5 and 30 μm. The threshold voltage of n-TFT is 0.4 V and -0.776 V for p-TFT.

Results and Discussions

The DC characteristics of the individual TFTs and hybrid complementary inverters were measured using an Agilent 4155B semiconductor parameter analyzer. Figures 3 show the transfer characteristics and current-voltage (I-V) curves of individual n- and p-TFTs with gate width to gate length W/L of 30 μm/5 μm. The channel of n-TFT is along Si (110)/<110> direction, while the p-TFT is along (001)/<110>. The threshold voltages V_T of n- and p-TFTs are 0.4 V and -0.776V, respectively. The maximum transconductance g_m reach 16.26 μS for n-TFT and 0.875 nS for p-TFT, at a drain voltage $|V_D|$= 100 mV. Note that, the p-TFTs have less current drive force than that of n-TFTs, which is believed to be due to the un-optimized boron doping and poor interface between the semiconductor and the metal.

Conclusion

We demonstrate the first hybrid-oriented complementary thin-film transistors and inverters made on flexible plastic substrate employing both single-crystal Si (001) and (110) together released from two SOI source wafers. The fabrication process requires tow-time transfer to align the SiNMs from tow different source SOI wafers side-by-side. The hybrid inverter is formed of n-TFT with channel direction along Si (110)/<110>, and p-TFT along Si (001)/<110>. The gate lengths of both devices are 5 μm, and the gate widths are 15 μm for n- and 30 μm for p-TFTs. DC performances of individual n- and p-TFTs were investigated, and typical field-effect transistor characteristics were achieved. It is very promising to further improve the switching speed of the inverters by combining n-Si (001)/<110> and p-Si (110)/<110> with the HOT and transfer approach described in this work.

Acknowledgments

This work was partially supported by NSF under grant DMR0520527 and AFOSR. The Program Officer at AFOSR is Dr. Gernot Pomrenke.

References

1. R. H. Reuss, B. R. Chalamala, A. Moussessian, M. G. Kane, A. Kumar, D. C. Zhang, J. A. Rogers, M. Hatalis, D. Temple, G. Moddel, B. J. Eliasson, M. J. Estes, J. Kunze, E. S. Handy, E. S. Harmon, D. B. Salzman, J. M. Woodall, M. A. Alam, J. Y. Murthy, S. C. Jacobsen, M. Olivier, D. Markus, P. M. Campbell, and E. Snow, "Macroelectronics: perspectives on technology and applications," *IEEE Proc.*, Vol. 93, No. 7, pp. 1239-1256, 2005.
2. K. Bock, "Polymer electronics systems: polytronics," *IEEE Proc.*, Vol. 93, No. 8, pp. 1400-1405, 2005.
3. T. W. Kelley, and P. F. Baude, C. Gerlach, D. E. Ender, D. Muyres, M. A. Haase, D. E. Vogel, and S. D. Theiss, "Recent progress in organic electronics: materials, devices and processes," *Chem. Mater.*, 16, pp.4413-4422, 2004.
4. E. Menard, R. G. Nuzzo, and J. A. Rogers, "Bendable single crystal silicon thin film transistors formed by printing on plastic substrates," *Appl. Phys. Lett.*, **86**, 093507, 2005.
5. J. Yoon, A. J. Baca, S.-I. Park, P. Elvikis, J. B. Geddes, L. Li, R. H. Kim, J. Xiao, S. Wang, T.-H. Kim, M. J. Motala, B. Y. Ahn, E. B. Duoss, J. A. Lewis, R. G. Nuzzo, P. M. Ferreira, Y. Huang, A. Rockett, and J. A. Rogers, "Ultrathin silicon solar microcells for semitransparent, mechanically flexible and microconcentrator module designs," *Nat Mater.*, vol. 7, pp. 907-915, 2008.
6. H. C. Yuan, G. K. Celler, and Z. Q. Ma, "7.8-GHz flexible thin-film transistors on a low-temperature plastic substrate," *Journal of Applied Physics*, **102**, 034501, 2007.
7. D. H. Kim, J. H. Ahn, W. M. Choi, H. S. Kim, T. H. Kim, J. Z. Song, Y. G. Y. Huang, Z. J. Liu, C. Lu, and J. A. Rogers, "Stretchable and foldable silicon integrated circuits," *Science*, vol. 320, pp. 507-511, 2008.
8. M. Yang, V. W. C. Chan, K. K. Chan, L. Shi, D. M. Fried, J. H. Stathis, A. I. Chou, E. Gusev, J. A. Ott, L. E. Burns, M. V. Fischetti, M. Ieong, "Hybrid-

orientation technology (HOT): opportunities and challenges," *IEEE Trans. Electron Device*, **53**, 965, 2006.

9. H.-C. Yuan, Z. Ma, C. S. Ritz, D. E. Savage, M G. Lagally, and G. K. Celler, "Complementary single-crystal silicon TFTs on plastic," *ECS Trans.*, 6(4), 139-144, 2007.

10. R. M. de Oliviera, M. Dalponte and H. Boudinov, "Electrical activation of arsenic implanted in silicon on insulator (SOI)," *J. Phys. D: Appl. Phys.*, 40, pp. 5227-5231, 2007.

11. P.A. Stolk, H.-J. Gossmann, D. J. Eaglesham, D. C. Jacobson, C. S. Rafferty, G. H. Gilmer, M. Jaraiz and J. M. Poate, "Phsical mechanisms of transient enhanced dopant diffusion in ion-implanted silicon," *J. Appl. Phys.*, **81**, 034501, 1997.

CHAPTER 4

MATERIALS AND CHARACTERIZATION II

SOI And Other Semiconductor-On-Insulator Substrates Characterization

A. Abbadie, Y.M. Le Vaillant, C. Figuet, E. Guiot

SOITEC R§D, Parc Technologique des Fontaines, 39126 Bernin, FRANCE

We focus in this paper on SOI and new engineered on-insulator substrates characterization. After a short review of on-line characterization techniques, we will detail off-line monitoring techniques such as preferential etching and Raman spectroscopy. We will especially highlight their complementarities on strained layers and the need for a proper evaluation of structural quality of on-insulator substrates.

Introduction

In order to improve device performances, one need to innovate more than just down-scaling. SOI wafers are amongst the engineered substrates that could be candidates for the 22nm CMOS technology node and beyond. The quality of SOI wafers did not stop being improved these last years, approaching the one of bulk Si products. Since conventional SOI process, numerous engineered substrates such as SiGeOI (1), GeOI (2), strained-SOI (3) etc. have been developed based on the Smart Cut^{TM} technology. Thin SOI wafers are used as templates for the formation of SiGe/Si dual channels (4) and of SiGeOI wafers (3) thanks to the Ge condensation technique. Efficient characterization techniques are needed, both on-line techniques for wafer processing and off-line ones for the monitoring of crystalline quality, electrical properties, lattice disorder etc.

After a short review of in-line monitoring techniques on SOI, we will focus on off-line techniques, especially preferential defect etching and Raman spectroscopy. We will illustrate for the first time their complementarities in evaluating the crystalline quality of strained dual channel substrates after relaxation.

In-line monitoring techniques

Standard silicon metrology enables in-line monitoring of geometrical and structural parameters on SOI such as nanotopology, flatness, doping, thickness, metallic contamination, thickness determination etc. There are also techniques that require adjustment and development on SOI. It has for example been demonstrated that, for laser scattering inspection systems on SOI wafers, reflectivity is an additional contributor to wafer noise (5-6), impacting also sizing accuracy on the wafer. Such reflectivity changes could lead to severe threshold limitations during inspection and also to variations within the wafer and wafer to wafer. The development of the SP1^{DLS} optical configuration enables a minimization of wafer background levels. More recently the development of SP2 inspection tools improves significantly the light interference issue between the buried oxide layer (Box) and thin Si films below 20nm.

Metallic contamination measurements have been much improved recently with the introduction of automated Inductively Coupled Plasma Mass Spectrometry and Total X-Ray Fluorescence tools etc. The non-contact Surface Potential Difference Imaging technique introduced in 2005 (also named Chemetriq) is very useful for analysing traces of metals, particle residues left after wet-cleaning and organics (7-8). Fast to use on Cz

Silicon (n and p-type), it has also been tested on standard SOI wafers (9). It is based on the Work Function (WF) difference between a probe material and the silicon surface, what can also be expressed as a Surface Potential Difference. Naturally, given the technique sensitivity to film surfaces and interfaces, developments have to be planned to adapt it on thin SOI wafers.

Finally, lifetime measurements directly reflect the SOI quality film thanks to the lifetime sensitivity on contaminants, dopants, cleanliness etc. Already tested on SIMOX and bonded-etched back SOI wafers (BESOI), it is a promising technique on SOI devices (10-12). However, due to both interfaces, sub-surface splitting and also thin Si films, lifetime predictions might be more difficult on hydrogen-induced splitting (or Smart Cut^{TM}) SOI wafers.

Off-line monitoring techniques

Destructive techniques are currently widely used in SOI wafer manufacturing in order to monitor the final SOI parameters. The pseudo-MOSFET technique can be used to assess the electrical properties of the top silicon layer and its interface with the buried oxide (13-14). It is a useful technique for a quick wafer inspection before device processing, as it provides the electrons and holes mobility, the interface state densities etc. Four-point probing of the film resistivity, spreading resistance profiling of the resistivity as a function of depth etc. are complementary techniques for a proper electrical evaluation of un-processed SOI wafers.

As far as crystalline quality analysis of silicon (and Ge-based semiconductor on-insulator substrates) is concerned, preferential etching techniques (15), that we will detail below, are mandatory for a fast and efficient monitoring of structural quality. Other less destructive techniques, such as Raman or X-Ray Diffraction, are essential to determine lattice disorder, strain profile or even Ge composition in Ge-containing substrates (16-17).

Preferential etching

Wet-chemical etching is widely used for crystal defect delineation in engineered silicon substrates. Different standard solutions containing hexavalent chromium in the form of $K_2Cr_2O_7$ or chromium-oxide (CrO_3) are available today for selective etching of different types of materials (18-19). The Wright-etch solution remains a standard for delineation of oxygen precipitates and bulk Stacking Faults. Physical techniques in development today on Si and SOI, such as Laser Scattering Tomography, are currently benchmarked with Wright-etch etching (20). Schimmel etch is also widely known to decorate dislocations or pile-ups on SiGe relaxed substrates (15). The most universal of these Cr-based solutions is diluted Secco etching solution (21) The recipe has been improved these last few years and a two-step etching has been developed, especially on SOI and sSOI, to overcome some limitations such as thin films etching, introduction of strained layers etc. (22-23).

These solutions are sensitive enough to differentiate between the potential energy at defect sites (caused by the so-called "strain field") and the potential energy in the perfect silicon lattice. However, their main drawback is the presence of toxic hexavalent chromium (Cr VI). New Chromium-free solutions have recently been developed to (i) get rid of toxic chromium and (ii) potentially improve the delineation of defects having a different potential energy than standard Secco etch pits, such as misfit dislocations and Stacking Faults (24-26). These solutions are based on the use of a halogen compound in the solution in combination with the typical oxidizing agent, nitric acid, usually present in such solutions. This halogen compound (bromine, iodine) is really necessary to initiate the oxidation reaction on SOI. Its presence and its concentration are most important. Indeed, without this additional agent, no oxidation or etching of SOI is observed. The mechanism can briefly be summarized using the two electrochemical equations below: oxidation of Si using the halogen agent X and nitric acid and dissolution of silicon dioxide using HF.

$$3Si\text{-}X + 2NO_3^- + 3H^+ \leftrightarrow 3Si\text{-}O_2 + 3HX + H^+ \qquad [1]$$
$$SiO_2 + 6HF \leftrightarrow SiF_6^{2-} + 2H_2O + 2H^+ \qquad [2]$$

After the etching, a subsequent HF dip is also crucial for defect delineation. Figure 1 shows a schematic defect delineation process of a defect on SOI substrates (26).

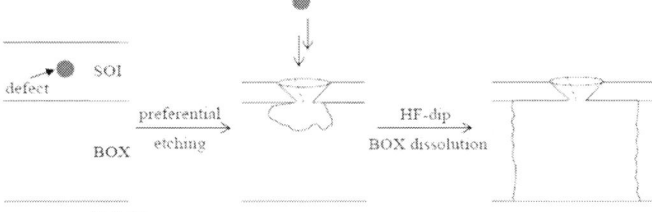

Figure 1 (26): Defect delineation mechanism of a defect on SOI substrate. Left: location of the original defect; center: selective etching producing an halo at defect site; right: subsequent HF dip dissolving the Box below the defect.

The etching behaviour depends on many parameters, the main ones being the composition of the solution, the temperature, the initial thickness of the etched material, the etching depth etc. The last two determine the selectivity of the solution (i.e. the ratio between a defect size and the minimum etched material thickness necessary to observe the defect). Figure 2 shows a typical Scanning Electron Microscopy image of a Oxidation-Induced Stacking Fault (OISF) - type defect resulting from non adapted thermal treatments and delineated using a Cr-free SOI solution (26).

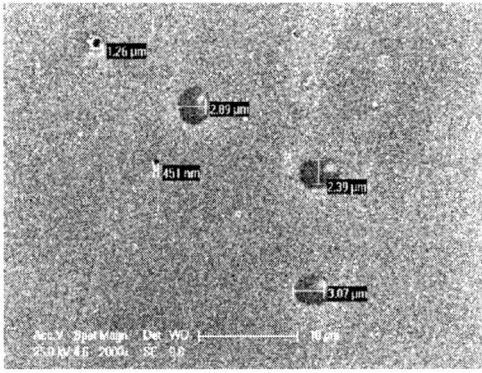

Figure 2 : Scanning Electron Microscope image of OISF - type defects resulting from non adapted thermal treatments and decorated using a Cr-free SOI solution.

Recently another etching technique has been proposed, which is based on the use of gaseous HCl in an epitaxy reactor (27). This etching technique has been now widely applied on Si, SiGe, Ge and on the corresponding on-insulator substrates such as SOI, sSOI, GeOI. Though the mechanism is completely different from chemical etches, the HCl etch pits densities and distribution have been found to be equivalent with the ones observed after (i) Schimmel etch on SiGe virtual substrates (15), (ii) Secco on strained silicon layers (23), (iii) Secco and Cr-free on pure epitaxial Ge and GeOI substrates (28), (iv) Secco and Cr-free on SOI substrates. Typical pits shapes, illustrating also the rather high selectivity of this technique, have been evidenced on the various materials used for decoration. Examples of square-based pits obtained after HCl etching on SOI and inverted pyramid pits obtained on GeOI substrates are shown figures 3 and 4.

Figure 3 : Optical microscope image (corresponding to a 2960µm^2 field size) representing OISF – type defects created on SOI wafers and decorated using HCl etch.

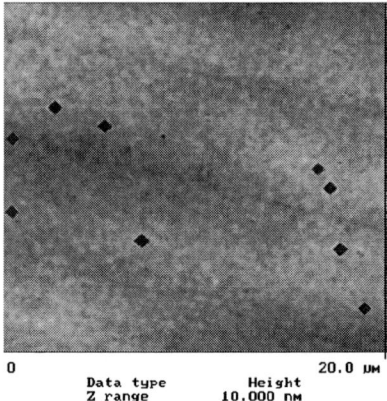

Figure 4 : Atomic Force Microscopy image (20x20µm^2) representing inverted pyramid pits obtained after a ~ 20nm HCl etch on a GeOI substrate.

Each technique is characterized by its own selectivity. It is also defined by an activation energy, which enables a comparison between different etching behaviours and determines the etching mechanism: diffusion-limited associated to low activation energies, reaction-limited linked to high activation energies (usually the case for preferential etching) and thermally activated (in the case of HCl etch). Figure 5 compares the different selectivities of Secco, Cr-free and HCl etches and their associated activation energies.

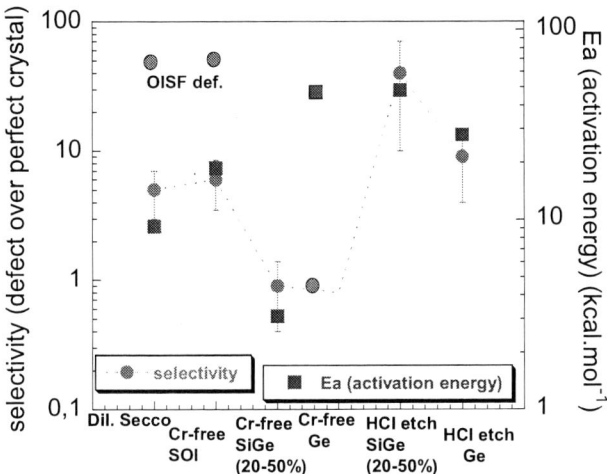

Figure 5: Selectivity of different etching techniques reported versus their activation energy (kcal.mol⁻¹).

The activation energy, calculated from the Arrhenius equation,

$$k = A \cdot e^{\frac{-E_a}{RT}} \qquad [3]$$

is defined as the energy (commonly expressed in term of surface potential) that have to be overcome for a chemical reaction to occur. In general, the activation energy for a diffusion-controlled process is less than that for a reaction-controlled one. However, in a reaction-controlled process, the strain field at defect sites affect the local reaction and thus the removal rate. The activation energy might be then an important parameter in the comparison of preferential etching characteristics of defect etching solutions.

The selectivity (or sensitivity to defect delineation) depends on the competition between defect etching and surrounding bulk etching. It can reasonably be defined as the ratio of the defect etch rate over the substrate etch rate (29).

An etching solution with higher activation energy may then be expected to show an enhanced selectivity for low strain fields. It is the case for HCl etch, which shows the highest selectivity on standard defects (>10) together with rather high activation energies (> 40kcal.mol⁻¹) in the low temperature regime (~ 400°C to 900°C) used for defect delineation. Meanwhile, Cr-free SOI and Secco show similar etching behaviours. Only Ge epitaxial substrates do not fit to the assumed connection between selectivity and activation energy (also dependant on the nature of material).

"Strained" layers : combination of Raman spectroscopy with defect etching

A preferential etching efficiency (chemical or gaseous) does not only depend on "chemical" composition and temperature. It is also influenced by metallic impurities (segregating at defect sites) and the strain field. For thin strained layers (< 200 Å), the etching reaches its physical limits notably because the thickness that can be etched is really small.

Raman spectroscopy is mandatory to evaluate strain relaxation (together with High-Resolution X-Ray Diffraction (HR-XRD)). We have recently combined Raman and defect etching in order to evaluate critical thickness on strained Si layers (sSi) grown on SiGe relaxed substrates (20% to 50%) with different sSi thicknesses up to 35nm (30). No change in surface morphology or roughness and no significant loss of strain occurred as the sSi layer thickness increased. This remarkable stability of tensile strained layers has also recently been outlined by Parsons et al (31-32). A level of strain relaxation for tensile strained silicon much lower than for compressively strained SiGe layers on Si has indeed been reported (32-34). It was explained by the "glide dominated" mechanism and the strong influence of Stacking Faults that pin the strain-relieving misfit dislocations and inhibit gliding. Such a mechanism was reported as a source of stability (35). However, if the combined use of defect etching and Raman spectroscopy is quite well optimized and established on tensile strained silicon, it is not so straightforward on compressive strained SiGe layers.

Dual channel structures (sSi grown on compressive strained SiGe layers, themselves grown on a template substrate which can be relaxed SiGe, SOI or sSOI) yield vastly improved n-MOS and p-MOS channel MOSFETs (4), (36). Reliable characterization techniques have however to be developed on such hetero-structures. Indeed, there is still a great deal of discrepancy and some "inconsistencies" in the literature concerning the Raman analysis of c-SiGe. For example, the values of phonon deformation potentials for intermediate Ge contents and of the strain shift coefficients in the whole Ge concentration range etc. are different depending on the authors (37-38). The low intensity of the peaks coming from the SiGe transition mode do not facilitate either the quantitative analysis.

<u>Before annealing</u>

Figure 6 represents a typical Raman spectrum of a dual channel sSi cap / c-SiGe stack, grown on a eXtra Strained Silicon-On-Insulator (XsSOI) substrate (39)

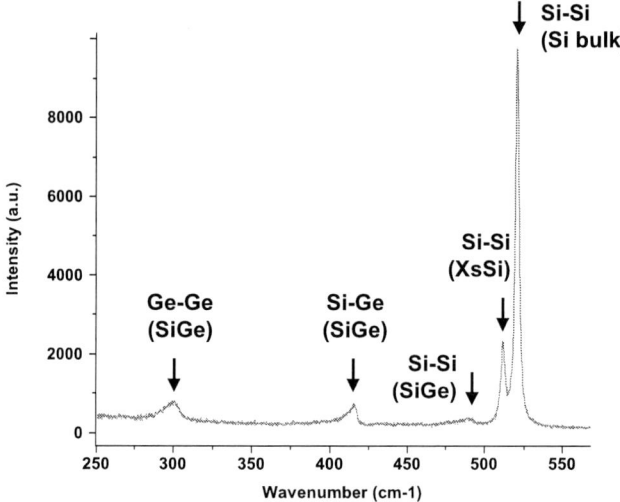

Figure 6: Typical Raman spectrum for a Dual Channel epitaxial structure sSi / $Si_{0.4}Ge_{0.6}$ / sSi / SiO_2 /bulk Si. Three peaks arise from the SiGe layer (Ge-Ge, Si-Ge and Si-Si); strained-Si layer and bulk Si layers

provide separate peaks due to the strain effect. All Raman modes are Longitudinal Optics except Si-Ge mode which is Transverse Optics.

As far as defect etching is concerned, the poor contrast between initial thin layers and very shallow trenches, resulting from misfit dislocations gliding, make defect detection and therefore their observation difficult. Figure 7 illustrates the compressive etched SiGe surface after a few ten of nm Cr-free etch. Only a few Threading Dislocations (TD) are visible, making the counting relative, meanwhile a small Stacking Fault is delineated. The TD Density (TDD) is estimated close to $1x10^4 cm^{-2}$, i.e. slightly lower than the expected TDD after a strained silicon layer etching.

Using a Secco etching and a long HF dip or the use of a $HF/H_2O_2/CH_3COOH$ solution after Secco etch (40), no better delineation was obtained under an optical miscroscope, confirming the empiric nature and the lack of reproducibility of etching on those strained layers.

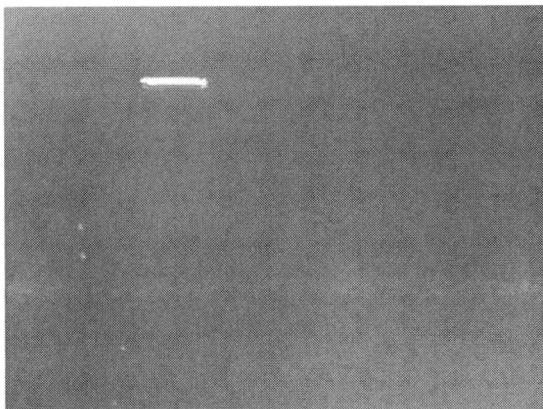

Figure 7: Optical microscope image ($23302.4\mu m^2$) after a Secco or Cr-free etch of a sSi / SiGe / XsSOI dual channel sample. White dots are Threading Dislocations, the small white line is a Stacking Fault.

After contrast adjustment and a new Cr-free etch, faint lines (though poorly delineated) can barely be distinguished, confirming the presence of Misfit Dislocations and the low density of decorated threading dislocations (figure 8).

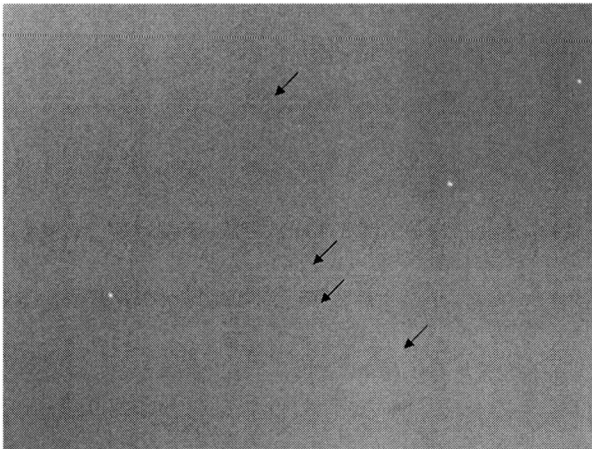

Figure 8: Optical microscope image (23302.4μm²) after a Cr-free etch of a sSi / SiGe / XsSOI dual channel sample. Faint lines labelled by small arrows are most likely Misft Dislocations.

After annealing

In order i) to check the efficiency and reliability of selective etching solutions and ii) make in relation defect formation and Raman, we have proceeded with some annealing experiments. We have annealed several pieces of the dual channel sSi cap / c-SiGe / eXtra Strained Silicon-On-Insulator (XsSOI) substrate, in a N_2 atmosphere during 1h at different temperatures >800°C, and out of the temperature range used for the substrate process (41).

Using the support of modeling, we have schematically summarized in Fig. 9 the different steps which are likely to be involved during the annealing.

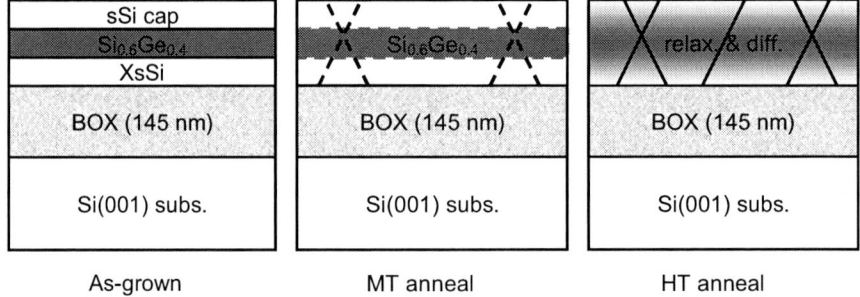

Figure 9: (left) as-grown structure; (middle) after Medium Temperature (800 & 850°C) annealing, defects are created and interdiffusion at the SiGe/sSi interfaces begins to occur; (right) after High Temperature (T > 850°C) annealing, numerous defects are present in the stack, which is interdiffused.

We have plotted in figure 10 the evolution of the root mean square (rms) surface roughness at small and higher scan sizes as a function of the annealing temperature. A slight increase of the rms roughness is observed for small scan sizes (5x5μm²) compared to the reference samples (and to a less extent for 20x20μm² scans). This change in topology most likely translates the plastic strain relaxation occurring for temperatures superior or equal to 800°C.

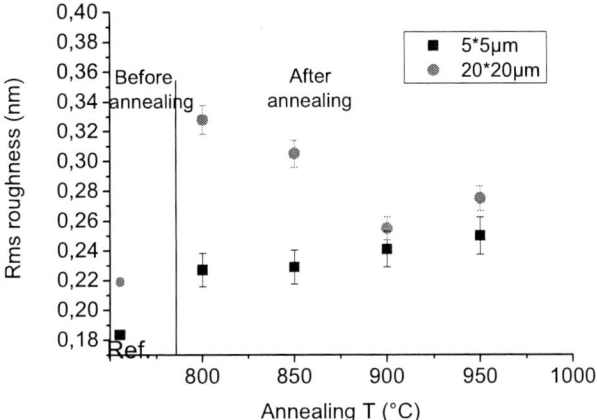

Figure 10: Evolution of surface rms roughness for small and large scan sizes (5x5µm² and 20x20µm²) as a function of the annealing temperature.

After annealing at 800°C, defects quantification becomes far easier, allowing the delineation of the threading arms of 60° dislocations (1st stage of the strain relaxation). These dislocations glide on (111) planes leaving in their wake misfit segments along the <110> directions. For higher annealing temperatures, a network of Misfit Dislocations (MD) (2nd stage of the strain relaxation) and presumably Stacking Faults (SF) appearing both as lines but with different contrasts) are also evidenced. Images after Secco etch on samples annealed at 800°C and 900°C are shown in figures 11 and 12, respectively. It is known that the cross-slip of misfit dislocations segments in the (111) plane can produce a 90° change in the direction of expanding interfacial misfit dislocations (42-43).

The cross-slip and blocking features associated with the observed truncated lines indicate they might be interfacial MD. Such a MD network is associated to high temperature annealing (≥ 900°C) with a strong surface cross-hatch. Such a contrast enhancement prevents the detection of isolated TD. That is why defect quantification becomes more and more approximate for samples annealed at high temperatures.

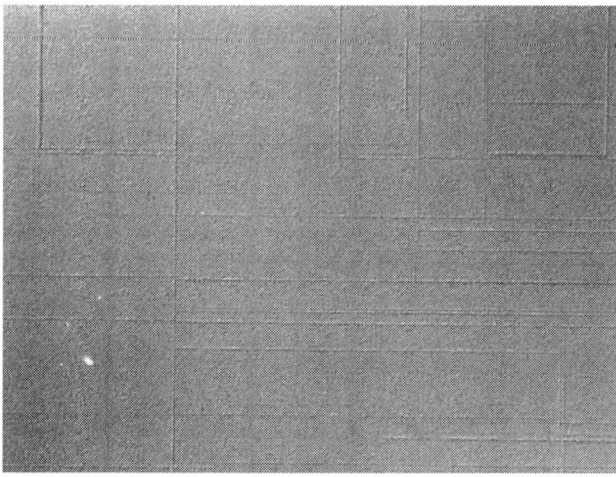

Figure 11: Optical microscope image (23302.4μm^2) after the Secco etching of a 800°C *annealed* sSi/SiGe/XsSOI dual channel sample. White dots are TD, while lines are MD (or SF, eventually).

Figure 12: Optical microscope image (23302.4μm^2) after the Secco etching of a 900°C *annealed* sSi/SiGe/XsSOI dual channel sample.

The TDD is estimated close to 1×10^5cm^{-2} whatever the annealing temperature. Meanwhile, the LLD remains at least constant, close to 1×10^3cm^{-1}, i.e values higher than the ones previously reported on single strained-silicon layers. We have also taken into account the cross-hatch reappearance, which prevents an accurate quantification of defects generated during relaxation.

We have used visible light (473 nm) Raman spectroscopy and backscattering geometry to estimate the strain relaxation of the XsSi layer as a function of the annealing temperature. It is important at that stage to remind that the experimental values reported below have to be considered with a pinch of salt, given that the Raman spectrum is coming from a tri-layer stack (i.e. sSi/c-SiGe/sSi) and from the Si bulk (not from a single layer).

Compressive strain values lower than 100% have been found for the c-SiGe layer on XsSOI (*compared to the same layer grown on unstrained SOI*). SiGe has indeed less to expand in the vertical direction to have the in-plane lattice parameter of a XsSOI substrate than that of an unstrained Si one (hence the lower compressive strain: in-between 30 and 45%). Similarly, the sSi layers in a XsSOI / c-SiGe / sSi stacks are slightly stretched in the growth plane by the compressive SiGe layer, resulting in a slightly increased tensile strain (101%) *compared to a reference XsSOI wafer.*

The strain variations measured in the annealed SiGe layers did not enable a meaningful conclusion. We have then focused on the evolution of the Raman spectra signal due to the XsSi layers (Fig. 13). The tensile strain in the XsSi layers leads to a peak at 511 cm^{-1}, i.e. at lower wave number than the one due to (relaxed) Si bulk (peak at 521 cm^{-1}). After annealing at 800°C or 850°C, the Raman shift is unchanged compared to the reference spectrum. The tensile strain in the XsSi layers is thus fairly stable with the annealing temperature up to 900°C. For T = 950°C, no Si-Si peak from the XsSi layers is visible anymore.

Beyond 850°C, strain relaxation is not the only mechanism to be considered. Ge diffusion into the XsSi layers (in which the SiGe layer is embedded) would readily explain the definite intensity reduction of the Si-Si peak (from XsSi), then its disappearance at 950°C (no more signal around 511 cm^{-1} at 950°C).

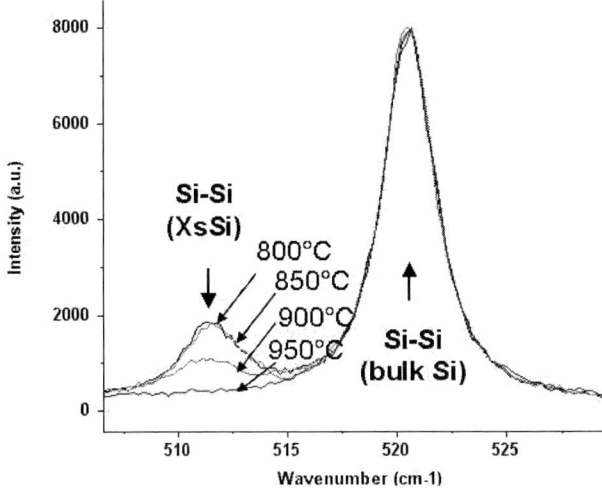

Fig. 13: Evolution of the Raman Si-Si (XsSi) transition mode, represented as an insert in fig.10, with the annealing temperature. Two Si-Si transitions modes are visible, coming from the strain Si layers (~511 cm^{-1}; 1.1% biaxial strain) and from the Si bulk (521 cm^{-1}; relaxed state).

The *measured* Ge concentration in the stack (~ 65% in the c-SiGe layer of the reference, un-annealed sample) fluctuates a lot (in-between 35 and 80%) with the annealing temperature, with no scientifically meaningful trends. The Ge concentration profile as a function of depth is however not box-like in annealed samples: the c-SiGe layer is sandwiched in-between XsSi layers, a periodic strain field is present in the XsSOI starting substrate (and that a second one occurs through the plastic relaxation of the c-SiGe layer) etc. All this complicates a lot the quantitative data extraction.

We have finally concluded that the strain field (its value and nature) might influence considerably defect selective etching (etching rate, defect decoration, etching sensitivity). There is also a difference in strain relaxation between tensile and compressive strained layers. Therefore the relation between defect formation and strain relaxation is not straightforward. Within the sensitivity of Raman technique, if a qualitative relationship is evident with crystalline quality, a quantitative relation with defect etching seems hardly possible on such dual channel layers.

Conclusions

Characterization techniques have to be adjusted when used on SOI and other semiconductor-on-insulator structures fabricated thanks to the Smart-Cut[TM] process, this because of the presence of a buried oxide layer and the interferences it creates. It is the case for light interference-based techniques or inspection systems, or also for off-line monitoring techniques, such as Pseudo-MOSFET or defect decoration methods.

Secco etching is a well known and established technique to delineate defects in thick layers. Its use on thin, strained layers and on substrates with a Box (needed in future CMOS technology nodes) is far from being straightforward, however. Their structural characterization requires adapted techniques which have to be continually improved.

We have developed etching techniques (chemical, gaseous or physical techniques) that could replace conventional Secco and Wright etches in order to decorate defects and oxygen precipitates in SOI substrates. The use of Cr-free etches solutions (based on a mix of different acids in specific ratios) and gaseous HCl etch for defect delineation has successfully been extended to new materials such as sSOI, GeOI etc. The defect decoration efficiencies of the various techniques have been compared. Cr-free solutions (especially on SOI) and HCl etches are sensitive defect decoration techniques which can be used to track peculiar and critical defects in the process (such as Oxidation Induced Stacking Faults).

We have finally dealt with the development and reliability of both characterization techniques, defect etching and Raman, on strained dual channel layers. To check the techniques efficiency and illustrate the complementarities between defect etching and Raman characterization, we have annealed dual channel samples consisting of sSi / c-SiGe / XsSOI substrate. Defects most likely due to plastic relaxation are clearly evidenced for high annealing temperatures. Their formation comes hand in hand with some surface roughening and penetration of the Ge atoms in the c-SiGe layer into the surrounding sSi layers. Such a mechanism could be confirmed by Raman analysis. However no quantitative relation was possible at this stage between defect etching and Raman.

Acknowledgments

The author would like to thank Pr. B. O. Kolbesen and J. Maelisse from Frankfort university and all persons from CEA-Leti and Soitec having participated to this work.

References

1. Y. Bogumilowicz, J. M. Hartmann, J. F. Damlencourt, B. Vandelle, A. Abbadie, A. M. Papon, G. Rolland, P. Holliger, C. Di Nardo, P. Besson, T. Ernst, T. Billon, C. Aulnette, N. Daval, O. Rayssac, B. Ghyselen, N. Cherkashin, A. Calverie, B. Cluzet and V. Calvo, Electrochemical Society Proceedings Vol. **2004-07**, p. 665 (2004).

2. L. Clavelier, C. Deguet, C. Le Royer, B. Vincent, J.-F. Damlencourt, J. M. Hartmann, O. Kermarrec, T. Signamarcheix, B. Depuydt, A. Theuwis, C. Quaeyhaegens, N. Cherkashin, Y. Morand, P. Rivallin, C. Tabone, S. Lagrasta, Y. Campidelli, S. Descombes, L. Sanchez, T. Akatsu, A. Rigny, D. Bensahel, T. Billon, N. Kernevez and S. Deleonibus, ECS Trans. Vol. **3**, no. 7, 789 (2006).

3. B. Ghyselen, J. M. Hartmann, T. Ernst, C. Aulnette, B. Osternaud, Y. Bogumilowicz, A. Abbadie, P. Besson, O. Rayssac, A. Tiberj, N. Daval, I. Cayrefourq, F. Fournel, H. Moriceau, C. Di Nardo, F. Andrieu, V. Paillard, M. Cabié, L. Vincent, E. Snoeck, F. Cristiano, A. Rocher, A. Ponchet, A. Claverie, P. Boucaud, M. N. Semeria, D. Bensahel, N. Kernevez and C. Mazure, Sol. State Electron. **48**, 1285 (2004).

4. F. Andrieu, T. Ernst, O. Faynot, Y. Bogumilowicz, J. M. Hartmann, J. Eymery, D. Lafond, Y. M. Levaillant, C. Dupré, R. Powers, F. Fournel, C. Fenouillet-Béranger, A. Vandooren, B. Ghyselen, C. Mazuré, N. Kernevez, G. Ghibaudo and S. Delonibus, 2005 IEEE International SOI Conference p. 223 (2005).

5. C. Malleville, E. Neyret, L. Ecarnot, T. Barge and A. J. Auberton-Herve, SOI conference, IEEE International, p.19 (2000).

6. C. Malleville, C. Moulin and E. Neyret, SOI conference, IEEE International, p.194 (2002).

7. C. Yang, Micro Magazine **23**, 55 (2005).

8. R. Bryant, Solid State Phenomena **134**, 289 (2007).

9. A. Danel, S. Sage, J.P. Barnes, D. Peters, R. Spicer, R. Bryant and R. Newcomb, Solid State Phenomena **134**, 289 (2007).

10. H. Shin, M. Racanelli, W. M. Huang, J. Foerstner, T. Hwang and D. K. Shroder, Solid State Electronics **43**, 349 (1999).

11. S. H. Renn, J. L. Pelloie and F. Balestra, Electron Devices, IEEE Trans. **45**, 2335 (1998).

12. S. Cristoloveanu and F. Balestra, Semiconductor Conference, IEEE **1**, 3 (1996).

13. S. Cristoloveanu, D. Munteanu and M. S. T. Liu, IEEE Transactions on Electron Devices **47**, 1018 (2000).

14. O. Kononchuck, F. Brunier and M. Kennard, ECS Trans. **6** (4), 225 (2007).

15. A. Abbadie, J. M. Hartmann and F. Brunier, ECS Trans. **10** (1), 3 (2007).

16. C. Villeneuve, K. K. Bourdelle, V. Paillard, X. Hebras and M. Kennard, J. Appl. Phys. **102**, 094905 (2007).

17. Y. Cayrefourcq, A. Boussagol and G. Celler, ECS Trans. **3** (7), 399 (2006).

18. G. A. Rozgonyi, in *Encyclopedia of materials: Science and Technology*, edited by S. Majahan, p. 8524, Elsevier Science Ltd, Amsterdam (2001).

19. B. O. Kolbesen, J. Maelisse and D. Possner, ECS Trans. **11** (3), 195 (2007).

20. K. Moriya, K. Nakashima, M. Sakai, K. Kashima, N. Inoue and R. Tadeka, the 5th International Symp. On Advanced Science and Technology of Silicon Materials (JSPS Si symposium), Hawai, p. 87 USA (2008).

21. F. Secco d'Aragona, J. Electrochem. Soc. **119**, 948 (1972).

22. J. Lu, R. Zhang and G. Rozgonyi, ECS Trans. **2** (2), 569 (2006).
23. A. Abbadie, S. W. Bedell, J. M. Hartmann, D. K. Sadana, F. Brunier, C. Figuet and I. Cayrefourcq, J. Electrochem. Soc. **154**, H713 (2007).
24. J. Maelisse, A. Abbadie and B. O. Kolbesen, ECS Trans. **6** (4), 271 (2007).
25. D. Possner, B. O. Kolbesen, V. Kluppel and H. Cerva, ECS Trans. **10** (1), 21 (2007).
26. J. Maelisse, A. Abbadie, F. Brunier and B. O. Kolbesen, ECS Trans. **16** (6), 309 (2008).
27. Y. Bogumilowicz, J.-M. Hartmann, R. Truche, Y. Campidelli, G. Rolland and T. Billon, Semicond. Sci. Technol. **20**, 127 (2005).
28. A. Abbadie, J. M. Hartmann, N. Cherkashin, C. Deguet, L. Sanchez, F. Brunier and F. Letertre, ECS Trans. Vol. **6** (4), 263 (2007).
29. Y. Kashiwagi, R. Shimokawa and M. Yamanaka, J. Electrochem. Soc. **143** (12), 4079 (1996).
30. J. M. Hartmann, A. Abbadie, D. Rouchon, J. P. Barnes, M. Mermoux and T. Billon, Thin Solid Films **516**, 4238 (2008).
31. J. Parsons, E. H. C. Parker, D. R. Leadley, T. J. Grasby and A. D. Capewell, Appl. Phys. Lett. **91**, 063127 (2007).
32. J. Parsons, R. J. H. Morris, D. R. Leadley, E. H. C. Parker, D. J. F. Fulgoni and L. J. Nash, Appl. Phys. Lett. **93**, 072108 (2008).
33. J. C. Bean, J. C. Feldman, A. T. Fiory, S. Nakahara and I. K. Robinson, J. Vac. Sci. Techn. A **2**, 436 (1984).
34. E. Bugiel and P. Zamseil, Appl. Phys. Lett. **62**, 2051 (1993).
35. M. L. Lee, D. A. Antoniadas and E. A. Fitzgerald, Thin Solid Films **508**, 136 (2006).
36. S. H. Olsen, P. Dobrosz, E. Escobedo-Cousin, S. J. Bull and A. G. O Neill, Mat. Sci Eng. B **124**, 107 (2005).
37. S. Nakashima, T. Mitani, M. Nicomiya and K. Matsumoto, J. Appl. Phys. **99**, 053512 (2006).
38. J. S. Reparaz, A. Bernardi, A. R. Goni, M. I. Alonso and M. Garriga, Appl. Phys. Lett. **92**, 081909 (2008).
39. T. Akatsu, J. M. Hartmann, C. Aulnette, Y. M. Le Vaillant, D. Rouchon, A. Abbadie, Y. Bogumilowicz, L. Portigliatti, C. Colnat, N. Boudou, F. Lallement, F. Triolet, C. Figuet, M. Martinez, P. Nguyen, C. Delattre, K. Tsyganenko, C. Berne, F. Allibert and C. Deguet, ECS Trans. Vol. **3**, no. 6, 107 (2006).
40. J. Lu, R. Zhang, G. Rozgonyi, E. Yakimov, N. Yarykin and M. Seacrist, ECS Trans. **2** (2), 569 (2006).
41. W. P. Huang, H. Cheng, G. Sun, R. F. Lou, J. H. Yeh and T. M. Shen, Appl. Phys. Lett. **91**, 14102 (2007).
42. M. Putero, N. Burle and B. Pichaud, Philos. Mag. A **81**, 125 (2001).
43. J. Lu, G. Rozgonyi, M. Seacrist, M. Chaumont and A. Campion, J. Appl. Phys. **104**, 074904 (2008).

Morphological and Electrical Comparison of GeOI Enriched Structures Obtained from SOI and sSOI Substrates

JF Damlencourt[1], B. Vincent[1], L. Clavelier[1], C. Le Royer1, Y. Campidelli[2], S. Bernasconi[2], Y. Morand[2], T. Nguyen[3] and S. Cristoloveanu [3]

[1] CEA-LETI –Minatec, 17 avenue des Martyrs, 38054 Grenoble Cedex 9, France
[2] STMicroelectronics, 850 rue Jean Monnet, 38926, Crolles-Cedex, France
[3] IMEP-INP Grenoble-Minatec, BP 257, 38016 Grenoble Cedex 1, France
Email : jfdamlencourt@cea.fr

Fabrication of high quality Ge/SGOI bilayer by Ge enrichment technique has been investigated. Using a strain SOI (20%) wafer as a starting substrate leads to the delay of plastic relaxation and thus, the crystalline defects generation, during the Ge enrichment. These sGeOI substrates show a lower layer roughness and a high resistivity compared to the ones obtained on conventional GeOI. Unfortunately, plastic relaxation is evidenced during the Ge thickening (final thickness up to 30nm). This can be overcome by using extremely strain SOI (XsSOI) as starting substrates.

Introduction

The challenges imposed by the scaling of Si devices make mandatory the study of new materials to overcome the physical limitations of Si. Ge was actually used in the very first transistors but was then abandoned in favour of Si due to the instability of Ge oxide. However, the recent introduction of high-K gate dielectrics makes the use of Ge possible in an advanced technology (1, 2). Its benefits for MOSFET applications are important: better transport properties than silicon hence higher saturation currents; lower band gap enabling lower supply voltages and lower power dissipation; finally its lattice parameter is compatible with GaAs for mixed circuit. Germanium On Insulator (GeOI) substrates benefit from both Ge properties and "On Insulator" advantages for microelectronic applications: better electrostatic control i.e. less short channel effects, reduced junction capacitances and lower substrate coupling in RF circuits. Another main advantage of GeOI substrates is the limited quantity of Ge used to obtain such a substrate but also its mechanical behavior, which is close to the one of a Si wafer.

Two fabrication methods for GeOI and SGOI (Silicon Germanium On Insulator) substrates are used: the SMARTCUT^{TM} process and the Ge condensation method (3-6). The latter technique consists in a Si selective oxidation done after a SiGe layer epitaxial growth on a SOI wafer. The SiGe layer is initially grown with a low Ge content. Its thickness decreases during the oxidation due to the Si consumption. During this Si selective oxidation performed under dry O_2 at high temperature, Ge pills up at the oxide interface. Competitively due to the high temperature used, Ge diffuses through the top silicon layer of the SOI substrate and then through the remaining SGOI layer (7). Ge diffusion and SiGe thickness decrease involve a Ge content enrichment. Ultra thin SGOI layers with enrichments up to 100% can be achieved after long enough oxidation times.

Today, ultra thin GeOI (~10nm) are not adapted for MOSFET fabrication due to a Ge consumption which may occur during wet processes. Therefore a Ge epitaxial regrowth to thicken the SGOI is mandatory. Those Ge/SGOI bilayers have already been described and present a Rms roughness up to 2nm (8). This roughness may be attributed either to Ge epi growth conditions or to the stacking faults present within the underlying SGOI condensed substrate (9).

This paper proposes fabrication of Ge/SGOI bilayers starting with better quality SGOI templates. Those templates can be obtained by the Ge condensation technique performed on SiGe layers grown on strain SOI (sSOI) substrates. Starting with sSOI substrates allow indeed to delay defects generation during the Ge enrichment and thus, to decrease the surface roughness (10).

Experimental Details

The SGOI prestructure for condensation has been prepared by growing a $Si_{0.9}Ge_{0.1}$ layer on a SOI and sSOI 20% wafer with an initial 20nm top Si thickness. Growth has been performed at 650°C, by Reduced Pressure - Chemical Vapor Deposition under 20 Torr. The SiGe thickness target is 75nm. This thickness is lower than the critical thickness for plastic relaxation (11). A silicon cap, 2nm thick, is added to the structure to prevent any GeO_x formation during the early oxidation stages. The cap is oxidized at a low temperature (700°C) to form an initial SiO_2 layer, acting as an upper Ge diffusion barrier.

Oxidations have been done under dry oxygen using a two steps temperature process. The first temperature used is 1050°C involving a high oxidation rate. 1050°C being the melting temperature of the $Si_{0.35}Ge_{0.65}$ alloy, a lower temperature step at 900°C is added to reach Ge content higher than 65%. Enrichments up to 100% are obtained with the second step. In order to favor Ge diffusion and get highly homogeneous Ge profiles within the SGOI layers obtained, some annealing steps are added periodically all along the process (12, 13). Samples obtained after enrichment and epitaxial thickening starting from SOI are called in this paper GeOI while the ones starting from sSOI are called sGeOI. sGeOI does not mean that substrates obtained a totally strained.

The silicon dioxide formed during the oxidation is removed in a concentrated HF solution (10%). The wafers are then loaded in the RPCVD epi tool and miscellaneous Ge thickness is deposited on both SGOI and sSGOI substrates (called afterwards GeOI and sGeOI respectively). The RMS roughness of these samples is assessed by Atomic Force Microscopy (AFM) imaging. The layers resistivity is obtained by four point probes measurement and the layers mobility is measured by Pseudo-MOS technique. The different results obtained on sSGOI are compared to the ones obtained on conventional SGOI but also on SOI and sSOI 20%.

Results and discussion

The figure 1 is 5x5µm² AFM pictures of a 30nm and 90nm thick conventional GeOI substrate obtained by Ge enrichment and epitaxial thickening. A typical cross-hatch pattern, fingerprint of the strain relaxation, is evidenced on the surface. In the meantime, the surface of sGeOI (starting from a 20% sSOI) substrate does not evidence any cross-hatch pattern, whatever the thickness regrowth is: from 30nm to 90 nm thick (figure 2).

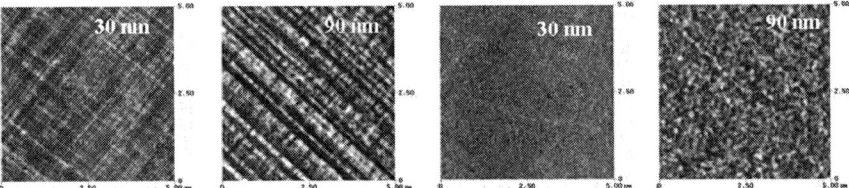

Figure 1: 5x5 μm² AFM picture of a 30nm and a 90nm thick conventional GeOI substrate obtained by condensation and epitaxial regrowth.

Figure 2: 5x5 μm² AFM picture of a 30nm and 90nm thick sGeOI substrate obtained by condensation and epitaxial regrowth.

Figure 3 depicts the Rms roughness of Ge surface extracted from 5x5 μm² AFM scans, as a function of the final Ge thickness (30 and 90 nm) for both GeOI and sGeOI. Whatever the starting substrate is (SOI or sSOI 20%) the Rms roughness increases with the finale Ge thickness. This can be explained by a strain relaxation during the epitaxial thickening of the SGOI.

However, a lower surface roughness is obtained, whatever the Ge thickness is, on sGeOI samples. The main conclusion which can be drawn is that using sSOI as starting substrate for Ge enrichment leads to a decrease of the surface roughness and delayed the plastic relaxation of the strain.

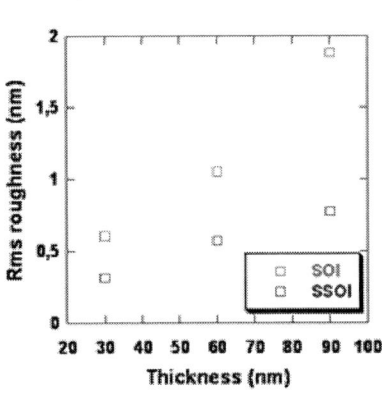

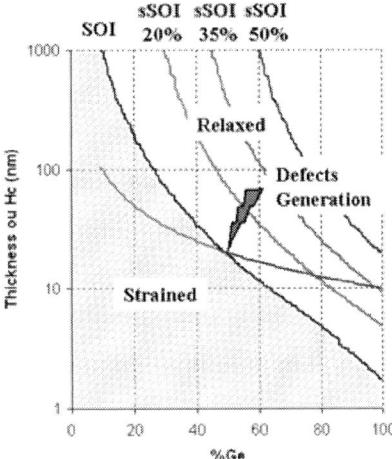

Figure 3: Rms roughness of GeOI surface starting from SOI (red symbols) and sSOI (blue symbols) substrates, as a function of the final Ge thickness.

Figure 4: Thickness of an enriched substrate (blue curve), critical thickness of plastic relaxation of a SiGe layer deposited on conventional SOI (purple curve) and on miscellaneous strain SOI (sSOI) substrates (20% :red curve, 35% green curve and 50% : brown curve) as a function of the Ge content.

This hypothesis is confirmed by figure 4 which depicts the thickness of an enriched substrate (pink curve) based on the Ge conservation, the critical thickness of plastic relaxation of a SiGe layer deposited on conventional SOI (blue curve) and on miscellaneous strain SOI (sSOI) substrates (20% : red curve, 35% green curve and 50% : brown curve) as a function of the Ge content. We highlight the existence of threshold enrichment for crystalline defects generation. This threshold is determined, as the intersection between the thickness of an enriched substrate and the critical thickness of plastic relaxation. The plastic relaxation of the strain is then delayed by using a sSOI (20%) as a starting substrate for Ge enrichment.

Moreover, this threshold enrichment can be increased until a Ge content close to 100% if the prestructure is performed on a sSOI 50%. It is an easy way to get perfectly crystalline strain-GeOI substrates by Ge enrichment.

Figure 5 depicts the resistivity measured by four point probes on GeOI starting from SOI (figure 5-red squares) and 20% sSOI substrates (figure 5-open blue squares).

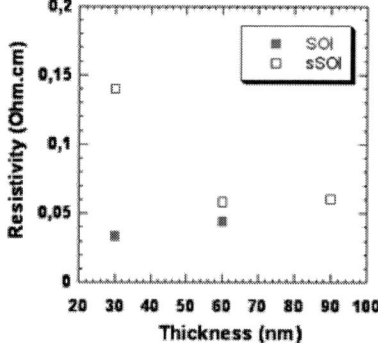

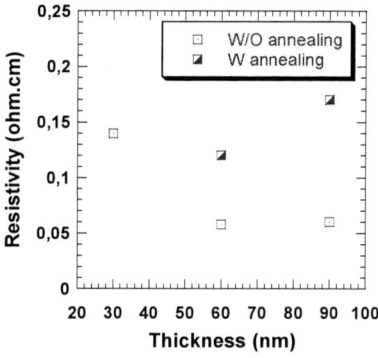

Figure 5: Resistivity of GeOI surface starting from SOI (red symbols) and sSOI (blue symbols) substrates, as a function of the final Ge thickness.

Figure 6 Resistivity of GeOI surface starting from sSOI before (open blue squares) and after (half blue squares) annealing at 650°Cas a function of the final Ge thickness

For GeOI starting from SOI substrates, the increase of the resistivity with the final Ge thickness is perfectly linear. Previous studies have shown that this resistivity is not due to an apparent doping level but to defects at the SiGe/Box interface (14, 15). Thus, the thicker the Ge layer is, the lower the impact of the interface. For sGeOI substrates, the resistivity decreases from $0.14\Omega.cm$ to $0.06\Omega.cm$ when the Ge thickness increases from 30nm to 60nm and then remains constant when the Ge thickness increases from 60nm to 90nm. This trend can not be only explained by interfacial defects. We assume that the decrease of the resistivity with the increase of the thickness is mainly due to the generation of crystalline defects into the layer (e.g. stacking faults) during the Ge thickening (i.e. plastic relaxation of the strain).

However, compared to the results obtained on conventional GeOI, the resistivity measured on sGeOI is increased by a factor of 4 for a Ge thickness of 30nm. For a 90nm thick sGeOI substrate, the resistivity measured by four point probes is the same than the one of a 90nm thick conventional GeOI.

The influence of post-epitaxy treatments on the layers resistivity has been also investigated. Figure 6 shows the evolution of the sGeOI layer resistivity as a function of the Ge thickness before (open blue square) and after annealing at 650°C under H_2 (half blue square). For thin layer, there is no benefit to perform an annealing treatment after thickening. For thicker layers (90nm), the resistivity is increasing by a factor of 3 after annealing treatment. We assume that H_2 annealing is able to cure the crystalline defects generated during the epitaxial regrowth and then to increase the layer resistivity.

Using sSOI as a starting substrates leads to a lower layer roughness and a higher layer resistivity. This is the fingerprint of a better material quality. Thus, cross sectional TEM pictures (not shown here) show a lower defects density for sGeOI compared to the one of conventional GeOI. These benefits are observed only on thin sGeOI substrates. Indeed, for thicker layer (up to 30nm), the plastic relaxation of the strain leads to the generation of crystalline defects and thus, to the decrease of the layer resistivity.

Electrical characterizations on GeOI and sGeOI were carried out by Pseudo-MOSFET (ψ-MOS), which is a very simple and efficient method to characterize the Ge-buried oxide interface (16, 17). This method uses the natural upside-down MOS configuration of the structure. The GeOI film serves as the transistor body and buried oxide (BOX) acts as gate oxide. The substrate bias V_G induces a conducting channel (inversion or accumulation) at the film-BOX interface. Two pressure-adjustable probes form the source and drain (Figure 7). The low-field mobility μ_0 is extracted by drawing $I_D/g_m^{0.5}$ as a function of V_G. These GeOI wafers (both conventional GeOI and sGeOI) were compared to unstrained and strained SOI materials.

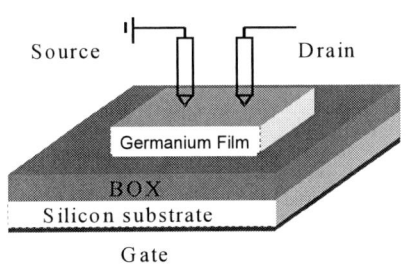

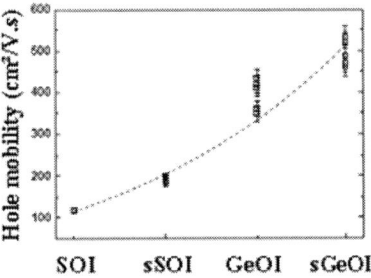

Figure 7: Schematic configuration of Pseudo-MOS transistor

Figure 8 : hole mobility extracted by the PseudoMOS technique on SOI, sSOI, GeOI and sGeOI substrates.

Figure 8 presents the hole mobility extracted by the PseudoMOS technique on SOI, sSOI, GeOI and sGeOI substrates. The sGeOI substrate shows an excellent hole mobility (~500 cm²/V.s) which exceeds recently reported values (18). The mobility enhancement over unstrained SOI, strained SOI and unstrained GeOI is 430%, 280% and 125%, respectively. Moreover, an electron mobility value can be measured on these layers (~100cm²/V.s) while on GeOI no electron mobility can be extracted.

Conclusion

Fabrication of high quality Ge/SGOI bilayers has been obtained by enrichment technique using sSOI 20% as starting substrate. Lower layer roughness and a higher layer resistivity is evidenced on 30nm thin sGeOI substrates compared to the value obtained on conventional ones (i.e. GeOI obtained by enrichment starting from conventional SOI substrates). This benefit is achieved thanks to the delay of strain relaxation during the Ge enrichment. However, during the epitaxial thickening, some plastic relaxation occurs, which leads to a decrease of the layer resistivity through the generation of stacking faults.

The use of 20% sSOI substrates, as starting substrates, for Ge enrichment, could be very interesting to fabricate strained SGOI substrates ([Ge] ~70%) and/or thin Ge/SGOI bilayers (t ≤ 30nm). Fabrication of defects free compressively strained GeOI layers by the Ge enrichment technique can only be obtained starting from XsSOI wafers (up to 30%).

References

1. L.Clavelier, C. Le Royer, C. Tabone, J.M. Hartmann, C. Deguet, V. Loup, C. Ducruet, C. Vizioz, M.Pala, T. Billon, F. Letertre, C. Arvet, Y. Campidelli, V. Cosnier, Y. Morand, *Proceedings of SNW Symposium*, p18 (2005).
2. N.Wu, Q. Zhang, C. Zhu, C. Shen, M.F. Li, D. S. H. Chan, N. Balasubramanian, *Int. Electron Device Meet. Tech. Dig.*, p. 563 (2005).
3. M. Bruel, *IEEE Electron. Lett.*, **31**, 1201 (1995).
4. C. Deguet, J. Dechamp, C. Morales, A.M. Charvet, L. Clavelier, V. Loup, J. M. Hartmann, N. Kernevez, Y. Campidelli, F. Allibert, C. Richtarch, T. Akatsu, F. Letertre, *Proceedings of ECS Symposium*, p.78 (2005).
5. T.Tezuka, N. Sugiyama, S.Takagi, *Appl. Phys. Lett.*, **79**, 1798 (2001).
6. T.Tezuka, N. Sugiyama, T. Mizuno, M. Suzuki, S. Takagi, *Jpn. J. Appl. Phys*, **40**, 2866 (2001).
7. N. Sugiyama, T. Tezuka, T. Mizuno, M. Suzuki, *J. Appl. Phys.*, **95**, 8 (2004).
8. J.F. Damlencourt Y. Campidelli, M.C. Roure, B. Vincent, E. Martinez, F. Fillot, Y. Morand, B. Arrazat, T. Nguyen, S. Cristoloveanu and L. Clavelier, *ECS Trans.*, **6**(4), 315 (2007).
9. B. Vincent, J.F. Damlencourt, V. Delaye, R. Gassilloud, L. Clavelier and Y. Morand., *Appl. Phys. Lett.*, **90**, 074101 (2007)
10. L. Clavelier, J.F. Damlencourt, C. Le Royer, Patent n° EN 08 354017
11. J.Huang, Z.Ye, H.Lu and D.Que, *J. Appl. Phys.*, **83**, 171 (1998).
12. J.F. Damlencourt, R. Costa, Patent n° E.N. 0601850.
13. B. Vincent, J-F. Damlencourt, P. Rivallin, E. Nolot, C. Licitra, Y. Morand and L. Clavelier, *Semicond. Sci. Technol.*, **22** (2007).
14. J.F. Damlencourt, Y. Campidelli, T. Nguyen, B. Vincent, C. Le Royer, Y. Morand, S. Cristoloveanu and L. Clavelier, *ECS Trans.*, **6**(1), 65 (2007).
15. W. Van Den Daele, E. Augendre, K. Romanjek, S. Cristoloveanu, C. Le Royer, L. Clavelier, J.F. Damlencourt, E. Guiot and B. Ghyselen, accepted for presentation at ECS 2009
16. S. Cristoloveanu, D. Munteanu, M.S.T. Liu, *IEEE Transaction of Electronic Devices*, **47**(5), p. 1018 (2000).
17. S. Cristoloveanu and S. S. Li, Boston, MA, Kluwer, (1995).
18. T. Akatsu, C. Deguet, L. Sanchez, C. Richtarch, F. Allibert, F. Letertre, C. Mazure, N. Kernevez, L. Clavelier, C. Le Royer, J.M. Hartmann, V. Loup, M. Meuris, B. De Jaeger, G. Raskin, *Proceeding of IEEE International SOI Conference*, p. 137 (2005).

CHAPTER 5

DEVICE PHYSICS AND MODELING I

100

ECS Transactions, 19 (4) 101-112 (2009)
10.1149/1.3117397 ©The Electrochemical Society

COMPARISON OF SOI FINFETS AND BULK FINFETS

Jong-Ho Lee, Ju-Wan Lee, Han-A-Reum Jung, and Byung-Kil Choi

School of Electronic and Electrical Engineering, Kyungpook National University, Daegu
702-701, Korea

Bulk FinFETs and silicon-on-insulator (SOI) FinFETs were
compared in terms of device structure, fundamental characteristics,
speed characteristics, model, and application. Bulk FinFETs have
shown several advantages over SOI FinFETs while keeping nearly
the same scaling-down characteristics as those of SOI FinFETs.
Compared to bulk FinFETs, SOI FinFETs could suppress any
possible leakage between source and drain, and has low
source/drain to substrate capacitance so that speed characteristics
could be better. Since SOI FinFETs have two more corners in the
bottom of the fin body than bulk FinFETs, bulk FinFETs can have
less process variation. Both devices seem to have their specific
applications.

Introduction

The scaling-down of MOSFETs is strongly required to achieve high integration
density and performance. However, in the deep sub-100 nm regime, the scaling of
conventional planar MOSFETs has been facing problems such as threshold voltage (V_{th})
lowering, subthreshold swing (SS) degradation, drain-induced barrier lowering (DIBL),
fluctuation of device characteristics with random channel dopant, leakage increase due to
dielectric tunneling, and band-to-band tunneling at the junction.

To solve some of problems, several device structures have been proposed: thin-body
silicon-on-insulator (SOI) MOSFETs [1], double-gate MOSFETs [2], triple-gate
MOSFETs [3], and FinFETs [4],[5]. Among these structures, FinFET is considering a
promising candidate for ultimate CMOS device structure because the device has
robustness against short channel effect (SCE), higher current drivability, nearly ideal
subthreshold swing (SS), and mobility enhancement [6]. FinFETs are classified into two
types: SOI and bulk FinFETs. Bulk FinFETs [7]-[9] have been built on bulk silicon (Si)
wafers and SOI FinFETs [10]-[12] have been built on SOI wafers.

Several studies on the comparison between bulk FinFETs and SOI FinFETs have
already performed [13],[14]. In [13] and [14], both devices have been compared in terms
of fin body doping and corner effect. We think both devices need to be compared in
views of speed characteristics, RF parameters, and I-V model. More comprehensive
discussion on bulk and SOI FinFETs is also needed.

In this paper, we compare bulk FinFETs and SOI FinFETs in terms of device
structure, fundamental characteristics, model, speed characteristics, and geometry
variation. FinFETs have two side channels, which has a sort of double-gate nature, and a
top channel. Analysis on these channels will be given and *I-V* model also will be
introduced briefly. We also briefly address key factors in device fabrication, the dilemma
and prospect of FinFETs.

FinFET Structure and Schematic Views

Fig. 1 shows 2-D cross-sectional views of SOI FinFET and bulk FinFET, respectively. H_{fin}, W_{fin}, and T_{ox} represent the fin height, fin width, and gate oxide thickness, respectively, in Fig. 1.

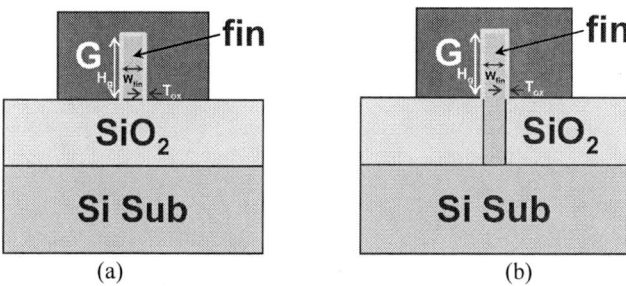

(a)   (b)

Fig. 1. 2-D cross-sectional view. (a) SOI FinFET. (b) Bulk FinFET.

FinFETs can be divided roughly by gate shape into two types: double-gate (DG) and tri-gate (TG). Fig. 2 shows 2-D cross-sectional views of (a) DG FinFET, (b) TG FinFET with 90° corner, and (c) TG FinFET with a half-circle top corner, respectively.

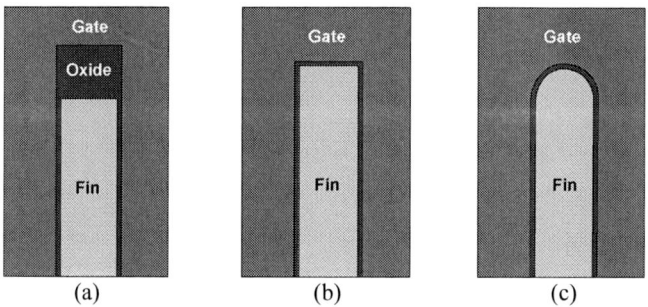

(a)   (b)   (c)

Fig. 2. 2-D cross-sectional views of FinFETs. (a) DG FinFET, (b) TG FinFET with 90° corner, and (c) TG FinFET with a half-circle top corner.

Fundamental Characteristics

Now we compare fundamental characteristics of both devices with the same body doping and body geometry. Fig. 3 shows the threshold voltage (V_{th}) and drain-induced barrier lowering (DIBL) of n-type FinFETs with L_g and H_g of 25 nm and 70 nm by changing the W_{fin}. The subthreshold slop (SS) characteristics are shown in Fig. 4. In Figs. 3 and 4, N_b, T_{ox}, and $x_{j,S/D}$ are 1×10^{19} cm^{-3}, 1.5 nm, and 66 nm, respectively. The n$^+$ poly silicon (Si) gate was applied for 3-D device simulation [15].The bulk FinFETs (solid circles) have nearly the same V_{th}, DIBL, and SS characteristics as those of SOI FinFETs (open circles).

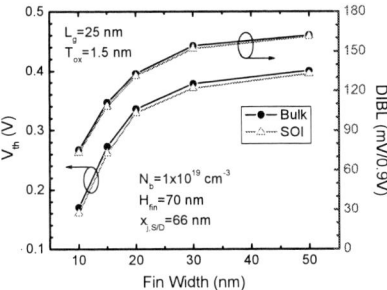

Fig. 3. V_{th} and DIBL chararcterstics of bulk and SOI FinFET versus W_{fin}. The n$^+$ poly Si gate was applied.

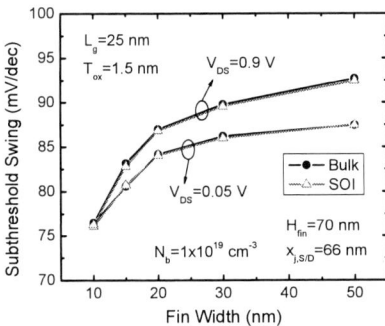

Fig. 4. SS characteristics versus W_{fin} for bulk and SOI FinFET as a function of V_{DS}.

Fig. 5 shows device temperature versus gate bias at a given V_{DS} of 0.9 V. L_g and T_{ox} are 30 nm and 1.5 nm, respectively. The bulk FinFET with W_{Fin} of 20 nm shows much lower device temperature than the 30 nm SOI FinFET since the body directly connected to the substrate has much higher heat transfer rate compared to that of the SiO$_2$ in SOI FinFET. At the V_{GS} of 0.9 V, bulk FinFET has a device temperature less by 130 °C than SOI FinFET.

Fig. 6 shows V_{th} versus back bias (V_{BS}) for planar MOSFET, bulk FinFET, and SOI FinFET with L_g=50 nm and N_b=2×10^{18} cm^{-3} at a given V_{DS}=0.05 V. The bottom oxide thickness under the fin body in the SOI FinFET is 200 nm. To compare the back bias effect of three devices, V_{th}s of three devices with the same channel width were made equal to 0.35 V at a low V_{DS} by controlling gate workfunction. Planar MOSFET has the largest back bias effect as shown in Fig. 8. But bulk and SOI FinFETs have an ignorable back bias effect for given V_{BS} from –0.4 V to 0.6 V.

Bulk FinFETs can be weak in a bulk punch-through in the fin body under gate when the source/drain extension junction depth (x_{jSDE}) is larger than H_g. Fig. 10 shows DIBL and SS characteristics of bulk FinFETs versus x_{jSDE} as a parameter of N_b. L_g, H_g ,W_{fin}, and T_{ox} of 50, 70, 20, and 1.5 nm, respectively. Here N_b is changed from 1×10^{16} cm^{-3} to 1×10^{17} cm^{-3}, and mid-gap workfunction is adopted. In Fig. 7, DIBL and SS are worse when x_{jSDE} is larger than H_g. The local boron doping (triangle symbol) located at x_p=70 nm is useful for the suppression of the DIBL because the local doping suppresses effectively the bulk punch-through. The local boron doping has a Gaussian doping profile with a peak concentration of 1×10^{18} cm^{-3} and a profile of 10 nm/dec.

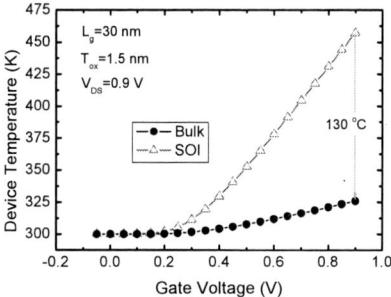

Fig. 5. Temperature characteristics of bulk and SOI n-type FinFET versus V_{GS}. The results were obtained through 3-D device lattice temperature simulation.

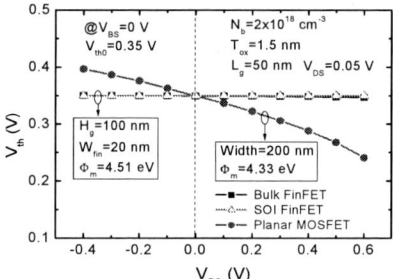

Fig. 6. V_{th} versus V_{BS} for planar MOSFET and bulk FinFET with L_g=50 nm and N_b=2×10^{18} cm^{-3} at a given V_{DS}=0.05 V. V_{th}s of two devices are 0.35 V at a V_{DS} of 0.05 V

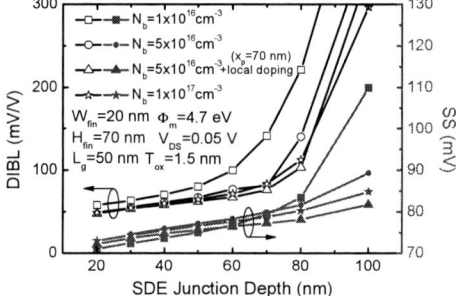

Fig. 7. DIBL and SS characteristics versus x_{jSDE} as a parameter of N_b. The local boron doping at x_p=70 nm is useful for the suppression of the DIBL.

Speed Characteristics

Fig. 8 shows change of full-down delay time and V_{th} versus V_{BS} as a parameter of device structure. The insert illustrates an inverter circuit. Here, supply voltage (V_{CC}) is 0.9 V. Gate length is 45 nm and a load capacitance is 20 fF. The delay time in both bulk and SOI FinFETs changes slightly with V_{BS}. Bulk FinFET shows slightly higher sensitivity of delay time with V_{BS} than SOI FinFET due to V_{th} change with the V_{BS} as shown in Fig. 8. Planar MOSFET shows significant increase of delay time with negatively increasing V_{BS}. For given negative V_{BS}, the delay of planar MOSFET degrades significantly due to the increased V_{th}. This means that the FinFETs are more effective device for the full-down circuits which consist of devices connected in series.

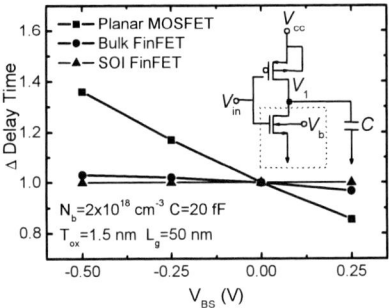

Fig. 8. Change of full-down delay time and V_{th} with V_{BS} as a parameter of device structure. Here, L_g and C are fixed at 45 nm and 20 fF, respectively. The insert illustrates an inverter circuit. The V_{CC} is 0.9 V.

RF characteristics of both bulk and SOI FinFETs were studied and compared in Fig. 9.

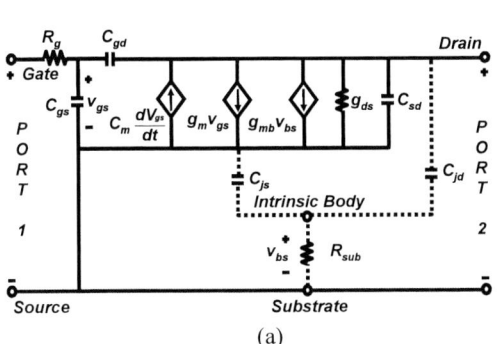

	Bulk DG FinFET	SOI DG FinFET
V_{th}	0.31 V	0.286 V
g_m	160 µS	166 µS
g_{ds}	0.427 µS	0.414 µS
C_{gs}	0.116 fF	0.113 fF
C_{dg}	0.0713 fF	0.073 fF
C_{gd}	0.037 fF	0.038 fF
R_g	461 Ω	472 Ω
C_{jd}	0.014 fF	– fF
R_{sub}	8200 Ω	– Ω
C_{js}	0.0151 fF	– fF
C_{sd}	−0.0037 fF	−0.0033 fF
f_T	170 GHz	177 GHz

(a)   (b)

Fig. 9. (a) Small signal equivalent circuit for RF device modeling [16]. (b) Comparison of RF parameters extracted from bulk and SOI DG FinFETs.

Both devices have an L_g of 60 nm, an H_g of 70 nm and a W_{fin} of 20 nm. The gate oxide thickness is 2 nm and the fin body doping concentration is 3.5×10^{18} cm^{-3}. n$^+$ poly-Si is applied. Fig. 9(a) shows small signal equivalent circuit of FinFET for RF device modeling at saturation region. The dashed line indicates substrate related parameters. The V_{th} of bulk FinFET is slightly larger than that of SOI FinFET since the bottom corner region in SOI FinFET has lower V_{th} due to electric field focusing. Fig. 9(b) shows comparison of RF parameters extracted from bulk and SOI FinFETs. DG SOI FinFET has no parameters related to the substrate due to its unique body structure. We compared RF parameters extracted from bulk and SOI DG FinFETs. RF parameters extracted from both devices are generally similar. Since the bulk FinFET has additional junction capacitance (C_{jd}), the cut-off frequency (fT) is slightly degraded by ~4% compared to that of SOI FinFET.

Corner and Width Effect

Corner effect together with the variation of fin body is very important point in FinFET technology. It is well known that the width of fin body needs to be smaller by ~70 % of a minimum channel length to keep DIBL less than 100 mV/V. We think it is very important to control body thickness uniformly over the wafer to make FinFETs practical. Without removing an appreciable variation in controlling uniform fin body width, FinFETs would not be practical, which seems to be a dilemma of FinFETs. It should be also addressed that the uniform control of the corner in the fin body is an important issue because the effect related to the corner gives a variation in V_{th} (or I_{off}), SS and DIBL and even device reliability. It is also well known that the a half circle (or a half cylindrical) shape of the top corner of FinFETs is optimum. Since the top corner effect in both FinFETs is the same, we tried to compare both devices by alleviating top corner effect with a thick top gate oxide. Fig. 10 shows cross-sectional views of bulk FinFET (a) and SOI FinFETs (b, c). In these views, the top oxide is set to 20 nm to alleviate the top corner effect. Now we can see more clearly the bottom corner effect in both FinFETs. SOI FinFET shown in Fig. 10 (b) shows the fin body formed on buried oxide. However the device shown in Fig. 10 (c) has an undercut under the fin body which can be formed by some process variation. This kind of variation is not observed in bulk FinFETs.

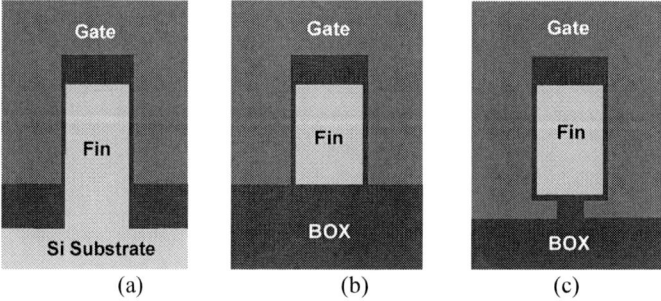

(a) (b) (c)

Fig. 10. Cross-sectional views of bulk FinFET (a) and SOI FinFEs (b, c). SOI FinFETs have 3 different body structures to show the bottom corner effect.

Fig. 11 compares I-V characteristics of FinFETs shown in Fig. 10. Here W_{fin} and H_{fin} are 20 nm and 70 nm, respectively. L_g of FinFETs is 100 nm and n$^+$ poly-Si is applied. N_b is 2×10^{18} cm^{-3}. As we can observe, SOI FinFETs show some V_{th} variation and resultant I_{off} variation. Compared to the I_{off} of bulk FinFET, SOI FinFET with the undercut shows several times higher I_{off}.

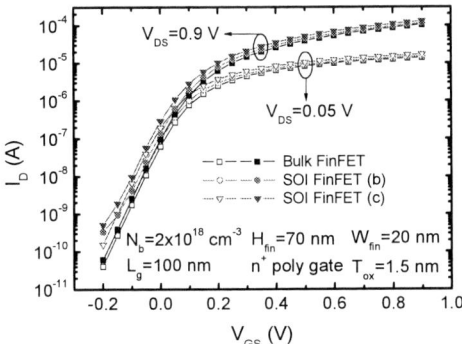

Fig. 11. I_D-V_{GS} curves of bulk and SOI FinFETs shown in Fig. 10 as a parameter of V_{DS}.

Fig. 12 shows V_{th} and DIBL of bulk FinFETs with a low body doping of 5×10^{16} cm^{-3} by changing W_{fin}. L_g and H_{fin} are 30 nm and 70 nm, respectively. Mid-gap gate workfunction was applied. This figure was prepared to show DIBL variation with the fin body with in FinFETs. Although we show the data in bulk FinFETs, nearly the same results could be found in SOI FinFETs. The insert shows a cross-sectional view of bulk FinFET used in the study. The V_{th} normally increases with increasing W_{fin} when the N_b is high (V_{th} is affected by N_b). However, the V_{th} in FinFETs with a low body doping (V_{th} is slightlty affected by N_b) decreases as the W_{fin} increases because of increasing DIBL shown on y2 axis of this figure. We can say that DIBL (and related V_{th} and I_{off}) in the FinFETs with a low doping is relatively sensitive with W_{fin}.

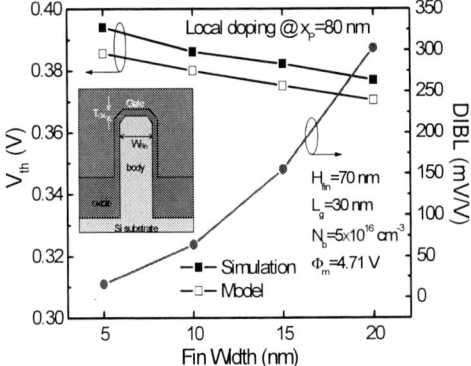

Fig. 12. V_{th} and DIBL characteristics of the device with 135°-corner. The body doping is 5×10^{16} cm^{-3} and midgap gate workfunction is applied. The local doping (Gaussian

profile: $N_{peak}=3\times10^{18}$ cm^{-3}) is located at $x_p=80$ nm from the fin top to suppress bulk punchthrough, and has no effect on the channel doping.

Model

DG nature is a key point in FinFET devices and needs to be understood well. The DC characteristics of DG MOSFETs are very important in accurate analysis, device design, and circuit design. To apply DG MOSFETs [17],[18] to integrated circuits, the DC behavior should be characterized and the parameters for DC characteristics should be extracted. Therefore, the DC models of DG MOSFETs have been studied extensively in case of undoped or doped channel.

V_{th}, SS, and I-V behaviors of DG MOSFETs with doped channel were parametrically modeled with the simple closed-form as follows [19]-[21]:

$$V_{th} = V_{FB} + 2\psi_B' + \frac{qN_b x_{dep}}{C_{ox}}\left(1 - \frac{x_h}{L_c} + \frac{\pi x_{df}}{4H_g}\right),$$

where

$$2\psi_B' = 2\psi_B + \delta_W + \delta_L + \delta_B . \qquad (1)$$

$$SS = \left[\frac{d\left(log_{10} I_{DS}\right)}{dV_{GS}}\right] = \frac{V_t \ln 10}{D_G},$$

where

$$D_G = 1 - a\left(\frac{e^{-0.5\Gamma_s}}{1 - e^{-2\Gamma_s}}\right)\left(\sqrt{\frac{c - aV_{GS}}{b - aV_{GS}}} + \sqrt{\frac{b - aV_{GS}}{c - aV_{GS}}}\right),$$

$$a = 1 - e^{-\Gamma_s} , \quad b = V_{bi} + \frac{qN_b W_{fin}}{2C_{ox}} + V_{FB} - \left(V_{bi} + V_{DS} + V_{FB} + \frac{qN_b W_{fin}}{2C_{ox}}\right)e^{-\Gamma_s} ,$$

$$c = V_{bi} + \frac{qN_b W_{fin}}{2C_{ox}} + V_{FB} + V_{DS} - \left(V_{bi} + V_{FB} + \frac{qN_b W_{fin}}{2C_{ox}}\right)e^{-\Gamma_s} . \qquad (2)$$

$$I_{D,diffusion} = 2\mu \frac{W}{L - L_{D1}} V_t^2 C_D \exp\left[\frac{D_G\left(V_{GS} - V_{th0}\right) + D_D V_{DS}}{V_t}\right]\left[1 - \exp\left(\frac{-V_{DS}}{V_t}\right)\right],$$

where

$$L_{D1} = \sqrt{\frac{2\varepsilon_{si}}{qN_b}}\left(\sqrt{V_{DS} + V_{bi}} - \sqrt{V_{bi}}\right) . \qquad (3)$$

$$I_{D,drift} = 2\mu C_{ox}' \frac{W}{L}\left(V_{GS} - V_{th0} - \frac{V_{DS}}{2}\right)V_{DS} \quad \text{for a low } V_{DS} \ (V_{DS} < V_{GS}-V_{th})$$

$$I_{D,drift} = \mu C_{ox}' \frac{W}{L - L_{D2}}\left(V_{GS} - V_{th}\right)^2 \quad \text{for a large } V_{DS} \ (V_{DS} \geq V_{GS}-V_{th}),$$

where

$$L_{D2} = l \cdot \ln\left[\frac{V_{DS} - V_{DS,sat}}{l \cdot E_{sat}} + \sqrt{\left(\frac{V_{DS} - V_{DS,sat}}{l \cdot E_{sat}}\right)^2 + 1}\right] . \qquad (4)$$

V_{th}, SS, and I-V behaviors of DG MOSFETs with undoped channel were analytical physically modeled as follows [22]-[24]:

$$V_{th} = V_{FB} + \frac{1}{1 - S_{gso}}\left[V_t \ln\left(\frac{Q_{TH}}{n_i W_{fin}}\right) - S_{dso}\right],$$

where

$$S_{gso} = \frac{1 + cos(\lambda)}{1 + sin(2\lambda)/2\lambda} e^{-4L\frac{\lambda}{W_{fin}}} \left[\frac{sinh(L\lambda/W_{fin})}{sinh(2L\lambda/W_{fin})} \right],$$

$$S_{gso} = \frac{1 + cos(\lambda)}{1 + sin(2\lambda)/2\lambda} e^{-4L\frac{\lambda}{W_{fin}}} \left[\frac{sinh(L\lambda/W_{fin})}{sinh(2L\lambda/W_{fin})} \right]. \tag{5}$$

$$SS = \frac{V_t \ln 10}{1 - S_{gs}},$$

where

$$S_{gs} = 2S_2 cos(0.5\lambda) e^{\frac{L\lambda}{W_{fin}}} cosh\left[(x_{min} - 0.5L)2\lambda/W_{fin} \right]. \tag{6}$$

$$I_D = \frac{\mu W}{L - L_D} \left[2V_t(Q_s - Q_d) + \frac{Q_s^2 - Q_d^2}{4C_{ox}} + 8V_t^2 C_{Si} \log\left(\frac{Q_d + 8V_t C_{Si}}{Q_s + 8V_t C_{Si}} \right) \right],$$

where

$$L_D = \lambda \cdot \ln \left[\frac{\varphi_d + \sqrt{\varphi_d^2 - \varphi_{sat}^2 + \left(\frac{\kappa v_{sat}}{\mu}\lambda\right)^2}}{\varphi_{sat} + \frac{\kappa v_{sat}}{\mu}\lambda} \right]. \tag{7}$$

Fig. 13(a) shows V_{th} versus L_g for DG MOSFET with W_{fin} of 10 nm. N_b is 8×10^{17}. The model (open circle symbols) is obtained from (1) and shows a good agreement with the simulation. Fig. 13(b) shows SS versus channel length for DG MOSFET with a W_{fin} of 15 nm. N_b and V_{DS} are 5×10^{18} cm^{-3} and 0.05 V, respectively. The dashed line represents ideal SS (60 mV/dec.). When D_G is 1 (maximum), SS is 60 mV/dec.. The model (open triangle symbols) obtained from (2) explains well the SS behavior calculated by numerical approach.

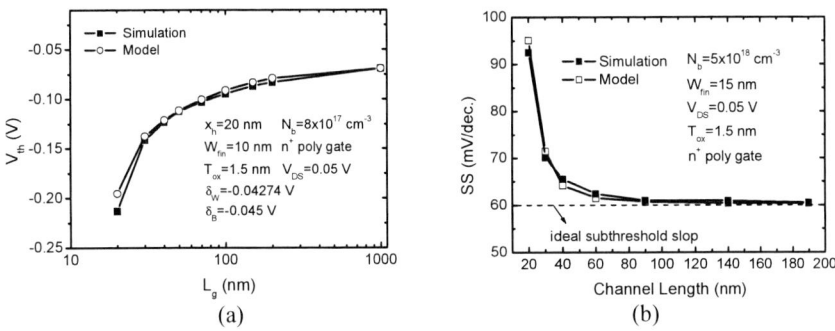

(a)   (b)

Fig. 13. (a) V_{th} versus L_g N_b for DG device with a W_{fin} of 10 nm. Model (open circle symbols) shows a good agreement with the simulation data. (b) SS versus channel length. W_{fin}, N_b, and V_{DS} are 15 nm, 5×10^{18} cm^{-3}, and 0.05 V, respectively. The ideal subthreshold slope is 60 mV/dec [20],[21].

Fig. 14(a) shows current versus gate bias as a function of channel length L for DG MOSFET with W_{fin} of 15 nm in all operational regions. N_b and V_{DS} are 5×10^{18} cm^{-3} and 0.05 V, respectively. The model obtained from (3) and (4) shows a good agreement with simulation in the viewpoint of both logarithmic and linear scales. Fig. 14(b) shows drift current ($I_{D,drift}$) versus V_{DS} as a function of V_{GS} for DG MOSFET with a W_{fin} of 15 nm at a fixed L of 30 nm. N_b and T_{ox} are 5×10^{18} cm^{-3} and 1.5 nm, respectively. The modeled data are consistent with those from simulation data.

Fig. 15(a) shows ΔV_{th} versus channel length for DG MOSFET with undoped channel as a function of W_{fin}. The model is obtained from (5) and shows a good agreement with the simulation. Fig. 15(b) shows SS versus channel length for DG MOSFET with undoped channel as a function of W_{fin}. The model obtained from (6) explains well the SS behavior calculated by numerical approach.

Fig. 16(a) shows I_D versus V_{GS} as function of V_{DS}. Fig. 16(b) shows I_D versus V_{DS} as a function of V_{GS}. In Fig. 17, the model is obtained from (7) and shows a good agreement with the simulation.

Thus DC I-V models for both FinFETs are good for device analysis and device design.

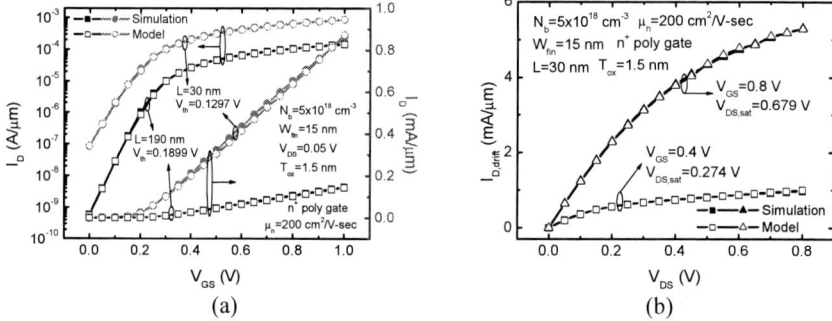

Fig. 14. (a) I_D versus V_{GS} as a parameter of channel length. W_{fin}, N_b, and V_{DS} are 15 nm, 5×10^{18} cm^{-3}, and 0.05 V, respectively. (b) $I_{D,drift}$ versus V_{DS} as a parameter of V_{GS}. L, W_{fin}, and N_b are 30 nm, 15 nm, and 5×10^{18} cm^{-3}, respectively [21].

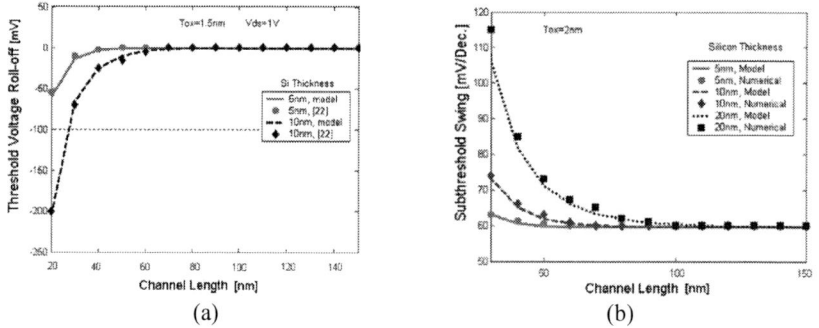

Fig. 15. (a) V_{th} roll-off versus channel length as a function of W_{fin} for DG MOSFET with undoped channel. T_{ox} and V_{DS} are 1.5 nm and 1 V, respectively. (b) SS versus channel as a function of W_{fin} for DG MOSFET with undoped channel. T_{ox} is 2 nm [22].

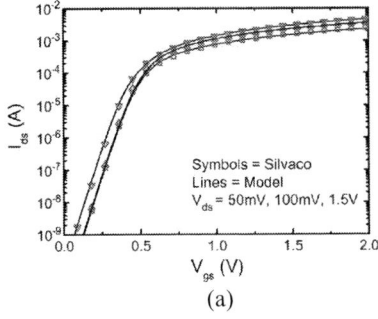 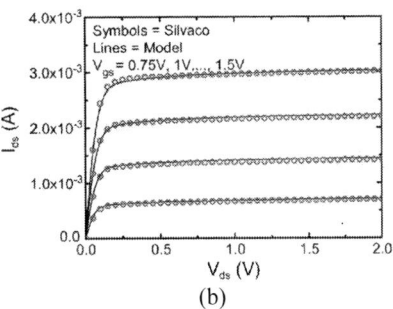

(a)   (b)

Fig. 16. (a) I_D versus V_{GS} as a parameter of V_{DS} for DG MOSFET with undoped channel. (b) I_D versus V_{DS} as a parameter of V_{GS}. In Fig. 17, T_{ox}, W_{fin}, and channel length are 2 nm, 15 nm, and 50 nm, respectively [23].

Discussion and Conclusion

We have compared SOI and bulk FinFETs in views of fundamental properties, speed characteristics, and models. The bulk FinFETs are built on bulk wafers which have low cost and low defect density compared to SOI wafers. Since the body of the bulk FinFETs is corrected directly to the substrate, a heat generated in the device under operation could be transferred to the substrate with a higher heat transfer rate compared to SOI FinFETs. For given the same body doping and channel geometry, both bulk and SOI FinFETs have nearly the same scalability. The bulk FinFETs have a bulk punch-through problem in the fin body when source/drain junction depth is larger than fin height. However, reasonable doping control in the fin body near the bottom of the gate electrode could suppress effectively the punch-through. Thus, compared to bulk FinFETs, SOI FinFETs could better suppress a leakage between source and drain. The SOI FinFETs have smaller parasitic capacitances (especially low source/drain to substrate capacitance), resulting in slightly faster delay time and slightly higher cut-off frequency in RF regime. More importantly, the SOI FinFETs shown higher corner effect originated from the bottom corner of the fin body, which gives a sort of a variation. If both FinFETs are to be practical, the control of uniform body width seems to be critical issue and would be a dilemma of FinFETs. Anyway FinFETs are expected to be utilized partly in sub-32 nm technology node if fabrication process is controlled well.

Acknowledgments

This work was supported by "the Korea Science and Engineering Foundation (KOSEF) grant funded by the Korea government (MEST) (No. R11-2008-105-01004-0) and the National Research Program for the 0.1 Tb Non-volatile Memory Development sponsored by Korea Ministry of Knowledge Economy" in 2009.

REFERENCES

1. Y.-K. Choi et al., in *IEDM Tech. Dig.* p. 919, (1999).

2. G. Kathawala, B. Winstead, and U. Ravaidi, *IEEE Trans. Electron Devices*, **50**, 2467 (2003).
3. B. Doyle et al., in *Symp. VLSI Tech. Dig.*, p. 133, (2003).
4. X. Huang *et al.*, *IEEE Trans. Electron Devices*, **48**, 880 (2003).
5. H.-Y. Chen et al., in *Symp. VLSI Tech. Dig.*, p. 6, (2003).
6. J. Kedzierski et al., in *IEDM Tech. Dig.* p. 437, (2001).
7. T. Park *et al.*, in *VLSI Technology Dig.*, p. 135, (2003).
8. T. Park *et al.*, in *IEDM Tech. Dig.*, p. 27, (2003).
9. T.-S. Park, E. Yoon, and J.-H. Lee, *Physica E*, **19**, 6 (2003).
10. A. Kranti and G. A. Armstrong, in *IEEE Int. SOI Conf.*, p. 96, (2005).
11. H.-C. Lin *et al.*, *IEEE Electron Device Lett.*, **24**, 102 (2003).
12. C.Y. Kang *et al.*, in *IEDM Tech. Dig.* p. 885, (2006).
13. M. Poljak, V. Jovanovic, and T. Suligoj, in *14th IEEE Mediterranean Electrotechnical Conf.*, p. 425, (2008).
14. C. R. Manoj, N. Meenakshi, V. Dhanya, and V. R. Rao, in *Int. Workshop on Physics of Semiconductor Devices*, p. 134, (2007).
15. ATLAS User's Manual [http://www.silvaco.com].
16. N.-K. Tak and J.-H. Lee , in *Topical Meeting on Silicon Monolithic Integrated Circuits in RF System Tech. Dig.*, p. 266, (2004).
17. G. A. Kathawala and U. Ravaioli, in *IEDM Tech. Dig.*, p. 683, (2003).
18. S. Harrison *et al.*, in *IEDM Tech. Dig.* p. 449, (2003).
19. B.-K. Choi *et al.*, *IEEE Trans. Electron Devices*, **54**, 537 (2007).
20. B.-K. Choi and J.-H. Lee, *Jpn. J. Appl. Phys.* **47**, 3396 (2008).
21. B.-K. Choi *et al.*, *Jpn. J. Appl. Phys.* **47**, 8253 (2008).
22. H. A. E. Hamid, J. R. Guitart, and B. Iniguez, *IEEE Trans. Electron Devices*, **54**, 1402 (2007).
23. F. Lime, B. Iniguez, and O. Moldovan, *IEEE Trans. Electron Devices*, **55**, 1441 (2008).
24. Q. Chen, E. M. Harrell, II, and J. D. Meindl, *IEEE Trans. Electron Devices*, **50**, 1631 (2003).

ECS Transactions, 19 (4) 113-118 (2009)
10.1149/1.3117398 ©The Electrochemical Society

Mechanisms that Explain Short-Channel Effects of Sub-50-nm-Channel Ultra-Thin Symmetric Double-Gate SOI MOSFET

Y. Omura and Y. Tahara

ORDIST, Department of Electronics, Kansai University, Yamate-cho, Suita, Osaka 564-8680, Japan

> Threshold voltage behaviors of the short-channel ultra thin symmetric double-gate (SDG) SOI MOSFET having a thin SOI body are simulated using the hydrodynamic transport model. We first propose models of threshold voltage and surface potential at the threshold that triggers the conventional drain-induced barrier lowering (*DIBL*) effect. From detailed analyses of device operations, it is demonstrated that the drain-induced potential rising (*DIPR*) effect at the middle of the SOI body plays an important role in determining the short-channel effects of SDG SOI MOSFET having a thin and low-doping body.

Introduction

In this study, we conducted simulations to estimate the influence of the physical parameters of the ultra thin short-channel symmetric double-gate (SDG) SOI MOSFET on threshold voltage (V_{th}). Since the main physics of the SOI FinFET [1] and omega-gate SOI MOSFET [2] can be taken to follow that of SDG SOI MOSFET, we focus on the SDG SOI MOSFET for simulation simplicity. We focus on the influence of *DIBL* and a new phenomenon on V_{th} behavior of SDG SOI MOSFET having a thin and low-doping SOI body and propose a comprehensive mechanism of V_{th} behavior appearing in the short-channel SDG SOI MOSFET. It is already known that the V_{th} of the ultra thin SOI MOSFET is raised by quantum effects when the SOI layer thickness (t_{SOI}) is very small [3]. It was shown recently that the threshold voltage of the SDG SOI MOSFET rises as t_{SOI} is reduced without any discernible quantum effects [4]. So, we propose an empirical model for threshold voltage of short-channel SDG SOI MOSFET in this article.

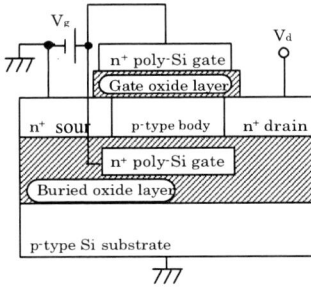

Figure 1. Schematic device structure considered.

Device structure and device parameters assumed

A schematic structure of the n-channel SDG SOI MOSFET assumed is shown in Fig. 1. Device parameters are summarized in Table 1. The drain voltage (V_d) ranges from 0.1 V to 1.0 V following the ITRS roadmap [5]. Basically, we used a two-dimensional device simulator

(*DESSIS* [6]) with a hydrodynamic transport model to analyze threshold voltage (V_{th}) characteristics; V_{th} is defined by extrapolating from of the I_d-V_g characteristic at V_d of 0.1 V and the $I_d^{1/2}$-V_g characteristics at V_d of 1.0 V. To simplify the discussion, in the simulations, we did not introduce a distinct quantum effect such as that of the density gradient model [7]. So, we limit the SOI layer thickness. In addition, to deepen our insight into the surface potential behavior, we also numerically solve Poisson's equation semi classically and discuss the physical background of the phenomenon observed in the V_{th} behavior.

Table. 1. Device parameters assumed

Parameters	Values [unit]
Gate oxide thickness, t_{ox}	3 [nm]
Buried oxide layer thickness, t_{box}	100 [nm]
Substrate doping conc., N_{sub}	1×10^{15} [cm^{-3}]
Source/drain doping conc., $N_{S/D}$	1×10^{20} [cm^{-3}]
n$^+$-gate poly-Si doping conc., N_G	1×10^{20} [cm^{-3}]
SOI layer thickness, t_{SOI}	5.0 -100[nm]
SOI doping conc., N_A	1×10^{15} - 1×10^{17} [cm^3]
Channel length, L	20 - 1000 [nm]

Results and discussion

We propose threshold voltage and surface potential models that reproduce the short-channel effects of the SDG SOI MOSFET as follows:

$$V_{th} = V_{FB} + \phi_{SURF0}(t_{SOI}, N_A) + \Delta\phi_{SURF}(t_{SOI}, N_A, L, V_d) + \frac{qt_{SOI}N_A/2 + qN_{inv}(\phi_{SURF0})}{C_{OX}}, \qquad [1]$$

$$\phi_{SURF0} = 2\phi_F + (\frac{kT}{q})\ln\left\{\frac{B - E\cdot N_A}{2C\cdot N_A} + \frac{1}{2C\cdot N_A}\sqrt{F}\right\}, \qquad [2]$$

$$\Delta\phi_{SURF} = \alpha L^{-\beta} + \gamma W_d \exp(\eta V_d^2) + \varsigma, \qquad [3]$$

where N_{inv} is the electron density of the inversion layer, ϕ_{SURF0} is the surface potential at the threshold of a 1-μm-channel device at $V_d = 0.1$ V [4], $\Delta\phi_{SURF}$ reveals he short-channel effects at the surface of SOI body and it is the shift of ϕ_{SURF} due to the *DIBL* effect that is generally observed in SOI MOSFET having a thick SOI body; parameters B to F are given as [4]

$$A = \frac{2q\beta}{\varepsilon_S}, \qquad [4a]$$

$$B = \frac{2kT}{\varepsilon_S}\alpha, \qquad [4b]$$

$$C = \left(\frac{q}{C_{OX}}\right)^2 \alpha^2 \left(1 + \frac{1}{\gamma}\right)^2, \qquad [4c]$$

$$D = \left(\frac{q\beta}{C_{OX}\gamma}\right)^2, \qquad [4d]$$

$$E = \left(\frac{q}{C_{OX}}\right)^2 \frac{2\alpha\beta}{\gamma}\left(1 + \frac{1}{\gamma}\right), \qquad [4e]$$

where parameters α to ζ are empirical parameters given in the Appendix. Other notations correspond to the conventional meanings. Since the purpose of this article is not to show a completely analytical expression for the threshold voltage, but to demonstrate a new threshold-voltage behavior of short-channel SDG SOI MOSFET, we follow the threshold voltage expression that has been already published in [4, 5]. In eq. (1), the term of $qN_{inv}(\phi_{SURF0})/C_{OX}$ that expresses the contribution to effective potential appears. This term plays a significant role in the threshold voltage determination when t_{SOI} and N_A are both small. In practice, it ranges from 5×10^{-2} V to 1.7×10^{-1} V for $t_{SOI}= 5$ nm and 10^{15} cm^{-3} $<N_A<10^{17}$ cm^{-3}, while qN_At_{SOI}/C_{OX} takes 7×10^{-5} V to 7×10^{-3} V. The primary physics (excluding short-channel effects) included in eq. (1) are already discussed in [4].

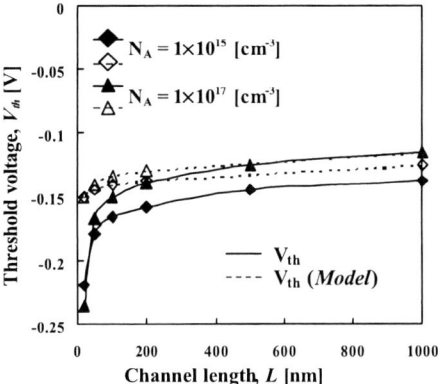

Figure 2. Threshold voltage (V_{th}) dependence on L for various N_A values. It is assumed $t_{SOI}= 10$ nm and $V_d= 1.0$ V.

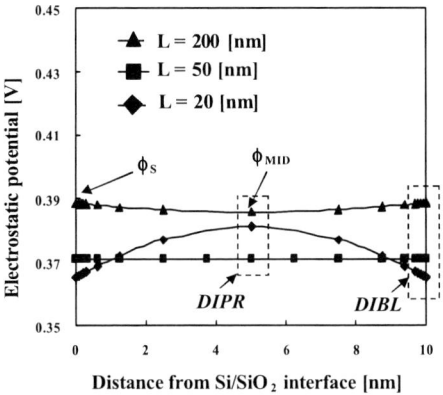

Figure 3. Electrostatic-potential depth profiles at the threshold of the SGD SOI MOSFET having various values of L. It is assumed $t_{SOI} = 10$ nm and $N_A = 1\times10^{17}$ cm^{-3}. It is assumed $V_g=V_{th}$ and $V_d= 1.0$ V. The depth profile of electrostatic potential is cut near the source junction.

Threshold voltage values simulated by *DESSIS* and those calculated by the proposed model are shown for $t_{SOI} = 10$ nm in Fig. 2 at $V_d = 1.0$ V. This figure shows that accuracy of the threshold voltage model is degraded as channel length (L) is reduced. On the other hand, the model successfully explains the threshold behavior of devices having a thick SOI body ($t_{SOI}>30$ nm) (not shown here). Therefore, this phenomenon is peculiar to the SGD SOI MOSFET having a thin SOI body ($t_{SOI}< 20$ nm). To clarify the reason for this, depth profiles of the electrostatic potential of SOI body are shown in Fig. 3 at the threshold for various channel length values. We can see that the difference between ϕ_S (surface potential) and ϕ_{MID} (mid potential) tentatively decrease as L is reduced, however, the heights of ϕ_S and ϕ_{MID} invert for $L = 20$ [nm]. The surface potential (ϕ_s) lowering that appears at the surface of SOI body is the conventional "*DIBL*" effect Since the gate-induced field at the middle of the SOI body is relatively weak, ϕ_{MID} has to be modulated by the drain voltage. We call this "*Drain-Induced Potential Rising* (*DIPR*)" effect; this phenomenon appears when the inversion layer carrier density (N_{inv}) is high at the threshold condition as clearly shown in Fig. 2. This phenomenon appears prominently for SDG SOI MOSFET with a thin body and a low body doping level, while SDG SOI MOSFET with a thick body suffers from the punch-through effect that is the potential lowering at the mid of SOI body. Therefore, the "*DIPR*" effect is quit different from the conventional "*DIBL*" effect. We think that *DIPR* should yield an additional subthreshold current at the middle of the SOI layer, resulting in additional roll-off in the threshold voltage.

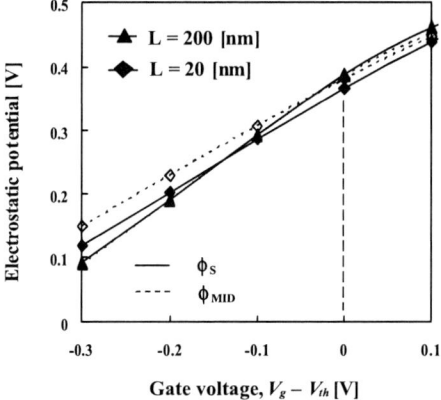

Figure 4. ϕ_S and ϕ_{MID} dependencies on gate voltage for $t_{SOI} = 10$ nm. It is assumed $N_A = 1{\times}10^{17}$ cm^{-3} and $V_d= 1.0$ V. The potential values are extracted from the depth profile of potential near the source junction.

We show, in Fig. 4, ϕ_S and ϕ_{MID} as functions of gate voltage (V_g) for various channel length values. It is seen that ϕ_{MID} is higher than ϕ_S for $L= 20$ nm at $V_g= V_{th}$. Here we introduce a new physical parameter ϕ_D ($= \phi_S - \phi_{MID}$) at the threshold to evaluate the impact of *DIPR* on the threshold voltage. ΔV_{th} and ϕ_D dependencies on channel length are shown in Fig. 5, where $\Delta V_{th}(L)= V_{th}(L) - V_{th}(1000\ nm)$. It can be seen that the behavior of ϕ_D basically matches that of the threshold voltage roll-off, which strongly suggests that *DIPR* contributes to the short-channel effects of the SDG SOI MOSFET.

According to the above consideration, we redefined the threshold voltage model as follows:

$$V_{th} = V_{FB} + \phi_{SURF0} + \Delta\phi_{SURF} + \theta\phi_D + \frac{qt_{SOI}N_A/2 + qN_{inv}(\phi_{SURF0})}{C_{OX}}, \qquad [5]$$

where θ is the empirical parameter. Since the purpose of this article is not to show a completely analytical expression for the threshold voltage, but to demonstrate a new phenomenon in SDG SOI NOSFET, here we simply define the threshold voltage. Threshold voltage dependency on channel length is shown in Fig. 6 as a function of channel length for $\theta = 5$; we assumed this value of θ because θ is insensitive to t_{SOI} for $t_{SOI} < 20$ nm. We can see that the new threshold voltage model basically reproduces the device simulation results since it takes account of the *DIPR* effect of ϕ_{MID}. However, the accuracy of the model should be improved in future by understanding the behavior of parameter θ in detail.

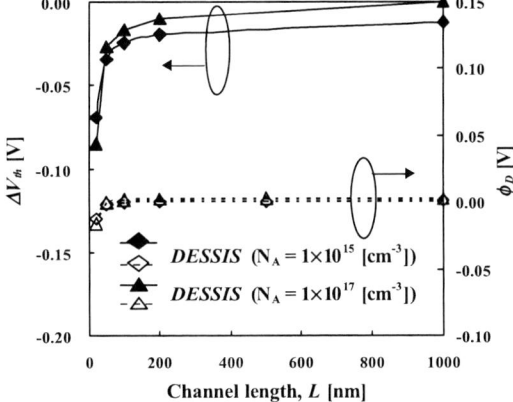

Figure 5. ΔV_{th} and ϕ_D dependencies on L for various N_A values and $t_{SOI} = 10$ nm at $V_d = 1.0$ V.

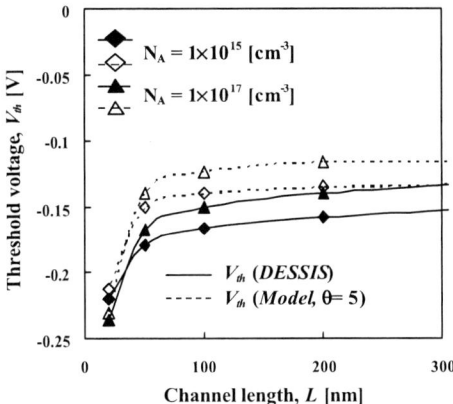

Fig. 6. Threshold voltage (V_{th}) dependence on L for various N_A values. It is assumed $t_{SOI} = 10$ nm.

References

1. D. Hisamoto, W.-C. Lee, J. Kedzierski, H. Takeuchi, K. Asano, C. Kuo, E. Anderson, T.-J. King, J. Bokor and C. Hu, *IEEE Trans. Electron Devices*, **47**, 2320 (2000).
2. F.-L. Yang, H.-Y. Chen, F.-C. Chen, C.-C. Huang, C.-Y. Chang, H.-K. Chiu, C.-C. Lee, C.-C. Chen, H.-T. Huang, C.-J. Chen, H.-J. Tao, Y.-C. Yeo, M.-S. Liang and C. Hu, *IEEE IEDM Tech. Dig.* (San Francisco, 2002) p. 255.
3. Y. Omura, S. Horiguchi, M. Tabe and K. Kishi, *IEEE Electron Device Lett.*, **14**, 569 (1993).
4. Y. Tahara and Y. Omura, *Jpn. J. Appl. Phys.*, **45**, 3074 (2006).
5. *International Technology Roadmap of Semiconductors*: http://public.itrs.net.
6. TCAD/*DESSIS* Operation Manual, ver.7.5 (Synopsis).
7. M. G. Ancona and G. J. Iafrate, *Phys. Rev.* B, **39**, 9536 (1989).

CHAPTER 6

DEVICE PHYSICS AND MODELING II

Advanced Design Methodology of High-Performance Sub-100-nm-Channel GAA MOSFET

S. Nakano[a], O. Hayashi[a], Y. Omura[a]
S. Yamakawa[b], and H. Wakabayashi[b]

[a] ORDIST, Department of Electronics, Kansai University, Yamate-cho, Suita, Osaka 564-8680, Japan
[b] Sony Corp, 4-14-1 Asahi-cho, Atsugi, Kanagawa 243-0014, Japan

This paper reconsiders the design methodology of the short-channel GAA SOI MOSFET and proposes an advanced concept to enhance its performance. The new ideas are based on gate field engineering and source and drain diffusion engineering. Validity of the proposal is demonstrated by device simulations.

Introduction

Ultra thin SOI MOSFET's and similar various advanced SOI devices are extensively studied to realize high-performance LSIs because they show high immunity to short-channel effects and high drivability [1]. The gate-all-around (GAA) SOI MOSFET [2] is one such promising device; its great potential is due to its electrostatic confinement of carriers in the Si body. Therefore it is expected to offer higher performance than the alternatives. However, no plausible design guideline has yet to be proposed because of the apparent drawback of parasitic resistance in the source and drain regions as well as mobility degradation due to the surface roughness effect and strong confinement effects.

This paper reconsiders the design methodology of the short-channel GAA SOI MOSFET and proposes an advanced concept to enhance its performance. The new ideas are based on gate field engineering and source and drain diffusion engineering. Their validity is demonstrated by various device simulation results.

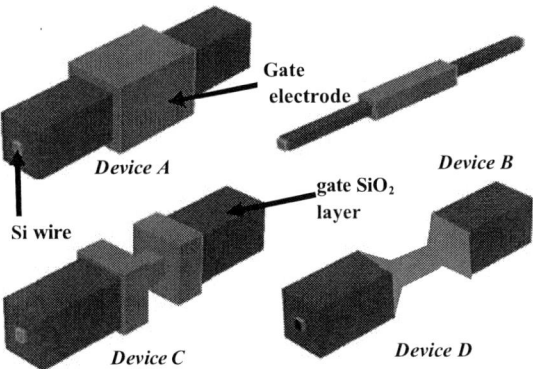

Figure 1. Schematic cross sections of the GAA SOI MOSFET's assumed in the simulations

Device Structures Assumed and Physical Parameters

This study simulates four different device structures as shown in Fig. 1. Figure 1 shows schematic bird's views of various 100-nm-long-channel GAA SOI MOSFET's with a Si body with 10 nm x10 nm cross-section: *device A* has a 20-nm-thick uniform gate SiO_2 layer, *device B* has a 1-nm-thick uniform gate SiO_2 layer, and *device C* has a 1-nm-thick gate SiO_2 layer that covers one half of the channel length. The basic structure of *device D* is similar to that of *device C*, but the SiO_2 layers near the junctions have a finite slope. The simulations use the hydrodynamic transport model and quantum-correction model of potential (density-gradient model) [3, 4].

All the devices have a 2-nm-long gate overlap at every junction edge and source and drain diffusion regions having 100-nm-long graded profiles to suppress short-channel effects. It is anticipated that such graded diffusion profiles yield high internal fields useful to the acceleration of carriers. However, such graded diffusion profiles usually yields undesired parasitic resistance in addition to the resistance stemming from the narrow wire dimension. Therefore, this disadvantage must be suppressed or eliminated by some additional techniques. We propose a new technique to exploit the advantages of the graded diffusion profile.

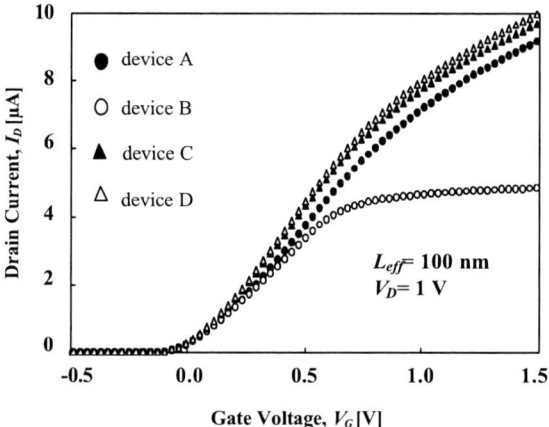

Figure 2. Current vs. voltage characteristics of GAA MOSFET's at V_D=1 V.

Simulation Results and Discussion

Figure 2 shows I_D-V_G characteristics of the four different GAA MOSFET's at V_D=1 V. It is clearly seen that *device A* has much higher drain current (I_D) than *device B*, which seems unexpected to us because this denies the simple expectation that a thinner gate oxide layer would raise the drain current of the MOSFET. We must consider whether short-channel effects of *device A*, which has a thick gate oxide layer, apparently raises its drain current. However, *devices A* and *B* have a sharp swing of 60 mV/dec at V_D of 1 V.

Since the cross-sectional area of the Si wire is 10 nm x 10 nm, electrostatic confinement of carriers is sufficient, and the thick gate SiO_2 layer does not allow short-channel effects even in *device A*[5, 6]. Then, we must investigate other possible mechanisms of high drain current of *device A*. It is seen in Fig. 2 that *devices C* and *D* have almost the same drain current as *device A*. Since the thin gate oxide layer of *devices C* and *D* does not result in a low drain current, it is suggested that the thick insulator covering the source and drain diffusions and junctions play an important role in maintaining high drivability.

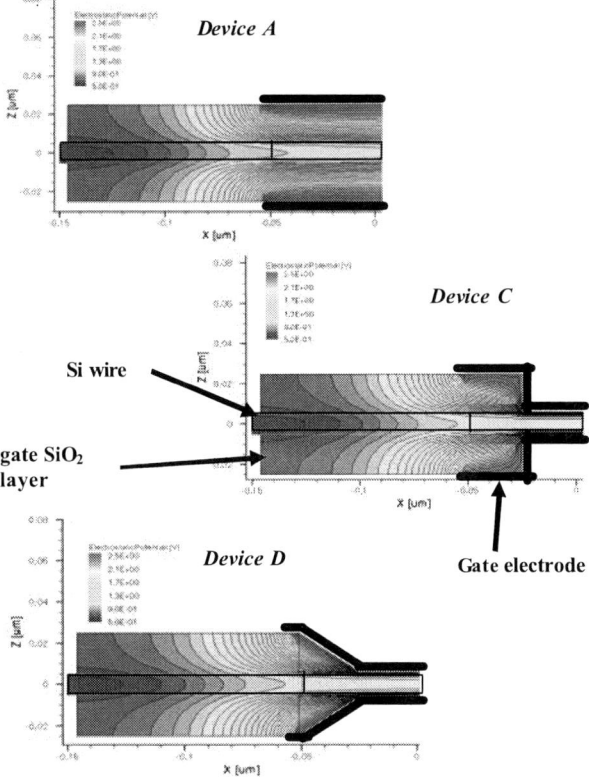

Figure 3. Potential contours of *devices A, C,* and *D* around the source junction at $V_D=1$ V and $V_G=2$ V.

The potential contours of *devices A, C,* and *D* around the source junction are shown in Fig. 3. The devices have a thick SiO_2 layer covering the source and drain regions, and they have a large potential gradient from the junction to the source contact; such a phenomenon clearly appears because the wire has a very small cross-sectional area and the electric field inside the oxide layer can change the field of the whole wire. Since the non-local effect of electrons along the diffusion plays an important physical role that should be accounted for by the hydrodynamic transport model, electrons inside the diffusion path are accelerated effectively, and their speed is higher than expected.

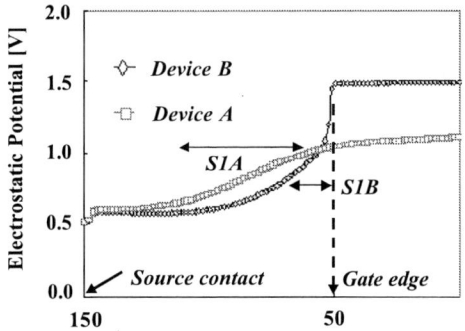

Figure 4. Potential profile of Si wire along the source-to-drain direction at $V_D=1$ V and $V_G=2$ V.

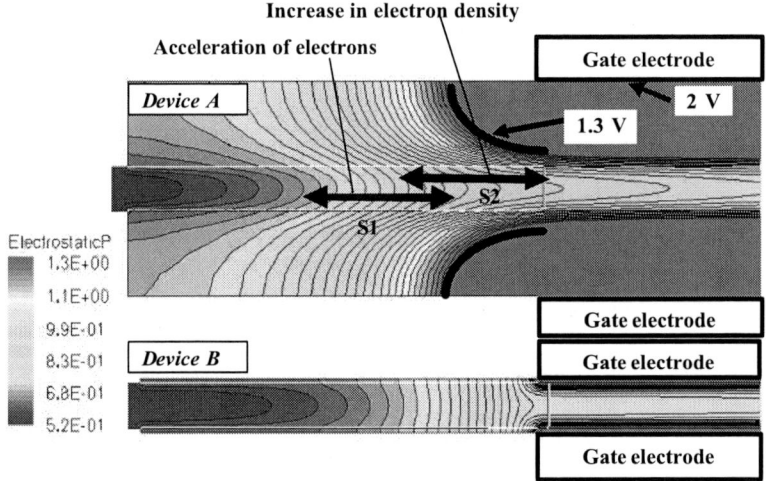

Figure 5. Potential contours of *devices A* and *B* around the source junction at $V_D=1$ V and $V_G=2$ V.

Figure 4 shows the electrostatic potential distributions along the source-to-drain direction for *devices A* and *B*. It should be noted that *device B* has a high field region (*S1B*) near the gate edge, but its width is narrow; such a narrow high-field region does not sufficiently contribute to carrier acceleration. On the other hand, *device A* has a wide high-field region (*S1A*) covering the diffusion path; this wide high-field region sufficiently contributes to the carrier acceleration as a result of non-local effect.

In addition, we must consider one more important effect appearing in *device A* Figure 5 shows the potential contours of *devices A* and *B*. In the illustration of the potential contours of *device A*, the potential contour of 1.3 V around the gate edge is shown as a bold black arch. The lateral extension of the 1.3-V contour raises the electron density of the Si wire over the width of *S2* because the lateral extension of contours creates the additional longitudinal field that impinges on the source diffusion; "Effective gate overlapping" length increases. This reduces the parasitic resistance of the 'low-doped' diffusion region of *S2*. In other words, the channel length of *device A* seems to be longer than the nominal length [7]. This yields suppression of the short-channel effects. In *device A*, the diffusion region of *S1* contributes to the acceleration of electrons as discussed later. On the other hand, in *device B*, the lateral extension of the gate field is very limited as seen in Fig. 5. Therefore, the increase in electron density around the gate edge is also limited.

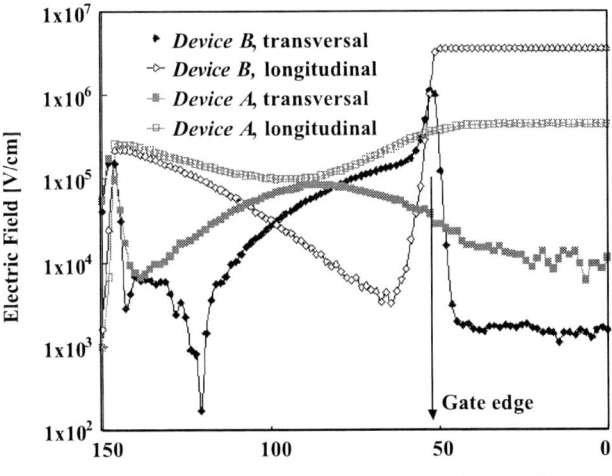

Figure 6. Electric field distributions of *devices A* and *B* along the source-to-drain direction around the source junction at V_D=1 V and V_G=2 V.

Finally, we show the longitudinal and transversal electric-field distributions along the source-to-drain direction for *devices A* and *B* in Fig. 6. The discussion of Figs. 4 and 5 in the above paragraph yielded two of the advantages of *device A*. The first merit of *device A* is its reduced parasitic resistance of its low-doped source diffusion region due to the longitudinal field near the gate edge. In Fig. 6, it is clearly seen that the longitudinal field of *device A* in the source diffusion region is much higher than that of *device B*. The second merit of *device A* is its effective acceleration of carriers in the diffusion path. In Fig. 6, it is seen that *device A* imposes a wide and high electric field on the diffusion path; this contributes to the acceleration of carriers due to the non-local effect.

Summary

This paper reconsidered the design methodology of the short-channel GAA SOI MOSFET and proposed an advanced technique to enhance its performance. The new ideas are based on gate field engineering and source and drain diffusion engineering. The simulation results suggest that a thick insulator covering the source and drain diffusions yields a wide lateral extension of the gate field which effectively increases carrier density in the low-doping diffusion region; this raises the drivability of the short-channel GAA MOSFET. In addition, the thick insulator covering the diffusion path creates a wide region over which a high transversal field is present; this region raises the transversal field in the graded diffusion region, so carriers in the region are accelerated.

Acknowledgments

This study was performed in collaboration of Sony Corp. and Kansai University, and financially supported by Sony Corp., Japan. The authors express their thanks to Drs. N. Nagashima and H. Ansai, Sony Corp. for extensive discussions and encouragement throughout the study.

References

1. *FinFETs and Other Multi-Gate Transistors*, edited by J.-P. Colinge, Springer Science, (2008).
2. J.-P. Colinge, H. M. Gao, A. Romano, H. Maes, C. Claeys, *IEEE IEDM* (San Francisco, 1990), p.595.
3. *Sentaurus*, Users Manual (Synopsys Inc.).
4. G. Paasch and H. Uebensee, *Phys. Stat. Sol.* (b), **113**, 165 (1982).
5. Y. Omura, the *10th Int. Symp. SOI Technol. & Dev.* (ECS, 2001) PV2001-3, p. 205.
6. Y. Omura, H. Konishi, and K. Yoshimoto, *J. Semicond. Tech. & Sci.,* **8**, 302 (2008).
7. H. Noda, F. Murai and S. Kimura, *IEEE IEDM* (Washington, D. C., 1993), p. 123.

Comparison of the Electrostatics of Bulk and SOI
Trigate MOSFETs

F. G. Ruiz, A. Godoy, I. M. Tienda-Luna and F. Gámiz

Department of Electronics, University of Granada, 18071 Granada (Spain)
Email: franruiz@ugr.es

In this work we compare the electrostatic performance of bulk and
SOI trigate MOSFETs. To do so, a self-consistent numerical
solution of the two-dimensional Schrödinger and Poisson
equations has been employed. The threshold voltage of bulk
trigates is higher than that of SOI devices, due to the presence of a
high-doping substrate, and this difference depends both on the
device size and the substrate doping.

Introduction

The use of multiple gate (MuG) architectures, such as FinFETs or Trigate
MOSFETs, allows a higher control of the Short Channel Effects (SCEs) that appear when
the channel lengths are scaled below 45nm (1). The first experimental steps have already
been given in that direction (2). Although MuG devices were originally designed on
Silicon on Insulator (SOI) wafers, there is a certain controversy in terms of performance
when these devices are fabricated on SOI or bulk substrates (3, 4). On one hand, the use
of insulator substrates allows a higher electrostatic control of the device (1). On the other
hand, multiple gate MOSFETs on bulk substrates reduce the fabrication cost, while
maintaining an acceptable gate control over the channel (4). First experimental results
showing the possibility of using bulk trigate MOSFETs have already been presented,
showing good characteristics compared with conventional planar transistors (5).

Bulk and SOI technologies have been experimentally compared in planar devices (6)
and FinFETs (3,4,7). However, a comprehensive study of the electrostatic behavior of
MuG devices fabricated on bulk and SOI wafers is still needed. In this work we have
carried out a comparative study of bulk and SOI trigate MOSFETs in order to gain
insight into the physical behavior of such devices and their scalability properties. Due to
the small size of the considered devices, quantum effects are dominant and thus a self-
consistent solver of the 2D Schrödinger-Poisson equations in a cross-section of the
structures has been used.

The paper is organized as follows. After this introduction, the numerical method used
to solve both equations will be introduced. Then, the geometry of the devices under
consideration (bulk and SOI) will be presented. Later, the main results of our simulations
will be shown, comparing bulk and SOI trigate MOSFETs. Finally, the main conclusions
will be drawn.

Numerical Method

The quantum-mechanical simulation of the MOSFETs under study in this work has been accomplished by using the effective mass approximation of the Schrödinger equation. This approximation uses an accurate description of the E–k relationship over only a limited range of energy near the valence or conduction band minima. To properly describe the electrostatic behaviour of our devices including the quantum effects, we also have to solve the Poisson equation turning the problem of studying the carrier concentration into the solution of the following coupled equation system:

$$-\frac{\hbar^2}{2}\nabla \cdot \left[\frac{1}{m^*}\nabla\Psi_n\right] + \left(V_h - V\right)\Psi_n = E_n\Psi_n \qquad [1]$$

$$\nabla\left(\varepsilon\nabla V\right) = -\rho \qquad [2]$$

where Ψ_n denotes the wavefunction, E_n its corresponding eigenvalue, V is the electrostatic potential, V_h is the heterojunction step potential that accounts for the Si–SiO$_2$ interface potential barrier, m^* is the tensor describing the effective electron mass and ρ is the net charge density where the fixed as well as the mobile charge are included.

Since typical (100) wafer orientation and <001> transport direction are considered, the solution of Eq. [1] can be accomplished by solving three times a two dimensional Schrödinger equation (one for each valley). Despite the previous simplification, the dependence of the charge density on the potential causes Eq. [2] to be non linear, making the system still too complex to be solved analytically. This is the reason why a discretization based on the Finite Elements is performed (8,9). This method offers the possibility of analyzing different geometries and dealing easily with interfaces between different materials. To reach a fast convergence, the Schrödinger and Poisson equations have been solved self-consistently using the predictor-corrector scheme proposed by Trellakis et al. (10), a reliable and robust algorithm that has been satisfactorily tested.

Geometry of the Devices under Consideration

Fig. 1 shows the geometry of the devices chosen for this study. SOI devices are simple from a geometrical point of view. They just consist of a silicon fin surrounded by the gate insulator at its top and lateral regions and by the buried insulator at its bottom. In this paper, both gate and buried insulators are SiO$_2$. The same insulator thickness is considered in top and lateral gates, which corresponds to the typical trigate configuration (11).

In bulk devices, the fin is grown over a silicon substrate, which has to be isolated from the gate. To do so, an oxide with thickness T_{ox2} is placed between them, as shown in Fig. 1. Thus, the channel height of a bulk device is $H_{Si}=H_{fin}-T_{ox2}$. In the fin region under the gate, the fringing electric field gives rise to parasitic capacitances that should be avoided. To do so, a higher doping concentration N_{A2} can be used, which helps the electrons to be confined into the channel region (4). If this higher doping concentration layer is extended underneath the fin, as shown in Fig. 1, it can also help to avoid the inversion of that region, reducing the insulator thickness necessary to achieve acceptable results. In the literature it has been shown that good results are obtained with $N_{A2}=10^{18}cm^{-3}$ and T_{ox2} in the range of 10 to 20nm (4), and therefore we will consider values close to those in our simulations.

Moreover, we will consider the same gate size in all the comparisons, as shown in Fig. 1, what means that we will not take into account the effective fin height reduction of bulk devices with respect to their SOI counterparts. Since the position of the interface (y_{int}) between the low-doping and the high-doping regions is not well defined, we will make a coherent comparison with SOI technology maintaining the same size of the fin region ($y_{int}= 0$).

In all the simulated devices, the channel is undoped, with $N_A=10^{14}cm^{-3}$. No corner rounding is assumed, so that the corner regions of the device present higher electron density than the lateral sides. However, an unique threshold voltage (V_{th}) is expected due to the small doping concentration (11). Midgap gate work function is considered, the gate insulator thickness is $T_{ox}=1nm$, and T_{box} and T_{bulk} have been chosen large enough to avoid any influence on the electron density.

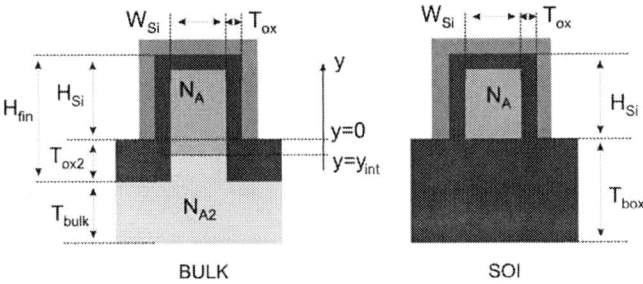

BULK   SOI

Figure 1. Geometry of the bulk and SOI trigate MOSFETs, along with their main geometrical and technological parameters.

Comparison between Bulk and SOI Trigate MOSFETs

In Fig. 2, we have compared the total electron charge per unit length, $N_t = \int n(x,y)\,dxdy$, as a function of the gate voltage (V_G), in square devices ($W_{Si} = H_{Si}$) with sizes varying from 5nm to 30nm. For bulk devices, the values considered for the substrate doping and the isolation oxide are $N_{A2}=10^{18}cm^{-3}$ and $T_{ox2}=10nm$. As can be seen in the figure, some differences appear between the electron densities obtained in bulk and SOI devices with the same size.

These differences are more evident when the threshold voltage (V_{th}) is depicted (inset of Fig. 2). Here, V_{th} is calculated as the maximum of the second derivative of N_t with respect to the gate voltage V_G $\left(d^2 N_t / dV_G^2 \right)$. The shrinking of the device size causes an increase of the threshold voltage in both, SOI and bulk devices, related to the quantum confinement effects and the corresponding increase of the subband energy levels. However, in the case of bulk trigates, their behavior cannot be explained only by the shrinking of the fin size, which is the same as in the SOI devices.

Fig. 3 depicts the 2D electron density of two SOI devices with 5nm and 20nm size, and their bulk counterparts, again with $N_{A2}=10^{18}cm^{-3}$ and $T_{ox2}=10nm$. As can be seen, the electron density in the top gate of the 20nm devices is quite similar, while noticeable differences are observed close to the bottom region (y=0). In that region, the effect of the space-charge region due to the junction created between the P+ substrate and the intrinsic channel is not negligible and it appreciably modifies the conduction band profile, as shown in Fig. 4 for the case of the 5nm device. In the strong inversion regime ($V_G=1V$), the potential profile in the top region of the devices is quite similar, which is coherent with the similar electron distribution shown in that region in Fig. 3. However, when V_G <V_{th} the effect of the junction made up by the channel and the substrate modifies the surface potential and therefore the threshold voltage of the device.

Effect of the Substrate Doping in the Threshold Voltage of Bulk Devices

From a theoretical point of view, the previously observed behavior can be explained from the threshold voltage expression of a bulk device with low-high step doping profile (12):

$$V_t \approx V_{fb} + 2\psi_B + \frac{qN_{A2}}{C_{ins}}\left(\sqrt{\frac{4\varepsilon_{Si}\psi_B}{qN_{A2}} + x_s^2} - x_s\right) \qquad [3]$$

where x_s is the low doping region length (in the case of the trigate corresponds to the fin height and therefore $x_s=H_{Si}$) and $2\psi_B$ is the surface potential needed to reach the channel inversion, that can be calculated as $2\psi_B = (2kT/q)\log(N_A/n_i)$.

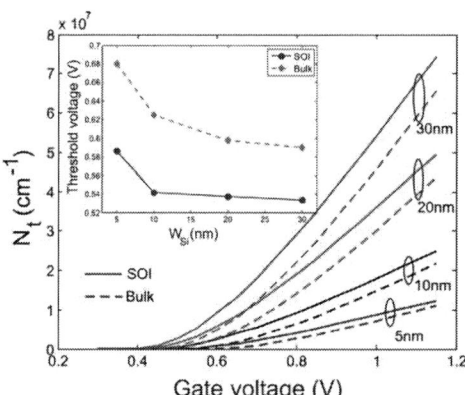

Figure 2. Total electron density per unit length as a function of the gate voltage. SOI and bulk square ($W_{Si}=H_{Si}$) trigate devices with different sizes are considered. The threshold voltage calculated for each device is shown in the inset.

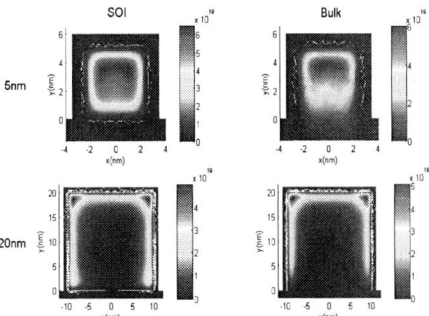

Figure 3. Comparison of the electron density of the 5nm and 20nm size SOI (left) and bulk (right) devices with $V_G=1V$.

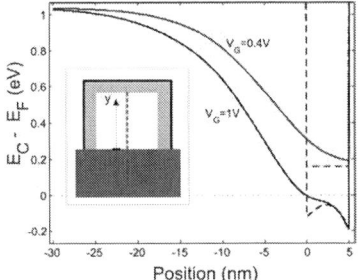

Figure 4. Conduction band profile of a bulk (solid lines) and a SOI trigate (dashed lines), both with fin size $W_{Si}=H_{Si}=5nm$, when two different gate voltages are applied: 0.4V and 1V. The inset shows the slice chosen for the plot.

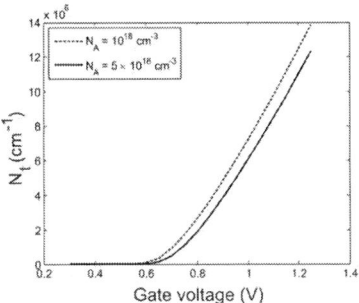

Figure 5. Comparison of the total electron density per unit length as a function of the gate voltage. Two devices are considered, both with fin size $W_{Si}=H_{Si}=5nm$, one with substrate doping (N_{A2}) of 10^{18} cm^{-3}, and the other of 5×10^{18} cm^{-3}.

Although the threshold voltage calculated with the former expression is not expected to fully agree with that achieved for the simulated trigate devices (since it is obtained for planar devices), it can qualitatively explain the increase of the threshold voltage due to the presence of a substrate layer with higher doping density N_{A2}. In fact, it can be determined from Eq. [3] that the threshold voltage depends on both the device height and also on the substrate doping density. When the silicon substrate doping varies, the space-charge region thickness is modified. If N_{A2} is increased, a higher gate voltage is needed to form the channel and thus the threshold voltage increases. An example of this behavior is shown in Fig. 5.

Conclusions

Important differences in the electron distribution of bulk and SOI trigates have been observed.. It has been shown that the threshold voltage of bulk trigates highly depends on their size by two different mechanisms: the effect of quantum-mechanical confinement, and of the underlying high-doping substrate. Furthermore, the influence of the substrate depends on the specific doping profile. Therefore, this variation of a fundamental parameter can significantly modify the performance of bulk devices compared to their SOI counterparts with similar geometrical characteristics. However, for a comprehensive benchmarking of both technologies it would be mandatory a deeper study that includes important features such as short-channel effects, cost and self-heating.

Acknowledgments

This work has been supported by the Spanish Government under project FIS-2008-05805 and the Junta de Andalucía under project P06-TIC-01899.

References

1. J. Park and J. P. Colinge, *IEEE Trans. Elec. Devices,* **49**, 2222-2229 (2002).
2. J. Kavalieros, B. Doyle, S. Datta, G. Dewey, M. Doczy, B. Jin, D. Lionberger, M. Metz, W. Rachmady, M. Radosavljevic, U. Shah, N. Zelick and R. Chau, 2006 Symp. on *VLSI Tech.,* 50-51 (2006).
3. M. Poljak, V. Jovanovic and T. Suligoj, *ISDRS* (2007).
4. C. R. Manoj, M. Nagpal, D. Varghese and V. R. Rao, *IEEE Trans. on Elec. Devices,* **55**, 609-614 (2008).
5. T. Park, S. Choi, D. Lee, U. Chung, J. T. Moon, E. Yoon, J. Lee, *Solid-State Elec.,* **49**, 377-383 (2005).
6. J. L. Pelloie, Y. Laplanche, T. F. Chen, Y. T. Huang, P. W. Liu, W. T. Chiang, M. Y. T. Huang, C. H. Tsai, Y. C. Cheng, C. T. Tsai and G. H. Ma, *Int. SOI Conf. Proc.,* 79-80 (2007).
7. Y. Li, *J. Comp. Electronics,* **5**, 371-376 (2006).
8. A. Godoy, A. Ruiz-Gallardo, F. Gámiz and C. Sampedro, *J. Comp. Electronics,* **6**, 1-3 (2006).
9. I. M. Tienda-Luna, F. G. Ruiz, L. Donetti, A. Godoy and F. Gámiz, *Solid-State Elec.,* **52**, 1854-1860 (2008).
10. A. Trellakis, A. T. Galick, A. Pacelli and U. Ravaioli, *J. Appl. Physics,* **81**, 12, 7880-7884 (1997).
11. F. G. Ruiz, A. Godoy, F. Gámiz, C. Sampedro and L. Donetti, *IEEE Trans. Elec. Devices,* **54**, 12, 3369-3377 (2007).
12. Y. Taur and T. H. Ning, *Fundamentals of modern VLSI devices,* Cambridge Press, 1998.

ECS Transactions, 19 (4) 133-138 (2009)
10.1149/1.3117401 ©The Electrochemical Society

Gate Stack Influence on GIFBE in nFinFETs

J. A. Martino[a], M. Rodrigues[a], A. Mercha[b], E. Simoen[b],
A. Veloso[b], N. Collaert[b], M. Jurczak[b] and C. Claeys[b,c]

[a] LSI/PSI/USP, University of Sao Paulo, Sao Paulo 05508-900, Brazil
[b] IMEC, Leuven B-3001, Belgium
[c] E.E. Dept., Katholieke Universiteit Leuven, Leuven B-3001, Belgium

The aim of this work is to study for the first time the gate induced floating body effect (GIFBE) of triple gate nFinFETs devices with different gate stack options for working function (WF) engineering. Based on electrical characterizations it is shown that the presence of a cap layer like Dy_2O_3 increases EOT and reduces the gate effective WF which decrease V_{FB} and V_T. As a consequence the transconductance also decreases and a positive shift of the onset of the GIFBE is observed both at room and at high temperature.

Introduction

Multiple gate FETs (MugFETs) have received considerable attention as the most scalable device, presenting excellent short-channel effects (SCEs) immunity and high current drivability [1,2]. FinFETs in combination with high-k dielectrics provide low leakage current devices [3]. However, their architecture offer limited possibilities in tuning the threshold voltage (V_T) for both nMOS and pMOS. Medium V_T values are set by a mid-gap work function (WF) metal gate electrode and a high-V_T can be achieved by increased doping concentration in wider fin devices. Low-Vt is particularly challenging to achieve in FinFETs due to their 3D-architecture and for targeted dense-pitch applications (e.g., scaled SRAMs)[4]. As a possibility, the insertion of a dielectric capping layer such as Dysprosium Oxide (Dy_2O_3) between the gate electrode/dielectric interface has been demonstrated as a way of obtaining low-V_T [5,6]. For the metal gate electrode engineering titanium nitride (TiN) has been most frequently used in FinFETs presenting low resistivity, compatible with standard CMOS flows from a thermal stability viewpoint and can be tuned by varying its thickness [7,8].

The so-called floating body effects (FBEs), originate from the non-grounded body and can degrade the device performance [9]. These effects are especially observed in partially depleted (PD) SOI MOSFETs, but they can also be observed in fully depleted (FD) ones when these are operated with an accumulating back gate bias [10]. For a reduced front-gate-oxide thickness, a new type of floating body effect (GIFBE) has been identified in linear operation, called the gate-induced floating body effect (GIFBE). It results from the body potential increase due to the electron valence band (EVB) tunneling that occurs for a large gate voltage (V_{GF}), giving rise to the injection of majority carriers in the film. As a result, a kink is induced in the linear I_D vs. V_{GF} characteristics and a second peak in the transconductance (g_m) is observed [11-13].

The aim of this study is to analyze the impact of the different gate stacks on some electrical parameters, especially on the GIFBE, of SOI FinFET architectures.

Experimental

MuGFET devices were fabricated on SOI wafers with a 65 nm Si film (H_{fin}) on 145nm BOX. The channel doping concentration is 1.10^{15} cm^{-3}. The reference gate stack consists of 5nm plasma enhanced-ALD (PE-ALD) TiN deposited on 2.3nm MOCVD HfSiO over a 1 nm RTO interfacial oxide layer (IL). The different stacking integration options are: Fig.1a) IL/HfSiO/Dy$_2$O$_3$/TiN; Fig.1b) IL/HfSiO/TiN/Dy$_2$O$_3$/TiN (with different thicknesses of TiN and also of Dy$_2$O$_3$ cap layer were considered for this case) and Fig.1c) the reference deposition sequence IL/HfSiO/TiN. The SEG used in most wafers prior to HDD I/I corresponds to a 750°C, 30min thermal budget. A 1050°C spike anneal in N$_2$ is used for junction activation. More process details are shown in ref (14).

The GIFBE is experimentally investigated at low drain bias V_{DS} for triple gate nFinFETs with a fin width of W_{fin}=2μm, with 10 fins and a channel length of L=10μm. For these devices dimensions the expected behavior should be similar the FD planar MOSFETs. The C-V curves were measured at 1 MHz using an Agilent 4284 A measurement system, and the flatband voltage (V_{FB}) and EOT were obtained using the NCSU CVC program [15]. All current–voltage measurements were measured using an HP-4156C parameter analyzer.

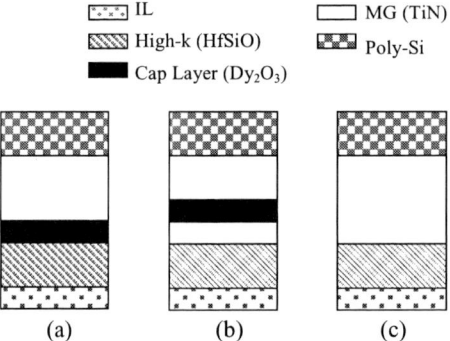

Fig. 1 – Schematics of gate stacks with different deposition sequences, changing the cap layer (Dy$_2$O$_3$) location: (a) IL/HfSiO/Dy$_2$O$_3$/TiN; (b) IL/HfSiO/TiN/Dy$_2$O$_3$/TiN; (c) IL/HfSiO/TiN.

Results and Discussions

From the gate to channel capacitance (C_{GC}) in function of the gate voltage curves shown in Fig. 2a one can see a shift in the flatband voltage (V_{FB}) to more negative values when compared to the reference. In addition, a reduction in the maximum capacitance (C_{GC}) is observed for devices with cap layer. The threshold voltage (V_T) was also analyzed through the drain current versus gate voltage for the different gate stacks (Fig. 2b). The V_T presents the same behavior than the V_{FB}, going to more negative values for the devices with cap layer.

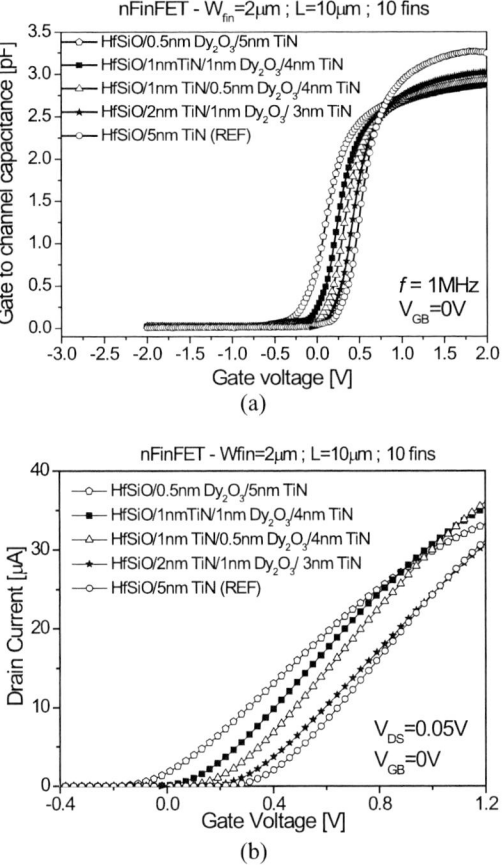

Fig. 2 – (a) Gate to channel capacitance and (b) drain current in function of the gate voltage for the different gate stacks.

Table I presents the values of EOT, V_{FB} and V_T extracted for the different gate stack combinations studied and one can see an increase in EOT and a reduction of V_{FB} and V_T when compared to the reference one. This variation in V_{FB} may be related to a reduction of the effective work function (eWF) due to the presence of a Dy_2O_3 cap layer.

TABLE I. EOT, V_{FB} and V_T extracted for the different gate stacks.

Combinations	EOT [nm]	V_{FB} [V]	V_T [V]
0.5 nm Dy_2O_3 / 5nm TiN	2.09	-0.90	-0.07
1nmTiN / 1nm Dy_2O_3/ 4nmTiN	2.02	-0.88	0.11
1nmTiN/0.5nm Dy_2O_3/4nmTiN	1.99	-0.87	0.19
2nmTiN / 1nm Dy_2O_3/ 3nmTiN	1.90	-0.84	0.27
5nm TiN (REF)	1.64	-0.80	0.35

The gate leakage current (I_g) as a function of the gate voltage is presented in Fig. 3. The highest leakage current level is observed for the reference device, which is in agreement with EOT values that are lower for the device without cap layer between HfSiO and TiN.

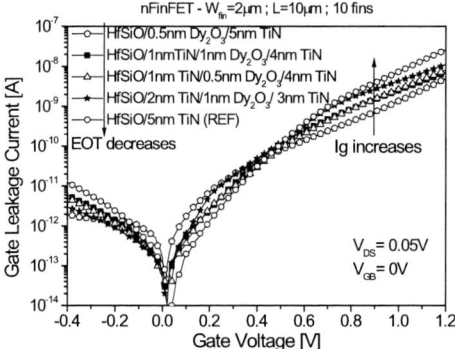

Fig. 3 – Gate leakage current in function of the gate voltage for the different gate stacks.

The GIFBE behavior (g_m second peak amplitude and horizontal shift) in devices with different gate stacks is presented in Fig. 4. The horizontal shift is better observed in Fig. 5 by the gate voltage corresponding with the maximum variation of the g_m near the second peak, V_{T2p} (obtained by the peak of the second derivative of $I_D \times V_{GF}$ in this region) as a function of V_T.

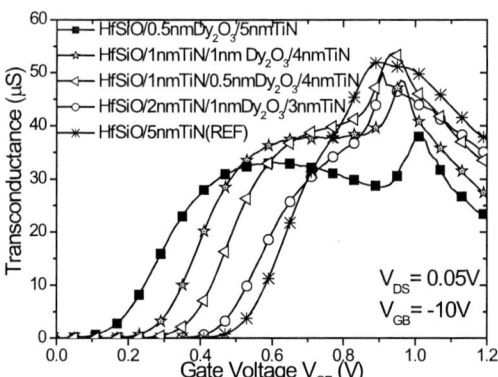

Fig. 4 – Transconductance for the different gate stacks for $V_{GB} = -10V$.

Fig. 5 shows that there is a good and clear correlation between V_{T2p} and V_T. When V_T decreases (which is related to the reduction of V_{FB}) an increase of V_{T2p} (g_m second peak amplitude decreases) is noticed thanks to the decrease of the gate tunneling current (I_g) due to the EOT increase. The GIFBE is also analyzed as a function of temperature till 100^0C. For increasing temperatures a reduction of both V_T and V_{T2p} is observed for all gate stacks, showing the same tendency, due to the Fermi level decrement.

136

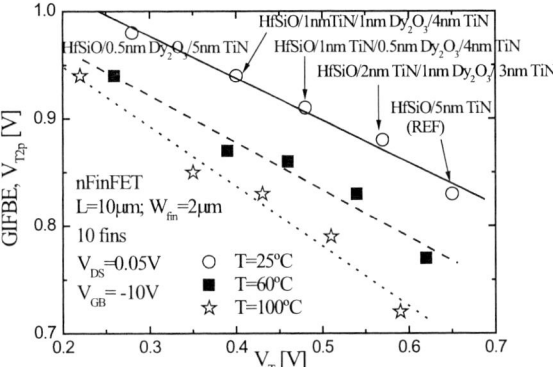

Fig. 5 – GIFBE onset (V_{T2p}) as a function of threshold voltage for different temperatures.

Fig. 6 presents the GIFBE variation with the V_{FB} that follows the same V_T tendency. With the subthreshold slope values smaller than 70 mV/dec trap oxide charges densities can be neglected. As a result, the V_{FB} shift for the cap layer devices when compared with the reference should really be related to the eWF variation.

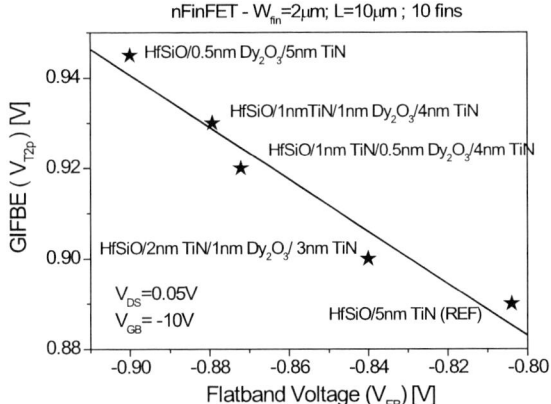

Fig. 6 – GIFBE onset (V_{T2p}) as a function of the flatband voltage for different gate stacks.

Conclusions

The influence of several gate stacks on the GIFBE in MuGFET devices is evaluated for the first time, showing that the insertion of a cap layer (Dy_2O_3) between the HfSiO and the TiN metal gate, increases EOT and decreases V_{FB} and V_T due to the eWF reduction. As a consequence, the I_g (EVB) decreases and the onset of GIFBE (V_{T2p}) increases. The same behavior was observed for different temperatures.

Acknowledgments

J.A. Martino and M. Rodrigues would like to acknowledge the Brazilian research-funding agencies of FAPESP, CNPq and CAPES for the support for developing this work.

References

1. D. Hisamoto, W. C. Lee, J. Kedzierski, H. Takeuchi, K. Asano, C. Kuo, E. Anderson, T. J. King, J. Bokor and C. Hu, *IEEE Trans. Electron Devices*, **47**, 2330 (2000).
2. E. J. Nowak, T. Ludwig, I. Aller, J. Kedzierski, M. Ieong, B. Rainey, M. Breitwisch, V. Gemhoefer, J. Keinert and D. M. Fried, *Proc. IEEE Custom Integr. Circuits Conf.*, p. 339 (2003).
3. G. Vellianitis, M. J. H. van Dal, L. Witters, G. Curatola, G. Doornbos, N. Collaert, C. Jonville, C. Torregiani, L. S. Lai, J. Petry, B.J. Pawlak, R. Duffy, M. Demand S. Beckx, S. Mertens, A. Delabie, T. Vandeweyer, C. Delvaux, F. Leys, A. Hikavyy, R. Rooyackers, M. Kaiser, R.G. Weemaes, F. Voogt, H. Roberts, D. Donnet, S. Biesemans, M. Jurczak and R.J.P. Lander, *IEDM Tech. Dig.*, p.684 (2007).
4. Y. Taur, *IEEE Transac. on Electron Devices*, **48**, 2861 (2001).
5. Tackhwi Lee; Se Jong Rhee; Chang Yong Kang; Feng Zhu; Hyoung-sub Kim; Changhwan Choi; Ok, I.; Manhong Zhang; Krishnan, S.; Thareja, G.; Lee, J.C., *IEEE Electron Devices Letters*, **27**, 640 (2006).
6. H.Y. Yu, S.Z. Chang, A. Veloso, A. Lauwers, C. Adelmann, B. Onsia, S. Van Elshocht, R. Singanamalla, M. Demand, R. Vos, T. Kauerauf, S. Brus, X. Shi, S. Kubicek, C. Vrancken, R. Mitsuhashi, P. Lehnen, J. Kitt, M. Niwa, K.M. Yin, T. Hoffmann, S. Degendt, M. Jurczak, P. Absil and S. Biesemans, *VLSI Symp. Dig. Techn. Papers*, p.18 (2007).
7. Y. Liu, S. Kijima, E. Sugimata, M. Masahara, K. Endo, T. Matsukawa, K. Ishii, K. Sakamoto, T. Sekigawa, H. Yamauchi, Y. Takanashi and E. Suzuki, *IEEE Transactions on Nanotec.*, **5**, 723 (2006).
8. K. Choi, H.-C. Wen, H. Alshareefa, R. Harrisb, P. Lysaght, H. Luanc, P. Majhid, and B. H. Lee, *ESSDERC/2005*, p. 101 (2005).
9. S. Cristoloveanu and SS. Li, Boston : Kluwer Academic Publishers, 1995.
10. J-P. Colinge, Boston : Kluwer Academic Publishers, 2004.
11. J. Pretet, T. Matsumoto, T. Poiroux, S. Cristoloveanu, R. Gwoziecki, C. Raynaud, A. Roveda and H. Brut, *ESSDERC*, p. 515 (2002).
12. A. Mercha, J. M. Rafí, E. Simoen, E. Augendre, and C. Claeys, *IEEE Trans. Electron Dev.*, **50**, 1675 (2003).
13. W. C. Lee and C. Hu, *IEEE Trans. Electron Dev.*, **48**, 1366 (2001).
14. A. Veloso, L. Witters, M. Demand, I. Ferain, N. J. Son, B. Kaczer, Ph. J. Roussel, E. Simoen, T. Kauerauf, C. Adelmann, S. Brus, O. Richard, H. Bender, T. Conard, R. Vos, R. Rooyackers, S. Van Elshocht, N. Collaert, K. De Meyer, S. Biesemans, and M. Jurczak, *VLSI Symp. Dig. Techn. Papers*, p. 14 (2008).
15. http://www.nnf.ncsu.edu/testing.

ECS Transactions, 19 (4) 139-144 (2009)
10.1149/1.3117402 ©The Electrochemical Society

Analytical Modeling of Double Gate Graded-Channel SOI Transistors for Analog Applications

F. A. L. P. Ferreira[a,*], A. Cerdeira[b], D. Flandre[c], M. A. Pavanello[a,d,**]

[a] Department of Electrical Engineering, Centro Universitário da FEI,
Av. Humberto de A. C. Branco nº 3972
09850-901 – Sao Bernardo do Campo – Brazil
* fferreira@fei.edu.br,** pavanello@fei.edu.br
[b] Sección de Electrónica del Estado Sólido (SEES), CINVESTAV, Mexico
[c] Laboratoire de Microélectronique, Université Catholique de Louvain, Belgium
[d] LSI/PSI/USP, University of Sao Paulo, Brazil

In this work we present the development of an analytical model for double gate (DG) Silicon-on-Insulator (SOI) nMOSFET transistor with graded-channel (GC), valid from weak inversion to strong inversion. Atlas numerical two-dimensional simulations and experimental results are used to validate the proposed model. Good agreement between simulated, modeled and experimental results is demonstrated.

Introduction

Double Gate (DG) transistors are considered an attractive option to improve the CMOS circuit performance, alleviating the occurrence of short-channel (SCEs) (1), thanks to the improved charge control in the channel, resulting in nearly ideal subthreshold slope, as well as reduced mobility degradation with respect to planar single gate devices (2-3).

Characteristics such as larger drain current, larger transconductance and reduced output conductance also make DG transistors attractive for analog circuits (4) in addition to their improved performance in digital circuits.

The graded-channel (GC) structure (Figure 1) consists in the use of asymmetrical doping profile in the channel region, obtained by adjusting the conventional threshold voltage (V_{th}) ion implantation to be performed at the source side only while the remaining channel, at the drain side, is masked preserving the natural wafer doping level (with length L_{LD}). This minimizes the electric field close to the drain (5). The GC structure was successfully implemented in double-gate configuration resulting in extremely improved analog performance, mainly an appreciable reduction of the drain output conductance leading to improved Early voltage ($\approx$1600V) and intrinsic gain (> 75 dB) for 3 µm long devices (4-5), only obtained recently in narrow FinFETs with fin width of 20 nm (6).

The DG structure used in this work is a Gate-All-Around (GAA), which has its channel surrounded by the gate oxide providing excellent analog behavior in extreme high-temperature and radiation environments. It can be considered as a DG due to the negligible current contribution from the sidewall gates.

139

This work presents the development of an analytical model for DG GC nMOSFET SOI transistors, valid from weak inversion to strong inversion. The model results were verified against Atlas two-dimensional numerical simulations and experimental data.

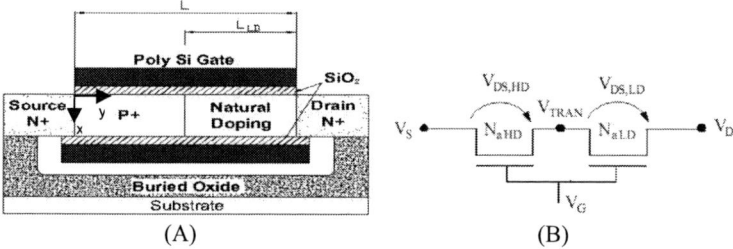

(A)   (B)

Figure 1. Cross-section of a Graded-Channel Gate-All-Around SOI nMOSFET (A) and series association of two GAA SOI transistors representing the electrical behavior of the GC GAA SOI structure (B).

Analytical Model Formulation

Considering a stepped variation in the doping level between the two channel regions of a DG GC transistor (figure 1A), the device equivalent circuit consists of a series association of two transistors (figure 1B) with different doping levels, one highly doped (HD) and other lightly doped (LD) representing the corresponding channel region, with common gates, as successfully used in the reference (7). Applying the current conservation in the structure, one can find the output drain current (I_{DS}) by calculating the current in one of the equivalent association transistors by using the voltage in the intermediate node V_{TRAN}, leading to a uniformly high doped transistor with channel length L_{HD} and drain voltage of V_{TRAN}. Thus the model strategy is to find an analytical solution for V_{TRAN} and apply the analytical model available in the literature for the HD uniformly doped transistor.

The strategy to obtain V_{TRAN} analytical solution is to split the two operational regimes, linear and saturation, modeling each regime using the equivalent circuit and then integrating it by using a fitting function.

In linear region, one can find transition voltage by using the classical triode current equation for each transistor, where the drain voltage in the HD transistor can be interpreted as $V_{TRAN,Triode}$. Neglecting the quadratic terms, one can solve V_{TRAN} for triode region as the following equation [1]:

$$V_{TRAN,TRIODE} = \frac{V_{DS}}{1 + \left[\dfrac{\mu_{nHD}}{\mu_{nLD}} \cdot \dfrac{L_{LD}}{L - L_{LD}} \cdot \dfrac{(V_{GS} - V_{th,HD})}{(V_{GS} - V_{th,LD})} \right]} \qquad [1]$$

In saturation region, the same current conservation technique is applied, but using existing DG models for each doping side. To calculate the drain current in the HD device, $I_{DS,HD}$, we used the model proposed in reference (8). This model is continuous in all operation regimes and needs no fitting parameters, as it is a physically charge-based model. The $I_{DS,HD}$ can be calculated as:

$$I_{DS,HD} = \frac{2W\mu_{HD}}{L-L_{LD}}\left[2UT\left(Q_{S,HD}-Q_{D,HD}\right)+\frac{Q_{S,HD}^2-Q_{D,HD}^2}{2C_{ox}}+UT.Q_{Dep}\ln\left[\frac{Q_{D,HD}+Q_{Dep}}{Q_{S,HD}+Q_{Dep}}\right]\right] \qquad [2]$$

where UT=kT/q (k is the Boltzmann constant and T is the absolute temperature), and $Q_{S,HD}$ and $Q_{D,HD}$ are the inversion charges at the source and drain for the HD transistor, that can be generically written as a function of the applied bias:

$$Q_{\chi,HD} = C_{ox}\left(-\frac{2C_{ox}UT^2}{Q_{Dep}}+\sqrt{\left(\frac{2C_{ox}UT^2}{Q_{Dep}}\right)^2+4UT^2\ln^2\left[1+\exp\left[\frac{V_{GS}-V_{th,HD}-V_{(y)}}{2UT}\right]\right]}\right) \qquad [3]$$

where χ can be substituted to S or D to calculate source or drain inversion charge, and $V_{th,HD}$, that also corresponds to the DG GC transistor threshold voltage. The remaining symbols have their usual meaning.

To calculate the drain current in the LD device, the model presented in reference (9) has been used. It is a complete analytical model for undoped body DG transistors, also continuous in all operational regimes. The $I_{DS,LD}$ can be solved as:

$$I_{DS,LD} = \frac{W\mu_{LD}}{L_{LD}}\left[2UT\left(Q_{S,LD}-Q_{D,LD}\right)+\frac{Q_{S,LD}^2-Q_{D,LD}^2}{4C_{ox}}+8UT^2C_{Si}\ln\left[\frac{Q_{D,LD}+2Q_0}{Q_{S,LD}+2Q_0}\right]\right] \qquad [4]$$

where $Q_0 = 4\beta C_{ox}$, and once again, $Q_{S,LD}$ and $Q_{D,LD}$ are the inversion charges at the source and drain, but for the LD transistor, that can be generically written as a function of the applied bias:

$$Q_{\chi,LD} = 2C_{ox}\left(-\frac{2C_{ox}UT^2}{Q_0}+\sqrt{\left(\frac{2C_{ox}UT^2}{Q_0}\right)^2+4UT^2\ln^2\left[1+\exp\left[\frac{V_{GS}-V_{th,LD}-V_{(y)}}{2UT}\right]\right]}\right) \qquad [5]$$

In order to reach an analytical solution for the transition voltage, equations [2] and [4] where used, considering $V_{(y=Leff)}=V_{TRAN,Sat}$ ($L_{eff}=L-L_{LD}$) in both models, with some simplifications: neglecting $Q_{D,HD}$ at the third sum term of [2], since it is only significant in triode, removing the unitary term in the logarithm of [3] and [5], as it is only important in subthreshold and neglecting first and third sum terms in [4], since this term is the dominating one when obtaining the current in the LD region. These approximations were checked with respect to the iterative solution obtained using both complete analytical models (7) and are valid for the saturation region. The $V_{TRAN,Sat}$ can be solved as:

$$V_{TRAN,Sat} = V_{GS} - CA\times\left(\frac{CB+\sqrt{CC}}{C_{ox}}\right) \qquad [6]$$

where CA, CB and CC are model parameters defined as:

$$CA = \frac{1}{2(2x+a)} \qquad [7]$$

$$CB = 2C_{ox}\left[a\times(V_{th,HD}-2UT-d)+2x\times(V_{th,LD}-z)\right] \qquad [8]$$

$$CC = 4x^2y^2 + 8a \times \left\{ \begin{array}{l} a \times (2UTC_{ox}b + 2C_{ox}{}^2UT^2 + C_{ox}.c + \dfrac{b^2}{2}) \\[4pt] + x \times (2C_{ox}c + b^2 + 4UTC_{ox}b + \dfrac{y^2}{4}) \\[4pt] + C_{ox}{}^2x \times \left[\begin{array}{l} 2z \times (V_{th,LD} - V_{th,HD} + d) \\ + 4UT(V_{th,HD} - d + z - V_{th,LD}) \\ + 2d(V_{th,HD} - V_{th,LD}) \\ - V_{th,LD}{}^2 - z^2 - V_{th,HD}{}^2 - d^2 \end{array} \right] \end{array} \right\} \qquad [9]$$

and a, b, c, x, y, z are:

$$a = 2\mu_{nHD}\frac{W}{L_{HD}}, b = Q_{S,HD}, c = UTQ_{Dep}\ln\left(\frac{Q_{Dep}}{Q_{Dep} + Q_{S,HD}}\right), d = \frac{-2C_{ox}UT^2}{Q_{Dep}} \qquad [10]$$

$$x = \mu_{nLD}\frac{W}{L_{LD}}, y = Q_{D,LD}, z = \frac{-2C_{ox}UT^2}{Q_0}$$

Then, using a fitting function to integrate [1] and [6], one can find the complete transition voltage as:

$$V_{TRAN} = V_{TRAN,SAT} - V_{TRAN,SAT} \cdot \left(\frac{\ln\left[1 + \exp(A_{TS} \cdot (1 - \frac{V_{TRAN,TRIODE}}{V_{TRAN,SAT}}))\right]}{\ln\left[1 + \exp(B_{TS})\right]} \right) \qquad [11]$$

where the parameters A_{TS} and B_{TS} control the linear and saturation region transition.

Thus, one can find the DG GC drain current by using eqn. [11] as the drain voltage into eqn. [2] calculating the drain current of the HD device.

Results and Discussion

ATLAS numerical two-dimensional simulations of GC GAA SOI nMOSFETs were used to validate this proposed equivalent model by comparing the equations solution with simulation results. The device characteristics used for the simulations are 2 nm gate oxide thick (t_{oxf}) and 50 nm silicon film thick (t_{Si}), with doping concentration of $N_{A,HD}=10^{17}$ cm^{-3} and $N_{A,LD}=10^{15}$ cm^{-3}. Devices with L=2 μm were used in the simulations. Figure 2A presents the drain current (I_{DS}) and the transconductance (g_m) as a function of the gate voltage (V_{GS}) curves extracted at $V_{DS}=1.5$ V, and figure 2B presents the results obtained for drain current (I_{DS}) and output conductance (g_d) as a function of the drain voltage (V_{DS}) with gate voltage overdrive ($V_{GT}=V_{GS}-V_{th}$) of 200 mV, both figures for L_{LD}/L ratio of 0.20, 0.30 and 0.50.

GC GAA SOI nMOSFETs were fabricated in a p-type wafer with an initial boron concentration of 10^{15} cm^{-3}. The 30 nm-thick gate oxide is grown and the threshold voltage ion implantation is performed, yielding a body doping level of about 10^{17} cm^{-3}. Three fingers connected in parallel with both channel length and channel width equal to 3 μm compose the measured devices fabricated at UCL (5). The final thickness of the silicon film is 80 nm. Transistors with L_{LD}/L=0.20, 0.35 and 0.50, compose the set of measured devices. Figure 3A presents I_{DS} and g_m as a functions of V_{GS} curves extracted at V_{DS}=0.1 V, and figure 3B presents the comparison between device measured and modeled I_{DS} as a function of V_{DS} with V_{GT}=200 mV.

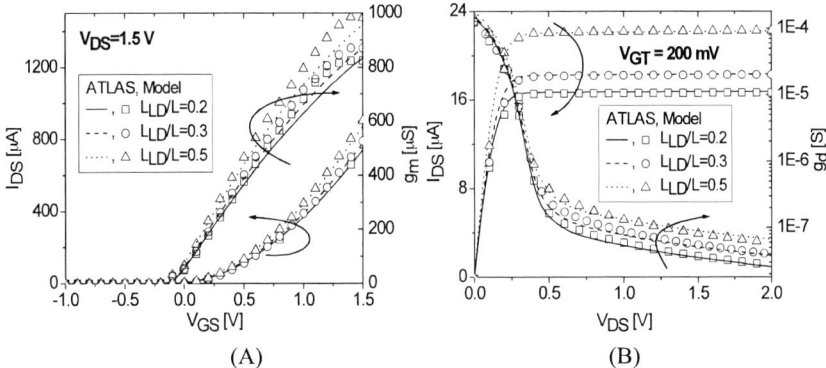

(A)   (B)

Figure 2. Simulated (lines) and modeled (symbols): (A) I_{DS} and g_m as a function of V_{GS} for V_{DS}=1.5 V and (B) I_{DS} and g_d as a function of V_{DS} for V_{GT}=200 mV.

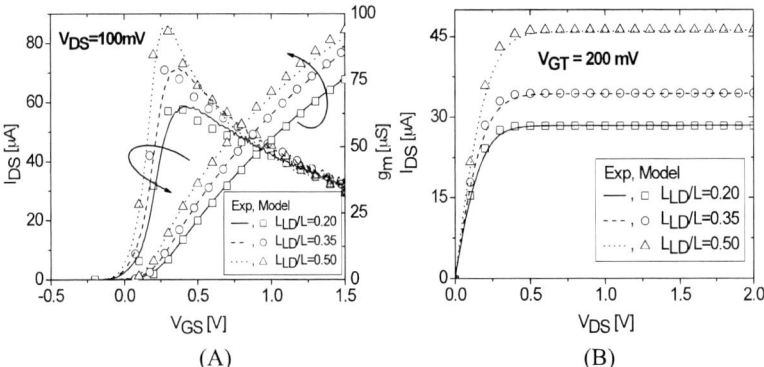

(A)   (B)

Figure 3. Experimental (lines) and modeled (symbols) (A) I_{DS} and g_m curves as a function of V_{GS} for V_{DS}=0.1 V, and (B) I_{DS} as a function of V_{DS} curves for V_{GT}=200 mV.

Table I presents calculated intrinsic voltage gain ($A_{V0}=g_m/g_D$) at V_{GT}=200 mV and V_{DS}=1.5 V for simulated and modeled results obtained with different L_{LD}/L ratios. Reasonable agreement with errors in the order of 1 dB where obtained.

TABLE I. Simulated and modeled values of intrinsic voltage gain (A_{V0}).

L_{LD}/L Ratio	Simulated A_{V0} (dB)	Modeled A_{V0} (db)
0.20	70.5	69.4
0.30	67.9	66.7
0.50	64.6	65.1

One important parameter for analog circuit design is the transconductance over drain current ratio (g_m/I_{DS}). Figure 4 plots the results of g_m/I_{DS} as a function of the normalized drain current ($I_{DS}/(W/L_{eff})$) for the 3μm GC GAA SOI with different L_{LD}/L ratios obtained from simulated and modeled results. Good agreement is found in weak and strong inversion regions whereas the error increases to 12 % in moderate inversion.

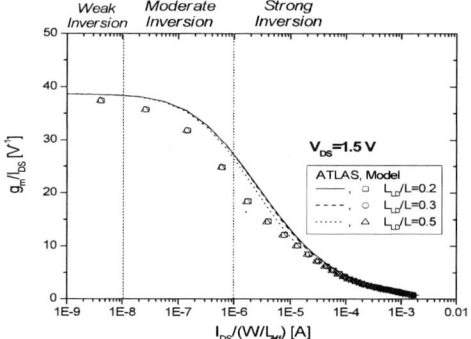

Figure 4. Simulated (lines) and modeled (symbols) g_m/I_{DS} as a function of the scaled drain current ($I_{DS}/(W/L_{eff})$) with $V_{DS}=1.5$ V, of the GC GAA SOI nMOSFET transistor.

Conclusion

The results demonstrate the good agreement between the proposed DG GC model, numerical simulations and experimental data, both in triode and saturation regions of operation, with errors smaller than 5%. The transconductance over drain current ratio was presented with good agreement in all regions of operation, although the error in moderate inversion increased to 12%. Also the intrinsic gain was studied with good agreement. The larger drain current caused by the L_{LD}/L ratio increase is well described by the model as well as the derivatives of I_{DS} with respect to V_{GS} and V_{DS}. The presented model describes satisfactorily the DG GC device characteristics and can be used for analog circuit design.

Acknowledgments

The authors would like to acknowledge the Brazilian research-funding agencies of CNPq and FAPESP for the support for developing this work.

References

1. M. Masahara, M. Liu, Y. Sakamoto, K. Endo, K. Matsukawa, Y. Ishii, K. Sekigawa, T. Yamauchi, H. Tanoue, H. Kanemaru, S. Koike and H. Suzuki, *IEEE Trans. on Electron Devices*, **9**, 52 (2005).
2. Y. Li and H. M. Chou, *IEEE Trans. on Nanotechnology*, **5**, 4 (2005).
3. D. Jiménez, B. Iñiguez, J. Suñé and J. J. Sáenz, *J. of Applied Physics*, **9**, 96 (2004).
4. A. Kranti, T. M. Chung, D. Flandre and J.-P. Raskin, *Solid-State Electronics*, **48**, 947 (2004).
5. M. A. Pavanello, J. A. Martino, J.-P. Raskin and D. Flandre, *Solid-State Electronics*, **49**, 10 (2005).
6. D. Lederer, V. Kilchytska, T. Rudenko, N. Collaert, D. Flandre, A. Dixit, K. De Meyer and J.-P. Raskin, *Solid-State Electronics*, **49**, 1488 (2005).
7. F. Ferreira, A. Cerdeira and M. A. Pavanello, *ECS Transactions*, **14**, 177 (2008).
8. B. Iñiguez, A. Lázaro, O. Moldovan, A. Cerdeira and T.A. Fjeldly, *NSTI-Nanotech*, **3**, 680 (2006).
9. O. Moldovan, D. Jiménez, J. Roig, F. A. Chaves and B. Iñiguez, *IEEE Trans. On Electron Devices*, **54**, 1718 (2007).

ECS Transactions, 19 (4) 145-152 (2009)
10.1149/1.3117403 ©The Electrochemical Society

Low-Temperature Measurements on Germanium-on-Insulator pMOSFETs: Evaluation of the Background Doping Level and Modeling of the Threshold Voltage Temperature Dependence

W. Van Den Daele[1], E. Augendre[2], K. Romanjek[2], C. Le Royer[2],
L. Clavelier[2], J.-F. Damlencourt[2], E. Guiot[3], B. Ghyselen[3], S. Cristoloveanu[1]

[1]IMEP-LAHC Minatec Grenoble-INP, 3 parvis Louis Neel, BP257, 38016 Grenoble cedex 1, France
[2]CEA-Leti MINATEC, F38054, Grenoble, France
[3]SOITEC S.A., Parc Technologique des Fontaines, Bernin, France
Email: vandendw@minatec.inpg.fr

Fully depleted GeOI (Germanium on Insulator) pMOSFETs with HfO$_2$/TiN gate stack and Si-passivation are studied at low temperature. The impact of the starting Ge material and N-type channel doping on threshold voltage is examined. As there is no evidence of a background doping in Ge films, the typical parasitic conduction and threshold voltage shift in p-channel GeOI MOSFETs are due to interface states. An extended model predicting the threshold voltage temperature dependence in GeOI transistors is proposed.

INTRODUCTION

The development of high-κ metal-gate stacks promoted germanium as a promising candidate to replace silicon for 'Beyond CMOS' technologies, mainly due to high hole mobility values in pMOSFETs [1]. High performance 70 nm GeOI pMOSFETs with pockets, Source/Drain extensions and V$_T$ adjustment have recently been demonstrated [2]: very good ON-state ($I_{ON} = 330$ μA/μm) and OFF-state ($I_{OFF} = 1$ μA/μm) currents were achieved. Fully-depleted GeOI structures are very attractive for excellent hole mobility and reduced leakage current. However, a critical issue is the positive V$_T$ shift for front and back channels, resulting in a parasitic OFF-state current in pMOSFETs. A strong positive back-gate voltage ($V_{BG} = + 60$ V) is needed to turn off the back channel and decouple front and back interfaces [2].

Several models and hypotheses [3, 4, 5] assume that the V$_T$ shift is due to unpassivated interface states in the Ge bandgap close to the valence band. Another hypothesis associates it to parasitic doping species introduced during the process. In this work, we explore, using systematic low-temperature measurements on various GeOI substrates, the doping problem as well as the threshold voltage behavior. We discuss the origin of the parasitic conduction and propose an extended model to predict the V$_T$ temperature dependence.

SUBTRATE AND DEVICES DESCRIPTION

200 mm GeOI wafers with a 40-80 nm thick germanium active layer were fabricated in CEA- Leti, using three processes: (i) Smart Cut^{TM} technology [6] with epitaxial Ge donor (Sample A), (ii) Smart-Cut using bulk Ge (Sample B), and (iii) Ge condensation technique (Sample C) [7, 8, 9]. First, Ge mesa structures were patterned, followed by silicon passivation and oxidation. 6 nm HfO$_2$ ALCVD, 10 nm TiN PVD, Poly-Si and a SiO$_2$ hard mask were deposited. The measured EOT is 2.4 nm. Source and drain were then implanted (using BF$_2$) and annealed (600°C), followed by a standard Ti/TiN/W

145

ECS Transactions, 19 (4) 145-152 (2009)

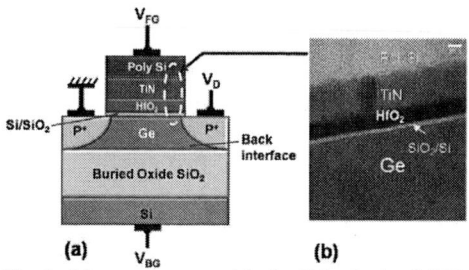

Sample	Non doped	N-type doped	t_{Ge} (nm)	t_{BOX} (nm)
Epitaxial	A	A'	65	320
Bulk	B	B'	75	390
Ge Con-densation	C		60	145

Fig. 1: Schematic structure (a) of a Fully-Depleted (FD) GeOI pMOSFET with a Si/SiO$_2$ passivation layer fabricated at the CEA-Leti; (b) High-resolution XTEM picture showing the Si/SiO$_2$/high-κ/TiN gate stack on Ge.

Fig. 2: Description of the characterized samples. Three different donors were used for the active Ge layer.

contact formation and Al metallization. N-type channel dopants have been implanted on epitaxial and bulk donor Smart Cut^{TM} GeOI wafers to compare the behaviour of non-doped and doped channels. The schematic structure of the isolated pMOSFETs is shown in Fig. 1(a) and a XTEM picture of the high-κ/TiN gate stack in Fig. 1(b). Note that the BOX (Buried Oxide) type and thickness are different for all samples. The samples description is shown in Fig. 2.

LOW TEMPERATURE MEASUREMENTS

Low-temperature $(77 - 300\ K)$ measurements have been performed by recording front-channel characteristics $I_D(V_{FG})$ as a function of back-gate bias V_{BG} (Fig. 3) on long transistors in linear regime ($L_G = 9\ \mu m$; $W = 10\ \mu m$; $V_D = -50\ mV$). Reciprocal measurements have been carried out for the back-channel MOSFET (Fig. 4). The threshold voltage was determined from the maximum of the drain current second derivative to

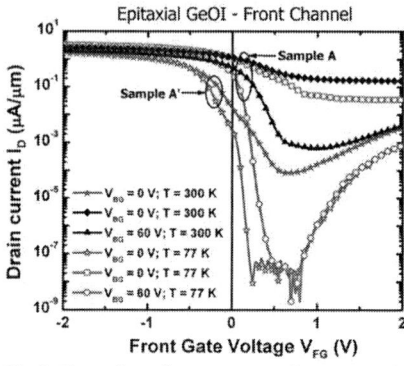

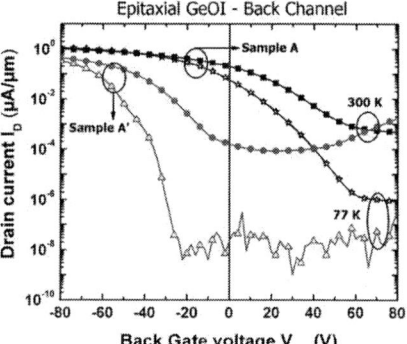

Fig. 3: Front-channel current versus front-gate voltage V_{FG} for samples A and A' at $T = 77\ K$ and $300\ K$. $V_{BG} = 0\ V$ or $+60\ V$ to turn the back interface into accumulation. $V_D = -50\ mV$ and $W/L_G = 10/9\mu m$.

Fig. 4: Back-channel current versus back-gate voltage V_{BG} for samples A and A' at $T = 77\ K$ and $300\ K$. Front interface is accumulated. For sample A, the back channel is activated at $V_{BG} = 0\ V$.

minimize series resistances [10]. We discuss the front and back channels parameters, independently extracted by turning the opposite interface into accumulation.

At room temperature ($T = 300\ K$) for $V_{BG} = 0\ V$, a high leakage current is observed on front channel characteristics (Fig.3) due to the parasitic conduction at the weakly inverted back interface. Therefore, a strong back-gate bias ($V_{BG} \simeq +60\ V$) is applied to decouple the two interfaces in undoped GeOI films. A practical solution to further reduce leakage and to adjust V_T for front/back channels is to implant the Ge film with N-type dopants (As in our case, see Figs. 3 and 4). The front and back drain currents shift towards the negative V_G values and the back-channel is turned off. Nevertheless, such N-type doping degrades the hole mobility in both channels [11]. At low temperature ($T = 77\ K$), the OFF-state current is lowered by reduction of TAT (Trap Assisted Tunneling) and SRH (Shockley Read Hall) mechanisms. The subthreshold behavior is improved because the swing S decreases, whereas the threshold voltage V_T becomes more negative. $I_D(V_{FG})$ characteristics of undoped and N-type doped MOSFETs, measured for $T = 77\ K$ and back-interface turned off , match for $V_{FG} \in [0.5\ V; 2\ V]$ highlighting that the BTBT (Band-to-Band Tunneling) is not influenced by the doping level.

EXTRACTION OF THE BACKGROUND DOPING

The high leakage current observed in Fig. 3 may originate from a parasitic doping. Our aim is to verify this scenario by evaluating the background doping. Film doping can be extracted from the front threshold voltage variations with temperature [12]. A first order model for fully depleted devices yields:

$$V_T(T) \simeq \phi_{MS}(T) + 2\phi_F(T) - \frac{Q_{OX}}{C_{OX}} \pm \frac{qNt_{Ge}}{C_{OX}} \quad \text{with} \tag{1}$$

$$\phi_{MS}(T) = \phi_{TiN} - \chi_{Ge} - \frac{E_G(T)}{2} - \phi_F(T) \quad \text{and} \quad \phi_F(T) = \pm \frac{kT}{q} \ln\left(\frac{N}{n_i(T)}\right)$$

ϕ_{TiN} is the work-function of the TiN metal gate, ϕ_F is the Fermi potential, and N is the doping level. Sign (+) corresponds to a P-type doping and sign (-) to a N-type doping. This approximation of V_T is valid when the film capacitance is much smaller than the oxide capacitance ($C_{Ge}/C_{OX} \ll 1$). We have evaluated $C_{Ge}/C_{OX} \approx 0.1$ for our devices.

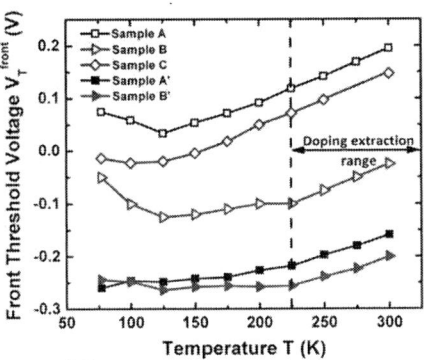

Fig. 5: Front-threshold voltage V_T^{front} versus temperature T for long pMOSFETs on GeOI. The V_T^{front} extraction is performed with the back interface in accumulation for all samples.

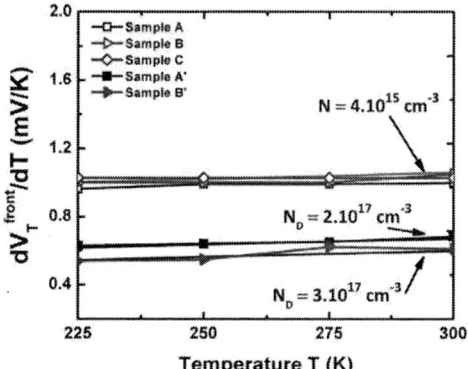

Fig. 6: Derivative of V_T^{front} and $\phi_F(T)$ with temperature for $T \in [225\ K;\ 300\ K]$. By adjusting the value of N in $\phi_F(T)$ it is possible to fit both derivatives and thus evaluate the background doping level.

After derivation, we have:

$$\frac{dV_T}{dT} \simeq \frac{d\phi_F}{dT} \tag{2}$$

Fig. 5 reproduces the variation of V_T with T. By fitting the V_T shift with the Fermi-level derivative (Fig. 6) for $T \in [225\ K;\ 300\ K]$, we evaluate the doping level for the different samples. The non-doped films (sample A, B & C) reveal a very small residual-doping level of $N = 4\times10^{15}\ cm^{-3}$. By contrast, in N-type doped wafers we obtain $N_D = 2\times10^{17}\ cm^{-3}$ for the epitaxial donor (sample A') and $N_D = 3\times10^{17}\ cm^{-3}$ for the bulk donor (sample B'). The doping concentrations in samples A' and B' correspond to the expected value introduced during the implantation step, which validates our extraction method for GeOI devices. Note that we confirmed the residual doping for epitaxial donor films to be smaller than $10^{16}\ cm^{-3}$ by independent Ψ-MOSFET measurements [10]. Such a low background doping definitely negates the possibility of unintentional P-type doping species introduced during the GeOI process. Similar threshold voltage extractions have been performed for back-channel $V_T^{back}(T)$ (see Fig. 7).

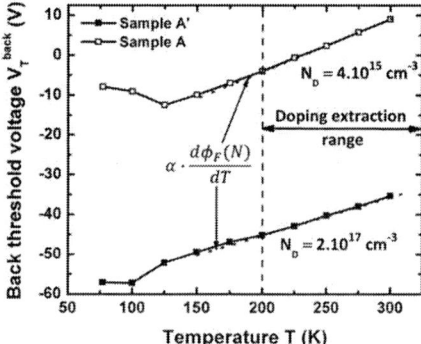

Fig. 7: Back-threshold voltage V_T^{back} versus temperature T for long channel pMOSFETs on GeOI (epi-wafers A & A'). The doping level evaluation method can be adaptated to back-channel by correcting the $\phi_F(T)$ derivative of $\alpha = C_{OX}/C_{BOX}$. Similar values of N_D are found for front gate analysis.

For both N-type doped and non-doped MOSFETs, it is possible to fit the dV_T^{back}/dT data with the derivative of $\phi_F(T)$ corrected by an α-factor, which accounts for the difference in gate oxide and BOX thickness: $\alpha = C_{OX}/C_{BOX}$. We deduce the same doping levels as in Fig. 6. It is concluded that this doping evaluation method is reliable for front and back GeOI channels in a temperature range of $T \in [225\ K;\ 300\ K]$. Nevertheless, the threshold voltage behavior at low temperatures (usually $T \leq 175\ K$) is more complicated and cannot be related to the sole Fermi-level variation. An appropriate model is proposed in the next section.

MODELING OF THE THRESHOLD VOLTAGE TEMPERATURE DEPENDENCE

The measurements above have demonstrated that the positive shift of threshold voltage at room temperature (Fig. 3-5) cannot be explained by a parasitic P-type doping. Recent studies [13] suggested that the P-type character of the Ge surface may be induced by native defects (dangling bonds mainly). We propose an extended model, valid in a wide temperature range. It is based on the negatively charged Ge surface description proposed in [13], which was efficent to describe the V_T shift for Ge pMOSFET but only at $T = 300\ K$. According to [13], we consider a CNL (Charge Neutrality Level) around $E_V + 0.08eV$ as well as charged donor and acceptor dangling bonds (DB) states, respectively at $E_V + 0.05eV$ and $E_V + 0.11eV$ (also predicted by [14]). The DB states are modeled by two gaussians:

$$D_{it}^{don}(E) = D_{it}^0 \exp\left(\frac{-(E-0.05)^2}{2\sigma^2}\right) \quad \text{and} \quad D_{it}^{acc}(E) = D_{it}^0 \exp\left(\frac{-(E-0.11)^2}{2\sigma^2}\right) \quad (3)$$

which are independent of temperature. Typically, the broadening is evaluated at $\sigma = 0.08\ eV$.

For $V_{FG} = 0\ V$ and the back-interface in accumulation, the surface potential Ψ_S^0 is determined by the charge neutrality condition : $Q_{it}^{don} + Q_{it}^{acc} + Q_{Ge} = 0$. At equilibrium, the CNL and Fermi level are aligned. Hence, Q_{it}^{acc} is calculated by integrating the D_{it}^{acc} distribution below the Fermi surface potential $\Psi_F^S = E_G - |\Psi_S^0| - kT/q \ln(N_C/N)$, and Q_{it}^{don} by integrating the D_{it}^{don} distribution above this value. Most of the Ge parameters vary with the temperature, such as the band gap E_G, the conduction/valence band density of states N_C/N_V, the Fermi-level ϕ_F, the intrinsic carrier concentration $n_i(T)$. Therefore to compute Ψ_S^0 as a function of temperature, these dependencies have to be implemented into the basic model. The surface potential Ψ_S^0 then depends on N, D_{it} and T. For each temperature, the surface Fermi potential is determined by:

$$\Psi_F^S(T) = E_G(T) - |\Psi_S^0| - \frac{kT}{q} \ln\left(\frac{N_C(T)}{N}\right) \quad (4)$$

with [15]: $\quad E_G(T) = 0.742 - 4.8 \times 10^{-4} \dfrac{T^2}{T+235}\ ; \quad N_C(T) = 1.989 \times 10^{15} T^{3/2}\ ;$

$$N_V(T) = 9.6 \times 10^{14} T^{3/2} \quad \text{and} \quad n_i(T) = \sqrt{N_C(T)N_V(T)} \exp\left(-\frac{qE_G(T)}{2kT}\right)$$

The surface potential Ψ_S^0, derived from the charge neutrality equation, is shown in Fig. 8 as a function of doping level N (assumed N-type), D_{it}^0, and for $T = 77K$ and $300K$. At fixed doping level ($N_D = 4.10^{15}\ cm^{-3}$), Ψ_S^0 is very sensitive to the temperature whatever

D_{it}, mainly due to the increasing value of Q_{it} when T decreases. Furthermore, for a given value of D_{it}^0 and if N_D is large enough, the Ψ_S^0 characteristics for $T = 77\ K$ and $300\ K$ can merge. This indicates that surface potential variation with T can be driven by N_D only. For high D_{it}^0 (e.g., $2.10^{13}\ eV^{-1}.cm^{-2}$), the merging area is reached only for heavy doping $(N_D \geq 10^{18} cm^{-3})$. The surface potential is hence very sensitive to temperature variations for high density of DBs states, if the doping level is low.

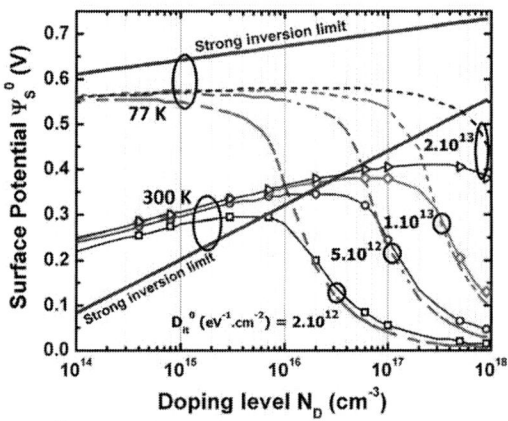

Fig. 8: Surface potential $|\psi_S^0(N, D_{it}, T)|$ as a function of doping level N_D for different D_{it}^0 at $T = 77\ K$ and $300\ K$. The thick lines correspond to the strong inversion limit for 77 K and 300 K.

The front threshold voltage definition for a FD pMOSFET with the back-interface in accumulation can be completed by the above T-dependent model:

$$V_T(T) = \alpha + |\psi_S^0(N, D_{it}, T)| - 2\left(1 + \frac{C_{Ge}}{C_{OX}}\right)\phi_F(T) + \frac{|Q_{it}(N, D_{it}, T)|}{C_{OX}} - \frac{qN t_{Ge}}{2C_{OX}} - \beta \quad (5)$$

$$\text{where} \quad \alpha = \phi_{TiN} - \chi_{Ge} - \frac{Q_{OX}}{C_{OX}}$$

We evaluated $\alpha = 0.13$, with $\phi_{TiN} \approx 4.6\ eV$, $\chi_{Ge} = 4.13\ eV$, and $Q_{OX} \approx 2.10^{12}\ cm^{-2}$ (which is a reasonable value for HfO$_2$/Ge gate stack). β is a correction factor linked to the back-interface passivation [16] and vary between 0 and 0.2 V. Fig. 9 shows the comparison between the computed V_T^{front} and the experimental values. The model can predict the characteristic elbow of $V_T(T)$ for low temperatures. Furthermore, an excellent correlation between measured and theoretical data is achieved for every samples by implementing the N_D and D_{it}^0 values displayed on the column diagram in Fig. 9. Note that the doping concentrations used in the model are equal to the experimental threshold voltage derivative extracted levels and D_{it}^0 values are comparable to literature data for Ge [5]. These results tend to validate our temperature-dependent model as well as the doping extraction method. The model emphasizes that negatively charged DB states, lying close to the valence band, are responsible for the positive V_T shift and the back-interface leakage issues in GeOI structures.

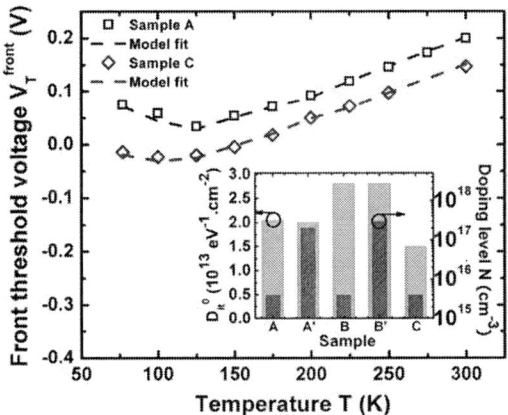

Fig. 9: Front threshold voltage V_T^{front} versus temperature T for samples A and C (non-doped). Experiments are represented by open symbols. Dash-lines correspond to our temperature-dependent model. Inset shows D_{it}^0 and N_D used for each samples.

CONCLUSION

Low-temperature measurements $(77 - 300\ K)$ on long channel FD GeOI pMOSFETs have been performed. An experimental evaluation method of the background doping has been validated on several samples corresponding to different GeOI wafers and doping levels (undoped and N-type doped). This evaluation is reliable in a temperature range commonly defined between 200 and 300 K. Doping level around $4.10^{15}\ cm^{-3}$ was extracted for non-doped samples, which negates the hypothesis of a P-type parasitic doping. Finally, we proposed a extended model to predict the evolution of the threshold voltage with temperature. The model provides excellent correlation with experimental data whatever the type of Ge layer and doping level. These results tend to confirm that V_T shift and back-interface leakage in GeOI pMOSFET are due to dangling-bonds states located close to the Ge valence band.

ACKNOWLEDGMENT

The authors would like to acknowledge B. Grandchamp, L. Benaissa, L. Sanchez, T. Signamarcheix and C. Tabone for competent help. This work has been carried out in the frame of Nanosmart and Nanosil projects.

REFERENCES

1. T. Yamamoto, Y. Yamashita, M. Harada, N. Taoka, K. Ikeda, K. Suzuki et al., *International Electron Devices Meeting, IEDM 2007*, 1041 (2007).
2. K. Romanjek, L. Hutin, C. Le Royer, A. Pouydebasque, M.-A. Jaud, C. Tabone et al., *ESSERC 2008 proceedings*, 75 (2008).
3. A. Dimoulas, P. Tsipas, A. Sotiropoulos and E.K. Evangelou, *Applied Physics Letters*, **89** , no. 25, 2110 (2006).
4. J. Weber, A. Janotti, P. Rinke, C.G. Van de Walle, *Applied Physics Letters*, **91** ,no.14, 2101 (2007).
5. K. Martens, J. Mitard, B. De Jaeger, M. Meuris, H. Maes, G. Groeseneken, F. Minucci, F. Crupi, *ESSERC 2008 proceedings*, 138 (2008).

6. C. Deguet, L. Sanchez, T. Akatsu, F. Allibert, J. Dechamp, F. Madeira et al. *IEEE Electronic Letters*, **42**, no. 7, 51 (2006).
7. T. Tezuka, N. Sugiyama, T. Mizuno, M. Suzuki, S. Takagi , *Applied Physics Letters*, **79**, 1798 (2001).
8. B. Vincent, J.-F. Damlencourt, P. Rivallin, E. Nollot, C. Licitra, Y. Morand, L. Clavelier, *Semiconductor Science and Technology*, **22**, no.3 237 (2007).
9. Q.T. Nguyen, J.-F. Damlencourt, B. Vincent, L. Clavelier, Y. Morand, P. Gentil, S. Cristoloveanu, *Solid State Electronics*, **51**, no.9, 1172 (2007).
10. S. Cristoloveanu and S.S. Li, *Electrical Characterization of Silicon-On-Insulator Materials and Davices*,p. 108 & 247, Kluwer Academic Publishers, Boston (1995).
11. W. Van Den Daele, E. Augendre, L. Clavelier, F. Allibert, E. Guiot, S. Cristoloveanu, *ESSERC 2008 fringe session proceedings*, 122 (2008).
12. G. Groeseneken, J.-P. Colinge, H.E. Maes, J.C. Alderman, S. Holt *IEEE Electron Device Letters*, **11**, no.8, 329 (1990).
13. P. Tsipas and A. Dimoulas, *Applied Physics Letters*, **94**, no.1, 2114 (2009).
14. P. Broqvist, A. Alkauskas, A. Pasquarello, *Physical Review B*, **78**, no.7, 5203 (2008).
15. NSM Archive website, *http://www.ioffe.rssi.ru/SVA/NSM/Semicond/Ge/index.html.*
16. K. Romanjek, E. Augendre, W. Van Den Daele, B. Grandchamp, L. sanchez,C. Le Royer et al., *accepted for presentation at INFOS 2009*, (2009).

ECS Transactions, 19 (4) 153-158 (2009)
10.1149/1.3117404 ©The Electrochemical Society

The Wave SOI MOSFET: A New Accuracy Transistor Layout to Improve Drain Current and Reduce Die Area for Current Drivers Applications

S. P. Gimenez

Department of Electrical Engineering, Centro Universitário da FEI, S. B. Campo, São Paulo, Brazil

This paper proposes a new transistor layout, called here simply as Wave, that can be used for any technology, to improve the current driver and enhanced layout packing with respect to Multifinger and Waffle structures, regarding the same geometric factor. Discussions about this novel layout approach are performed regarding matching, avalanche and electro static discharge. To verify the benefits of the Wave structure, a comparison with a Multifinger and Waffle is carried out. Defining a figure-of-merit as integration factor [(W/L)/A], the Wave features a better efficiency than Multifinger and Waffle layouts, as 35.9 % and 28.1%, respectively. The Wave approach allows a saving of 26.1 % and 21.8% in the power SOI MOSFET size as compared to Multifinger and Waffle layouts.

Introduction

The planar SOI MOSFET technology has been continually scaled down in order to improve the drive current, power consumption, devices integration capability and speed of analog and digital integrated circuits (1). Typically, MOSFETs with high drain current, i.e., with large geometric factor (W/L, where W and L are the channel width and length, respectively), are implemented by connecting in parallel several identical transistors with the same L, but with a small channel width (W'), given by W/N, where N is the number of individual device (2). This structure shares the drain and source contacts between adjacent transistors and consequently reduces the die area, parasitic junction capacitances and device mismatching (2). Different layouts have been studied, such as multifinger (interdigitated) (2) and waffle (2, 3), to improve the transistor electrical performance in general, focusing on analog and digital high accuracy integrated circuits. In Figure 1 are displayed the individuals drain currents (I_{DS}) that occur in only one direction [I_{DS-x} and I_{DSx} (Fig. 1.a) or I_{DS-y} and I_{DSy} (Fig. 1.b) or I_{DS-xy} and I_{DSxy} (Fig. 1.c)] of each transistor in these three Multifinger layouts. Thus, the Multifinger layout is not able to equalize the systematic and random errors effects of fabrication process that affect the drain current in both directions (x or y), reducing the device matching between devices and consequently degrading the electrical performance of analog and digital integrated circuits (4).

The Waffle (3), as presented in Figure 2, is another layout alternative to implement transistors with larger W/L ratio. It presents higher drain current (I_{DS}), smaller die area (A), lower parasitic junctions capacitances (C_j) and smaller influence of the systematic and random errors effects due to fabrication process, than multifinger structure. This occurs because Waffle layout presents two different drain currents for each direction (x and y), in contrast of Multifinger layout (2, 3, 4 and 5). And, in the Waffle, the fabrication process variations that affect I_{DS-x} and I_{DS-y}, can be equalized by the

153

variations that occur in I_{DSx} and I_{DSy}, respectively and therefore it tends further improve the transistors matching, in comparison with multifinger layout, and consequently improve the electrical performance of analog and digital integrated circuits (6).

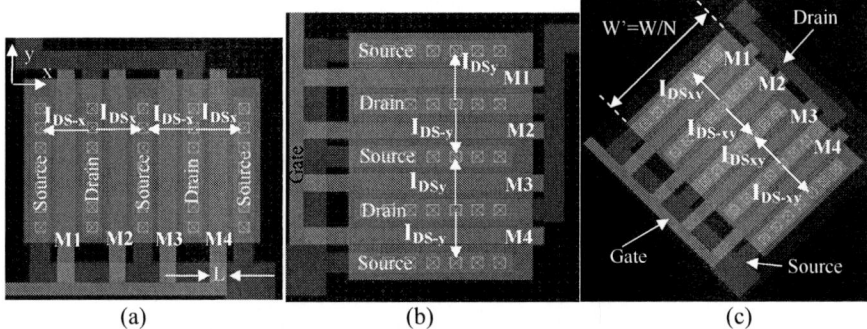

(a)   (b)   (c)

Figure 1. Example of Multifinger layouts, implemented with four SOI nMOSFETs (M1, M2, M3 and M4) connected in parallel, and positioned in different directions with I_{DS} in the x direction (a), y direction (b) and one direction different of x and y directions (c).

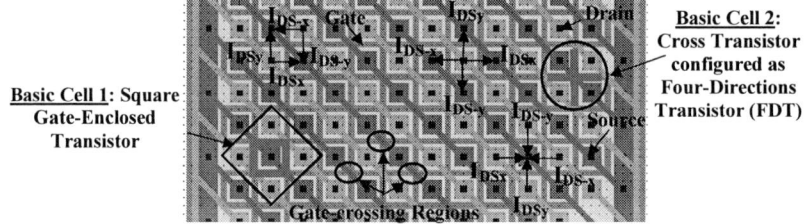

Figure 2. Example of SOI nMOSFET implemented with Waffle layout.

Note that, in Waffle layout, we can regard two different basic cells. The first one is known as Square Gate-Enclose Transistor (basic cell 1: square geometry) (5) and the second basic cell is known as Cross Transistor configured as FDT (basic cell 2: cross shape) (6). In Ref. (6), was performed 3D numerical simulations with second basic cell of Cross Transistor configured as FDT and it was showed that the resultant parallel (longitudinal) electric field ($\vec{\varepsilon}_{//}$) is null in the center of gate-crossing regions and consequently the total drain current density (J_T) is equal to zero, as can be seen in Figure 3.

Therefore, the gate-crossing regions do not operate as MOSFETs (6), and there are improvement opportunities to reduce further the Waffle layout die area. Additionally, in the waffle layout there is a larger numbers of bends in its channels, producing sharp corners ($90°$) in the source and drain regions, where the avalanche phenomenon occurs at lower voltages than multifinger (2). Avalanche phenomenon positioned in these regions reduces the power dissipation of the Waffle layout and consequently this structure is strongly affected the in terms of Electro Static Discharge (ESD) (2).

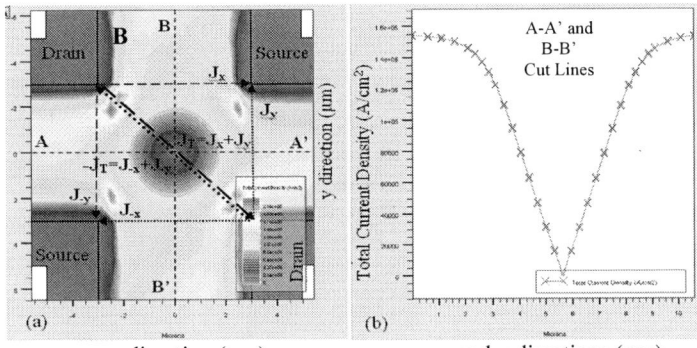

x direction (μm)   x and y directions (μm)

Figure 3. 3D numerical simulation of J_T in the Cross SOI nMOSFET configured as FDT: (a) superior view and (b) J_T as a function of x and y directions (A-A' and B-B' cut lines) (6), where J_x, J_{-x}, J_y, J_{-y} are the drain current density components in the x and y directions.

The Wave SOI nMOSFET ("S" shape layout) and Simulations Results

In order to reduce further die area of the Waffle layout (avoid gate-crossing regions), avalanche effects, ESD (2) and improve the transistor matching (4), a new layout structure is proposed, called Wave SOI MOSFET (filed in August 2008, reference number: 018080049797, Patent, INPI, Brazil) (Figure 4). This new Wave SOI nMOSFET was originated by dividing circular gate SOI nMOSFET (7) into two equal parts (semicircles) and after shifting one semicircle up to implement an "S" shape. Now, this new layout presents symmetrical drain/source dimensions, according with conventional SOI MOSFET and in contrast of circular gate SOI nMOSFET. Besides, this new structure maintains the radial direction of the drain current in each one of semicircles and also the same geometric factor $[f_g=(W/L)_{conv.}=2.\pi/\ln(R2/R1)$, R1 and R2 are initial and final radius that define the SOI MOSFET channel length (L), where R2=R1+L] of the circular gate SOI nMOSFET (7). Also, in Figure 4 is presented the drain current density vectors in inferior and superior semicircles of the Wave layout.

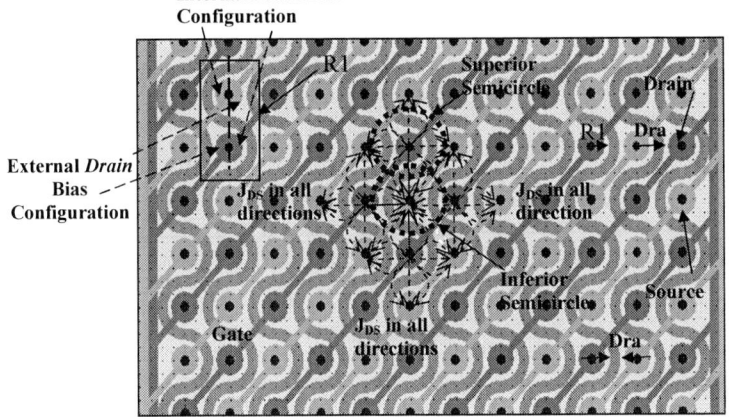

Figure 4. Drain current density vectors in the Wave SOI nMOSFET.

155

Note that, the superior semicircle of Wave SOI nMOSFET layout is configured to operate with internal (right site of drain contact) and external (left site of drain contact) drain bias (7) in the same time. The same occurs in the inferior semicircle of this new structure, regarding internal and external source bias. Therefore, we believe that the effects due to internal and external drain/source bias are smaller Wave than in circular gate layout. Additionally, as the drain current density occurs in all directions (radial direction) in the Wave structure, the process variations influence (systematic and random errors) in the drain current are smaller than in the Waffle layout, because its drain current density occurs in only four directions. Besides, it can observe that, the absence of corners in the Wave layout, where the edges are limited by reticulated pattern photomasks, can also reduce further the avalanche and ESD phenomenon, in comparison with Waffle structure (8). In Figure 5 is displayed the 3D numerical simulated I_{DS} x V_{GS} (Fig. 5.a) and g_m (Fig. 5.b) characteristics for partially depleted (PD) Wave SOI nMOSFET (Fig. 5.c) implemented with semiconductors devices editor DeviEdit3D (9), for V_{DS} equal to 25 mV and 0.1 V, respectively. n-well CMOS process parameters were attributed: gate-oxide (t_{ox}), silicon film (t_{si}) and buried-oxide (t_{box}) thickness are 3 nm, 100 nm and 390 nm, respectively, n^+ ($N_{drain/source}$) and P-substrate ($N_{P-Subs.}$) concentrations are equals to $1x10^{20}$ cm^{-3} and $5.5x10^{17}$cm^{-3}, respectively. Some ATLAS3D models were regarded in 3D-simulations (9): Lombardi's vertical and horizontal electric-field-dependent mobility, Fowler-Nordheim tunneling (electron and holes), Selberherr impact ionization, Shockley-Ready-Hall recombination are used. The threshold voltage (V_{TH}) of 0.12 V and subthreshold slope (S) of 69 mV/dec. were extracted for I_{DS}xV_{GS} curve for V_{DS}=10 mV.

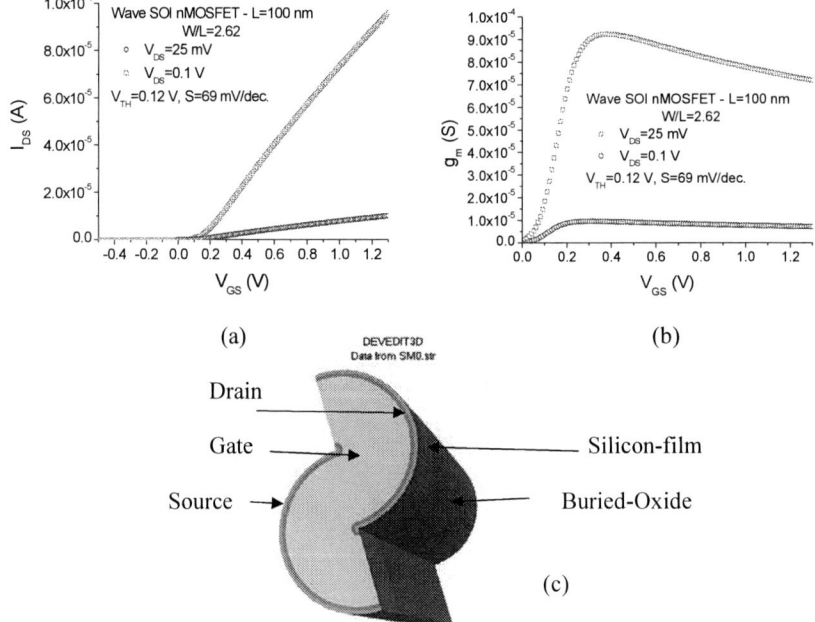

Figure 5. 3D Numerical Simulated I_{DS}xV_{GS} (a) and g_m (b) curves of PD Wave SOI nMOSFET implemented with semiconductors devices editor (c) for different V_{DS}.

Layout Comparisons

Three different layouts were done for the same SOI nMOSFET with L=0.8 μm and W/L=750. The design rules of 0.8 μm CMOS technology (lambda=0.4 μm) were used to implement these structures. The Waffle and Wave layouts were implemented with minimum-dimension transistors supported by this technology. The following layouts were implemented: **I.** Multifinger Layout: 10 conventional SOI nMOSFETs in parallel and each transistor with W/L=60μm/0.8μm=75; **II.** Waffle Layout: 47 Waffle SOI nMOSFETs in parallel, each one with W/L=12.8μm/0.8μm≈16) (Figure 1); **III:** Wave Layout [12 transistors in parallel implemented with 49 Wave SOI nMOSFETs ("S" shape), each one with W/L= $2.\pi/\ln(2.4/1.6) \approx 15.5$, where R1=1.6 μm and R2=2.4 μm] (Figure 5).

In order to perform the comparative study between these three layouts, two characteristics were taken in account: the die area (A) and the figure of merit defined by (W/L)/A ratio, here called as <u>integration factor</u>, that shows us the drain current driver capability per die area unit of the transistor. Table I show us the results of these comparisons. The die area gain (A Gain) and the integration factor gain [(W/L)/A Gain] are also presented in this Table I.

Table I. Die Area (A), die area gain in percentage (A gain), integration factor [(W/L)/A] and integration factor gain in percentage [(W/L)/A Gain] of the different layouts (Multifinger, Waffle and Wave SOI nMOSFETs), regarding L=0.8 μm and W/L=750.

Layout	A (μm²)	A Gain (%)	(W/L)/A (μm⁻²)	(W/L)/A Gain (%)
Multifinger	2544	-	0.295	-
Waffle	2396	5.8 (Multifinger)	0.313	6.1 (Multifinger)
Wave	1872	26.4 (Multifinger) 21.8 (Waffle)	0.401	35.9 (Multifinger) 28.1 (Waffle)

Analysing Table I, regarding a SOI nMOSFET with L=0.8 μm and W/L≈750, the Waffle layout presents a area gain and a integration factor gain around 5.8% and 6.1% respectively, larger than multifinger structures. The Wave layout presents a die area gain of 26.4% higher than multifinger and approximately 21.8% larger than Waffle layout. These results are due to the integration factor of Wave structure is higher than multifinger and Waffle layouts, around 35.9% and 28.1%, respectively.

Therefore, the Wave approach is an efficient alternative to implement analog and digital integrated circuits applications in order to reach large integration capability.

Conclusions

In this work, the main characteristics of Multifinger and Waffle layouts were discussed and it was observed by 3D numerical simulations that gate-crossing regions of Waffle structure do not operate as transistor, wasting die area. Thus, a novel Wave approach was introduced in order to further reduce occupied die area. Besides, several others Wave characteristics were discussed, such as the influence of internal and external drain/source bias configuration, matching, avalanche and ESD.

The Wave layout is originated based on circular gate SOI MOSFET. Several studies have already showed that circular gate present better performance than conventional counterpart has. Here in this paper we discussed that the new Wave structure maintain the same geometric factor than circular gate transistor and can improve others, such as matching, avalanche and ESD characteristics.

Comparisons between die area and integration factor of three different layouts (Multifinger, Waffle and Wave) regarding the same SOI nMOSFET (L=0.8 μm and W/L=750) were performed. The overall SOI MOSFET size implemented with Wave approach provides reductions of 26.1 % and 21.8% with respect to a Multifinger and Waffle, respectively.

Therefore, the Wave ("S" geometric shape) is a new alternative to implement planar SOI MOSFETs, resulting higher integration capacity, transistor matching and avalanche/ESD immunity than Multifinger and Waffle layouts. Therefore, the Wave layout can be used in order to reach high accuracy analog and digital integrated circuits.

Owing to its layout compactness, the Wave is also a good approach for the implementation of integrated circuits featuring larger current rates (current drivers and power stages), with important reduction on a smart-power die size and consequently integrated circuits cost.

References

1. J.-P. Colinge, *FinFETs and Other Multi-Gate Transistors*, p. 1, Springer, New York (2008).
2. A. Hastings, *The Art of Analog Layout*, p. 388, 416, 417, Prentice Hall, New Jersey (2001).
3. Sang L, P. K. T. Mok, W.-H. Ki, P. K. Ko, M. Chan, *Solid-State Electronics*, **47**, 785 (2003).
4. M. J. M. Pelgrom, A. C. J. Duinmaijer, and A. P. G. Welbers, *IEEE J. Solid-State Circuits*, **24**, 1433 (1989).
5. A. Giraldo, A. Paccagnella, and A. Minzoni, Solid-State Electronics, **44**, 981, (2000).
6. S. P. Gimenez and Marcello Bellodi, in Fifth Workshop of the Thematic Network on Silicon-on-Insulator Technology, Devices and Circuits, F. Gamiz, C. Claeys, J.-P Colinge, S. Cristoloveanu, O. Engstrom, O. Faynot, D. Flandre, A. Godoy Editors, PV 1, p. 71, EUROSOI, Goteborg (2009).
7. L. P. Dantas and S. P. Gimenez, in Analytical and Diagnostic Techniques for Semiconductor Materials, Devices, and Process 7, D. K. Schroder, L. Fabry, R. Hockett, H. Shimizu and A. Diebold, Editors, PV 11-3, p. 85, The Electrochemical Society Proceedings Series, Pennington, NJ (2007).
8. S. Dabral and T. J. Maloney, *Basic ESD and IO design*, p. 184, John Willey and Sons, New York (1998).
9. Atlas User Manual, Silvaco International (www.silvaco.com) (2007).

CHAPTER 7

DEVICE TECHNOLOGY II

ECS Transactions, 19 (4) 161-174 (2009)
10.1149/1.3117405 ©The Electrochemical Society

SOI as Platform for Transition from Micro to Nano

Francis Balestra

Sinano Institute, IMEP-Minatec (CNRS-Grenoble INP, UJF), 3 Parvis Louis Néel,
BP 257, 38016 Grenoble cedex 1, France

Silicon On Insulator (SOI)-based devices seem to be the best
candidates for the ultimate integration of ICs on silicon. The
performance and physical mechanisms are addressed in single- and
multi-gate thin film Si, Ge and III-V MOSFETs. The impact of tensile
or compressive uniaxial and biaxial strains in the channel, of high k
materials and metal gates as well as metallic Schottky source-drain
architectures are discussed. The interest of SOI-based emerging and
beyond-CMOS nanodevices for long term applications, based on
nanowires, carbon electronics or small slope switch structures is
presented. Finally, the possible 3D integration of multi-channels and
stacked nanowires is also shown.

1. Introduction

Since the 60's the shrinking of electronic components has been driven by the fabrication
of integrated circuits, which will continue for at least the next two decades. The critical
feature size of the elementary devices (physical gate length of the transistors) will drop
from 25nm in 2007 (65nm technology node) to 5nm in 2020 (14nm technology node).
In the sub-10nm range, Beyond-CMOS devices will certainly play an important role and
could be integrated on CMOS platforms in order to pursue integration down to nm
structures. Si will remain the main semiconductor material in a foreseeable future, but
the needed performance improvements for the end of the ITRS Roadmap [1,2] will lead
to a substantial enlargement of the number of materials, technologies and device
architectures.

Silicon On Insulator-based devices seem to be the best candidates for the ultimate
integration of Integrated Circuits on silicon [3,4]. An overview of the main advantages
of SOI for the Nanoelectronics of the next two or three decades is presented in this
paper. Nanoscale CMOS, emerging and beyond-CMOS nanodevices, based on
innovative concepts, technologies and device architectures, are addressed. The
flexibility of the SOI structure and the possibility to realize new architectures allow to
obtain optimum electrical properties for both low power and high performance circuits.
These transistors are also very interesting for high frequency and memory applications.
The performance and physical mechanisms are investigated in single- and multi-
gate/multi-channel thin film MOSFETs. A comparison between the performance of Si,
Ge and III-V semiconductors on insulator is given. The impact of tensile or compressive
strains in the channel, of high k materials and metal gates as well as metallic Schottky
source-drain architectures is discussed. The interest of advanced emerging and beyond-
CMOS nanodevices on SOI for long term applications, based on nanowires, carbon
electronics or small slope switch structures are also presented.

161

2. More Moore

a. New Channel Materials for Ultimate CMOS

As simple scaling of silicon CMOS becomes increasingly complex and expensive there is considerable interest in increasing performance by using strained channels, which reduces the valley degeneracy and lowers the carrier effective mass, thus improving carrier mobility and drive current in a device. The combination of the advantages of SOI structures for improving the electrostatic control and of strained semiconductors is very promising. An original method for strained silicon on insulator (sSOI) is based on thin SiGe buffer layers relaxed by the following technique. In the process, a 200 nm thick pseudomorphic SiGe layer grown on a Si wafer and capped with a few nanometers pure Si layer is implanted with He^+ ions below the SiGe/Si substrate interface. During post-implantation annealing, diffusive implanted He^+ ions form overpressurized He bubbles which punch out dislocations loops. Under the influence of the strain field, these dislocation loops move up to the interface forming misfit dislocations that lead to strain relaxation of the SiGe layer. If the Si cap layer thickness lies below a critical thickness (< 8 nm), an induced strain is transferred immediately during the relaxation of the SiGe layer. Finally, the thickness of the strained Si cap can be increased by a further epitaxial growth step. In the next fabrication step, the wafer is cleaved with hydrogen implantation according to the Smart-Cut® process and the stack layer is bonded to an oxidized handle wafer. After a final etch back step, the strained Si layer remains directly on the SiO_2. Low threading dislocation densities have been obtained with this method. Split-CV (capacitance-voltage) mobility measurements shown in figure 1 demonstrate the enhanced performance of the fabricated sSOI, with electron mobility values as high as 1200 cm^2/Vs [5].

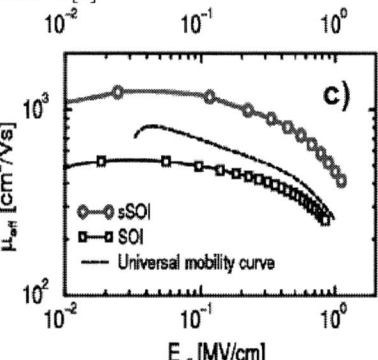

Fig. 1. Enhanced mobility curves of sSOI with respect to SOI [5]

Using 50nm sSOI with 1GPa biaxial tensile stress as starting material, uniaxial tensile strain can be obtained by lateral strain relaxation of patterned structures (Fig. 2a,b). Fig. 2c shows the transfer characteristics of a strained and an unstrained Nanowire-nFET. The NWs have a square cross-section of 40 × 40nm^2 and a length of 3μm. The subthreshold slope of both devices is between 70−80mV/dec. Both devices have a very low off-current and a high Ion/Ioff ratio. The on-current of the uniaxially strained device is enhanced on the average by a factor of 2.5 due to uniaxial tensile strain, while strain does not deteriorate the low off-current. The inset of Fig. 2c shows a plot of $Id/gm^{1/2}$ against Vg for a strained and unstrained device measured at a source drain voltage of 50mV. The slope of the linear part of the curves is related to the carrier mobility. The mobility enhancement due to uniaxial tensile strain is found to be x = 2.3. For maximum performance enhancement due to lateral strain relaxation sSOI devices must have a large length to width ratio [6].

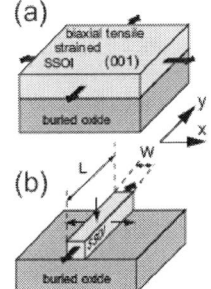

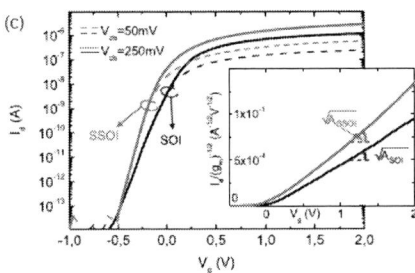

Fig. 2. (a) 50nm sSOI with 1GPa biaxial tensile stress is used as starting material. (b) Uniaxial tensile strain is obtained by lateral strain relaxation of patterned structures. (c) Transfer characteristicsof two NW-FETs, one fabricated on SOI and one on uniaxial sSOI. The channel length of both devices is L = 3μm and the gate oxide thickness tox = 5nm. The inset shows the $Id/gm^{1/2}$-plot for the devices. The slope of the linear region is related to carrier mobility [6]

Strain engineering is also useful for mobility enhancement for Si film thickness in the sub-10nm range [7], which will be needed for the ultimate integration of nanoMOSFETs. A similar enhancement of electron mobility in 3.5nm SOI devices under biaxial and uniaxial tensile strain has been obtained. The electron mobility is also enhanced in 2.3nm Si layer under uniaxial tensile strain (Fig. 3), and the hole mobility increases in 2.5nm film under uniaxial compressive strain.

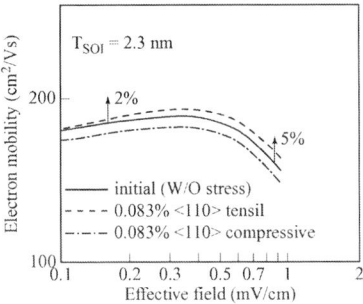

Fig. 3. Electron mobility in 2.3-nm ultra-thin-body MOSFET under <110> uniaxial strain [7]

b. Metallic Schottky Source/Drain MOSFETs

As CMOS technology is entering the decananometre era, the contact resistance associated with the silicide/silicon interface is identified as one of the biggest challenges to solve in order to preserve current drive capabilities. In that context, source/drain engineering takes an increasing importance in the development of leading edge CMOS generations because of the increasing impact of S/D series resistances on transistor performance. In order to further pursue down-scaling of MOSFETs in the sub-32 nm gate length range, novel devices that hierarchically combine alternative materials as well new architecture concepts such as multi-gated channel have been proposed. Considering that the aforementioned innovations are expected to contribute to a higher current drive at shallower junction depth and reduced silicide thickness, extremely severe constraints are placed on the junction and contact technologies.

In order to address this challenge, one alternative is to implement metallic S/D combined to a dopant segregation (DS) strategy at reduced thermal budget. The expected benefit is to considerably reduce the specific contact resistance of the metal/semiconductor junction while keeping activated dopants sharply localized at the interface. The efficiency of dopant segregated Schottky contacts has been demonstratd [8-10]. The implementation of dopant segregated contact is illustrated here by considering the following three distinctive features: i) implant-to-silicide (ITS), ii) band-edge low Schottky barrier (SB) to holes (PtSi) and iii) thermal budget limited to 500°C. It has been shown that the ITS scheme coupled to BF_2^+ provides a sub-100 meV barrier consistently with the boron pile-up observed at the PtSi/Si interface. A new state-of-the-art current drive performance has been established for SB-MOSFETs at 25 nm gate length: I_{on} of 530 $\mu A/\mu m$ at $V_g=V_d=-1.1V$. Fig.4 demonstrates that metallic S/D competes with best unstrained channel SOI p-FET technologies. A record RF performance for a 30-nm p-type unstrained thin-film fully depleted SOI SB MOSFET has been demonstrated with a f_T of 180 GHz [8]. The effect of strained semiconductor had also been studied: carrier injection from a metallic junction should benefit from band splitting and from the corresponding Schottky barrier height reduction.

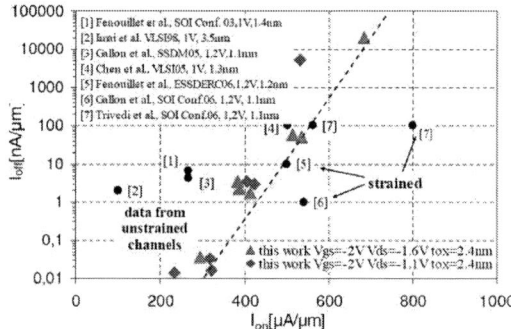

Fig. 4. Ion-Ioff state-of-the art of S/D p-MOSFET on SOI substrate indicating that Boron DS p-MOSFETs is leading the SOA of both SB and conventional unstrained thin-film SOI technologies [8]

c. High-k Materials

In the search for new insulator materials for the 22 nm CMOS node and beyond, the dielectric constant itself is not sufficient as a physical quantity to fulfil for technical device specification. The ultimate device property in this context is the gate leakage current for a given equivalent oxide thickness needed to achieve a high enough capacitive coupling between transistor gate and channel. For a perfectly amorphous or monocrystalline insulator material, this quantity is limited by tunnelling. Thus, the combination of dielectric constant, k, and the energy barrier height governed by the offset, ΔE, between the energy bands of the insulator and silicon is crucial from this viewpoint. For electron leakage, the relation between these two quantities can roughly be described as hyperbolic such that $\Delta E \times k = C_E$, where C_E is a constant, determined by the leakage and gate coupling properties required for a certain technology node [11]. For the "low standby power" (LSTP) 22nm bulk CMOS node, $C_E \approx 70$ eV, while the corresponding requirement for thin double gate SOI technology can be set to $C_E \approx 30$ eV. These relations are summarized in Fig. 5.

As future 22 nm and 16 nm node transistors probably will be produced in SOI technology, the lower C_E value mentioned above offers a large number of possibilities for gate dielectrics as seen in Fig. 5. This means that tunneling probably will not be the limiting factor. Instead the material choice will be governed by other properties like thermal stability and charge carrier traps.

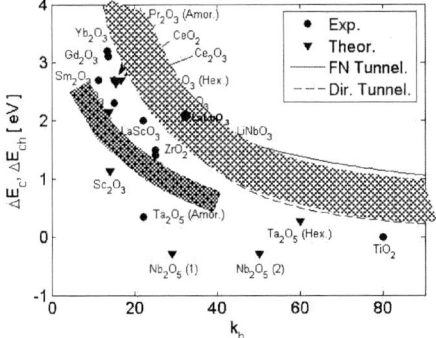

Fig. 5. Energy offset values between the conduction bands of various dielectrics and silicon versus dielectric constant. The upper shadowed area is a border for the 22 nm LSTP bulk CMOS node. The lower shadowed area represents the corresponding border for FD DG SOI technology [11]

d. Multi-Gate Devices

Multi-gate MOSFETs realized on thin films are the most promising devices for the ultimate integration of MOS structures due to the volume inversion or volume accumulation in the conductive layer (for enhancement- and depletion-type devices, respectively), leading to an increase of the number and the mobility of electrons and holes as well as driving current (additional gain in performance in a loaded environment), optimum subthreshold swing and the best control of short channel effects and off-state current, which is the main challenge for future nanodevices due to the power consumption crisis and the need to develop green/sustainable ICs [12].

Fig. 6 shows a comparison of short channel effect for bulk, single gate SOI, single gate SON with thin buried oxide and double gate MOSFETs. For sub-30nm gate length integration, the advantages of SOI structures, thin film and BOX as well as multi-gate architectures are clearly demonstrated [13].

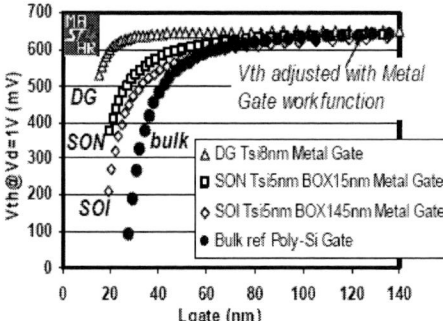

Fig. 6. Comparison of $V_t(L_g)$ for various MOSFET architectures from bulk to multi-gate devices [13]

The on-current Ion of the MOSFET is limited to a maximum value I_{BL} that is reached in the ballistic transport regime. Fig. 7 reports the self-consistent Monte Carlo simulation of the ballistic ratio BR=Ion/I_{BL} versus Drain-Induced-Barrier-Lowering, which is one of the main short channel effects, showing that one can increase the BR by scaling the gate length, thus increasing the longitudinal field at the source, but this comes at the cost of a larger DIBL. For a given DIBL, an increased ballisticity is obtained for low doping double gate SOI devices [14].

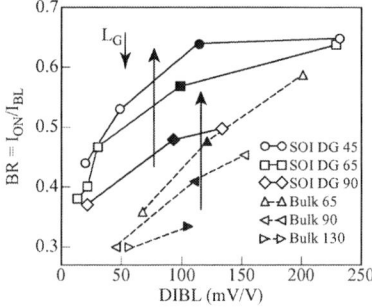

Fig. 7. Ballisticity ratio at Vg=Vd=Vdd vs. DIBL. Filled symbols represent transistors with the nominal gate length for the high-performance MOSFET of each technology node [14]

The transfer characteristics of several multiple-gate (1, 2, 3 and 4 gates) MOSFETs, calculated using the 3D Schrödinger-Poisson equation and the Non-Equilibrium Green's Function (NGEF) formalism for the ballistic transport or MC simulations, have shown similar trends. The best performance (drain current, subthreshold swing) is outlined for the 4-gates (QG-Quadruple Gate or GAA-Gate All Around) structure [15,16] (Fig. 8a).

However, Fig. 8b also demonstrates that the propagation delay in triple gate (TG) and quadruple gate MOSFETs are degraded due to a strong rise of the gate capacitance. A properly designed double-gate (DG) structure appears to be the best compromise at given off-sate current Ioff [16].

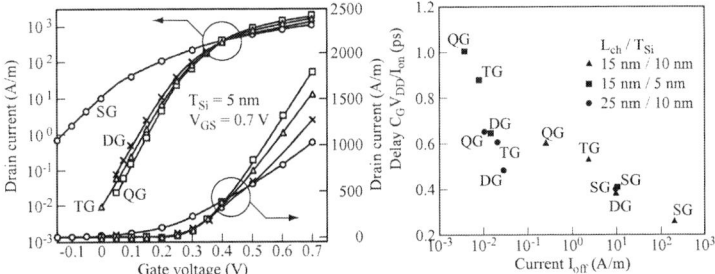

Fig. 8. (a) Id(Vgs) at Vds=0.7V in thin layers for different multi-gate architectures ; (b) Propagation delay versus Ioff for single-gate and multi-gate SOI devices [16]

In decananometer MOSFETs, gate underlap is also a promising solution in order to reduce the DIBL effect. Fig. 9 presents the variations of the driving current Ion, the subthreshold current Isub and the gate direct tunneling current Igdt versus gate underlap [17]. The on-current is almost not affected by the gate underlap whereas the leakage currents are substantially reduced due to a decrease in DIBL and drain to gate tunneling current. A reduction of the effective gate capacitance Cg for larger underlap values at iso Ion has also been shown. This reduction of Cg leads to a decrease in the propagation delay and power.

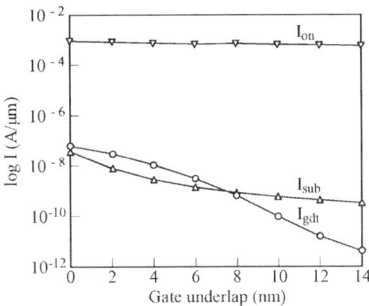

Fig. 9. Ion, subthreshold (I_{sub}) and gate direct tunneling (I_{gdt}) currents as a function of gate underlap [17]

Another important issue deals with the comparison of channel materials. Fig. 10 is a plot of the driving current as a function of gate length for Si, strain Si, Ge and GaAs n-channel double-gate MOSFETs, for a given Ioff according to ITRS for the next 3 generations of HP ICs. The four materials are chosen in their optimized orientation and short channel effects and access resistance are included. When neglecting source to drain tunneling, Ge and GaAs devices lead to the best Id for sub-10nm gate length. However, when source-drain tunneling is included, only 2G strain Si MOSFETs satisfy the need of ITRS in the sub-decananometre range [18].

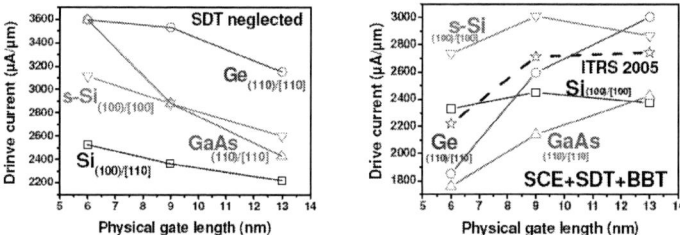

Fig. 10. Driving current of n-channels 2G MOSFETs obtained by simulations as a function of channel materials (Si, strain Si, Ge, GaAs) and source-drain tunneling, taking into account short channel effects and band-to-band tunneling [18]

In order to reach very high performance at the end of the roadmap, multi-bridge-channel MOSFETs (MBCFET) or multi-channel MOSFETs (MCFETs) present very high driving current larger than those of GAA devices and exceeding the ITRS requirements (Fig. 11) [19,20].

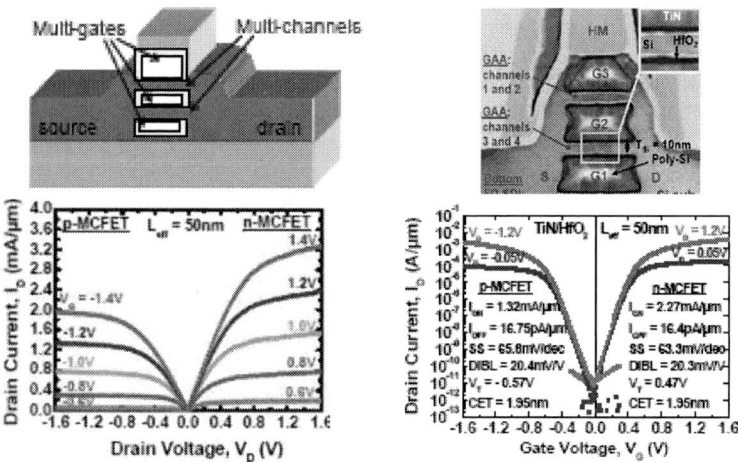

Fig. 11. Multi-channels realized using SON technology leading to very high drain currents with a very good control of leakage currents [20]

e. Variability

SOI devices can also be interesting for reducing the variability in decananometre MOSFETs, which also represents a major challenge at the end of the roadmap. Sources responsible for local and inter-die threshold voltage (Vt) variability in undoped ultra-thin FDSOI MOSFETs with a high-k/metal gate stack have been experimentally discriminated. Charges in the gate dielectric and/or TiN gate workfunction fluctuations are determined as major contributors to the local Vt variability and it is found that SOI thickness variations have a negligible impact down to t_{Si}=7nm. Moreover, t_{Si} scaling is shown to limit both local and inter-die Vt variability induced by gate length fluctuations. The highest matching performance ever reported for 25nm gate length MOSFETs has been achieved (Pelgrom coefficient Avt=0.95mV.μm), demonstrating the effectiveness of the undoped ultra-thin FDSOI architecture in terms of Vt variability control [21].

High immunity in Vt variability due to the possible use of undoped ultra-thin SOI has also been demonstrated in multi-gate SOI devices, which are also more tolerant to line edge roughness induced variability [22].

3. Emerging and Beyond-CMOS Nanodevices

The objectives in this domain are to explore the horizon beyond conventional CMOS, or beyond Moore, in order to overcome possible downscaling or performance limits of CMOS in the next two decades. In this field, SOI structures can also be considered as a platform for pursuing the integration in the nanoworld.

a. Nanowires and Carbon Electronics

It has been shown previously that multi-gate architectures based on the concept of volume inversion are very promising in order to overcome the number of challenges for CMOS integration (short channel effects, driving current, etc.) down to at least decananometre gate length devices [12]. Gate-All-Around semiconductor nanowires can be seen as the ultimate integration of these innovative nanodevices and present very interesting properties down to the sub-10nm range.

The combination of strain effects with these 1D structures can lead to very high performance ICs for the end of the roadmap. Top-down bended gate-all-around nanowires have been fabricated in order to improve carrier mobility and driving current [23]. A bending induced by thermal oxidation in suspended nanowires has been demonstrated. A maximum tensile strain is obtained in the middle of the wire. Tensile stresses from 200Mpa to 2Gpa can build-up in these suspended nanowires. The substantial enhancement of electron mobility in these structures has been demonstrated.

3D integration of Nanowires is a promising solution to boost the driving current and keep a very low off-state current. Using SON process, three stacked GAA sub-15nm nanowires (Fig. 12), with 100nm length and 1.8nm EOT (high k/metal gate stack) has shown extremely high driving current and very low leakage currents (Ion/Ioff : 6.5mA/μm-27nA/μm for nMOS - Ion/Ioff : 3.3mA/μm-0.5nA/μm for pMOS). A new optional independent gate nanowire with a FinFET-like structure named ΦFET has also been recently reported, leading to new design flexibility (Fig. 12). These 3D structures can be extended to a combined vertical and lateral integration for logic, memories and NEMS applications (Fig. 12). The 3D-NWFET and ΦFET, compared to a co-processed FinFET, relaxes the channel width requirement for a targeted DIBL and improves transport properties (Fig. 13). ΦFET also exhibits significant performance boosts compared to Independent-Gate FinFET (IG-FinFET): a 2-decade smaller Ioff current and a lower subthreshold slope (82mV/dec. instead of 95mV/dec.). This highlights the better scalability of 3D-NWFET and ΦFET compared to FinFET and IG-FinFET [24,25].

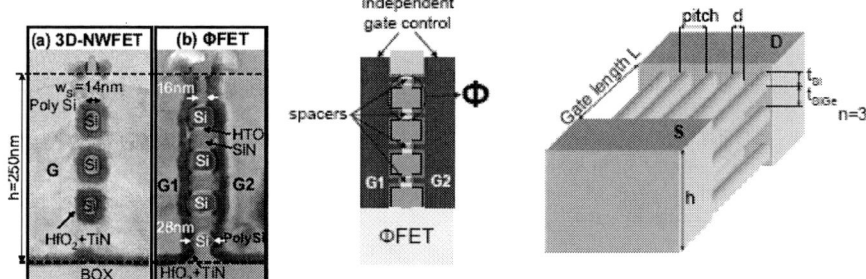

Fig. 12. (a), (b) and (c) Cross-sectional pictures of 3D-NWFET and ΦFET (spacers are introduced to obtain stacked nanowires with independent gate operation) ; (d) Lateral and vertical integration of 3D Nanowires [24,25]

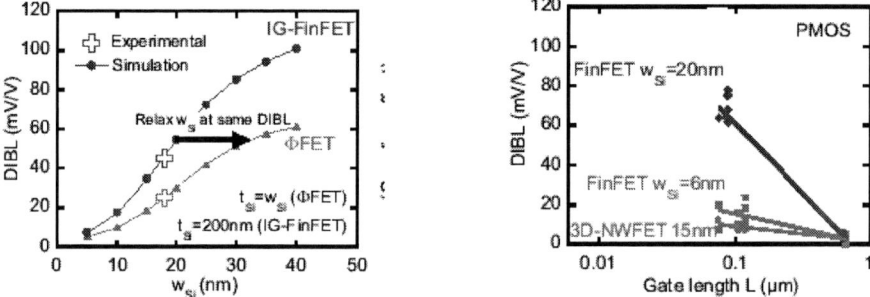

Fig. 13. (a) Simulated and measured DIBL vs. Si width wSi for ΦFET and IG-FinFET in the 3TFET configuration ; L=50nm ; t_{Si} is the Si thickness. (b) DIBL versus gate length L for 3D-NWFET compared with FinFET. These plots show the strong reduction of SCEs in 3D NWs and ΦFET [24]

Another promising future options to enhance Si-based ICs performance is the introduction of carbon electronics. These last years, many works have been devoted to carbon nanotubes, due to their remarkable electrical properties. One major challenge for the implementation of CNT-based electronic circuits is a controllable assembly of CNTs onto electrodes. Another very recent solution for carbon electronics deals with the development of two-dimensional carbon sheets, which have been demonstrated to be thermodynamically stable. Graphene monolayers lead to excellent electronic properties with extremely high carrier mobility and to the possibility of using top-down CMOS compatible process flows. A back- and top-gated graphene flake on SOI has recently been studied, with a total length from source to drain of about 7μm, a width of 265nm and a gate length of 500nm. An ambipolar conduction controlled by the front and back gates has been observed, confirming that the field effect can be applied to graphene devices. Preliminary results show the impressive potential of graphene, with electron and hole mobilities around 4800 cm^2/Vs in pseudo-MOS structures at 300K, and mobilities in the range 500-700 cm^2/Vs obtained in top-gated devices [26]. Band gap tuning by using for instance nanoribbons or bilayer graphene could be used to improve gate modulation and device characteristics.

b. Small Slope Switches

Even though the aggressive scaling will continue to play an important role in the future nanoelectronics, new technology drivers, such as *ultra-low power* and *new functionality* will open alternative ways for future high performance systems.
One interesting class of Beyond-CMOS devices are the small slope switches. A small slope electronic switch is defined as a solid-state semiconductor device showing a value of the subthreshold slope smaller than the 60mV/decade limit for a conventional MOSFET, set by the Boltzmann distribution at room temperature. The smaller the value of S, the more abrupt the transition from the off- to the on-state and closer the switch to the ideal case. Benefits of small slope switches are the ultra-low standby power due to a very low I_{off} but also the high-speed potential and dynamic power savings, since less power is drawn per transition when the subthreshold slope is more abrupt. Development of small slope switches requires exploration of new physical principles for very abrupt off-on transition, such as: (i) impact ionization, (ii) band-to-band tunneling, (iii) electro-mechanical instability, or iv) ferroelectric gate. It is worth noting that all these small slope device architectures can be implemented as extensions of advanced silicon CMOS or by hybridization of silicon CMOS with other compatible technologies (SiGe, nanowire, CNT, nano-electro-mechanical structure).

The Tunnel FET [27-29] is a gated p–i–n diode with a gate over the intrinsic region; it exploits the gate controlled electron tunneling from the valence band of the p-region to the conduction band of the i-region for reversed biases, resulting in very abrupt off-on transition. They have been reported on Si, III-V and CNT alternatives. Multi-gates TFETs have recently shown to present very interesting performance. Double-gate high-k Tunnel-FETs can be scaled to shorter lengths before important characteristics such as transconductance, Ion/Ioff, and subthreshold swing are degraded (Fig. 14). Optimized silicon body of 7-8 nm (for Lg=50nm) in DG tunnel FET with high-k dielectrics leads to improved characteristics with higher Ion (Ion/Ioff # 10^{11}) and smaller subthreshold swing (average: 57mV/dec − minimum point swing: 11mV/dec) [29].

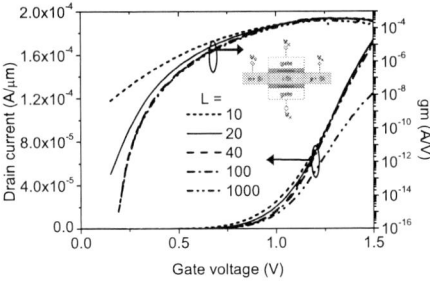

Fig. 14. Effect of gate length scaling from 1μm down to 10nm on I_D-V_G and g_m-V_G characteristics of high-K Double-Gate Tunnel FET [29]

Double gate TFET with strained-Ge heterostructure channel has recently shown record high drive currents (Ion~300μA/μm) and good swing~50mV/dec, due to small bandgap and tunnel resistance of s-Ge and DG electrostatics. Fig. 16 shows the corresponding experimental Id-Vg characteristic. Using a TFET simulator, the scalability of (A) symmetric and three asymmetric DG TFET configurations (Fig. 15), (B) Underlap drain, (C) Low drain-doping and (D) Lateral heterostructure has been studied with their ability to solve the ambipolar problem and achieve high ON and low OFF currents. Increasing the drain underlap and lowering the drain doping can kill the ambipolar behavior. These approaches are not fundamentally different and both lower leakage due to a long depletion width and low E-field on the drain side. Unfortunately, both these TFET structures need a very long non-scaling Drain of >30nm to achieve a S<60mV/dec. This puts a severe restriction on device scalability. The ambipolar problem can also be solved using a lateral heterostructure of a large Eg material (like Si) at the drain side (Fig. 15) to reduce the tunneling. Fig. 16 shows that a Source-Side heterostructure can effectively solve the ambipolar problem. The lateral heterostructure is the most scalable approach to solving the ambipolar problem [30].

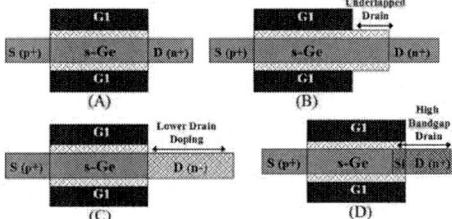

Fig. 15: Symmetric s-Ge DG TFET structure (A) exhibits ambipolar characteristics (see Fig. 16). In order to achieve high ON and low OFF currents, asymmetry can be introduced into the TFET by (B) using an under-lapped drain, or (C) using a lower doping on the drain side, or (D) using a lateral heterostructure of large bandgap material (like Si) to reduce the tunneling currents from the drain during OFF-state [30]

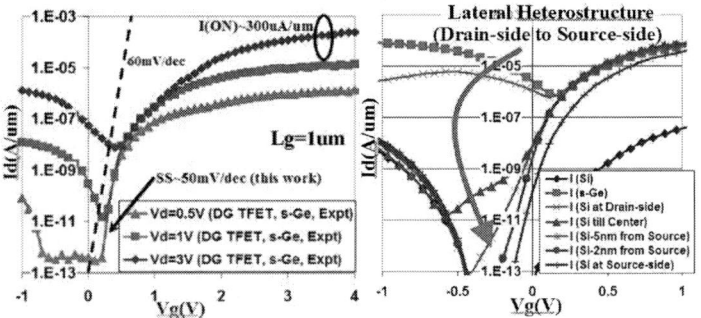

Fig. 16. (a) Experimental DG TFET with strained-Ge heterostructure channel shows record high drive currents (Ion~300μA/μm) and good swing~50mV/dec ; (b) The effect of the location of the s-Ge/Si interface on the leakage of the lateral heterostructure TFET is shown. Simulations show that a Source-Side heterostructure (interface between s-Ge and Si is at the source-side) is needed to achieve a S<60mV/dec [30]

A novel transistor design, based on a FinFET architecture, which utilizes positive feedback to achieve steep switching behavior has also been proposed (Fig. 17). The feedback (FB) FET exhibits very low subthreshold swing (~2 mV/dec) and high Ion/Ioff ratio (~10^8) to allow for significant reductions in gate voltage swing (to below 0.5V). It is a new candidate to replace the MOSFET for future low power electronic devices and can operate as a n- or p-channel device. Steep switching behavior is achieved by dynamically reducing Vt as the transistor is turned on, via barrier-height modulation by carrier accumulation in the drain- and source-offset regions inducing a positive feedback and an abrupt switch into the on-state (Fig. 18) [31].

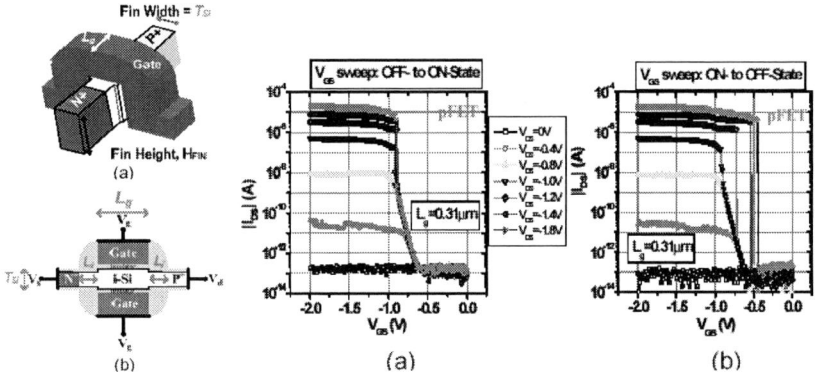

Fig. 17. Schematic illustration of the FB-FET structure

Fig. 18. Measured Ids-Vgs characteristics of a short-channel FB-FET operated as a p-channel device: (a) Forward sweep vs. (b) Reverse sweep Lg=0.31µm, W=0.2µm [31]

5. Conclusion

Silicon On Insulator-based devices seem to be the best candidates for the ultimate integration of Integrated Circuits on silicon down to nm structures. An overview of the performance of nanoscale CMOS, emerging and beyond-CMOS nanodevices, based on innovative concepts, technologies and device architectures, has been addressed. The flexibility of the SOI structure and the possibility to realize new architectures allow to obtain optimum electrical properties for both very low power, high performance circuits and memory applications. SOI platforms will thus enable to overcome the number of technological and physical challenges we are facing for ultimate CMOS and post-CMOS nanodevices in order to speed up technological innovation for the Nanoelectronics of the next 3 decades.

Acknowledgements

This work was partially supported by the European Networks of Excellence NANOSIL (FP7) and SINANO (FP6) devoted to Silicon-based Nanodevices.

References

1. ITRS Roadmap: http://www.itrs.net/
2. ENIAC Strategic Research Agenda : http://www.eniac.eu/web/SRA/local_index.php
3. F. Balestra, *SOI devices*, Wiley Encyclopedia of Electrical and Electronics Eng., 1999
4. S . Cristoloveanu and F. Balestra, in advanced Semiconductor and Organic Nano-techniques, H. Morkoc Ed., Academic Press, pp. 325-366, 2003
5. F. Driussi, D. Esseni, L.Selmi, M. Schmidt, M.C. Lemme, H. Kurz, ESSDERC 2007
6. S.F. Feste, S. Mantl, Proc. ULIS'2009
7. K. Uchida, R. Zednik, C.H. Lu, p. 229, IEDM'2004
8. G. Larrieu, E. Dubois, R. Valentin, N. Breil, F. Danneville, p. 147, IEDM'2007
9. Z. Zhang, A. Qiu, R. Liu, M. Ostling, S.L. Zhang, Electron Dev. Lett. 28, 565, 2007
10. Q.T. Zhao, U. Breuer, E. Rije, S. Lenk, S. Mantl, Appl. Phys. Lett. 86, 062108, 2005
11. O. Engström, B. Raeissi, S. Hall, O. Buiu, Solid State Electronics, 51, 622, 2007
12. F. Balestra, S. Cristoloveanu, M. Benachir, J. Brini, IEEE El. Dev. Let. 8, p. 410, 1987
13. T. Skotnicki, International Summer School MIGAS'2008, Grenoble, July 2008
14. S. Eminente, D. Esseni, P. Palestri, p. 609, IEDM'2004
15. M. Bescond, K. Nehari, J.L. Autran, p. 617, IEDM'2004
16. J. Saint-Martin, A. Bournel, P. Dollfus, p. 61, ULIS'2005
17. A. Bansal, B.C. Paul, K. Roy, Proc. IEEE Intern. SOI Conf., p. 94, 2004
18. Q. Rafhay, R. Clerc, Solid-State Electronics, 52, p. 1474, 2008
19. E.-J. Yoon, S.Y. Lee, S.M. Kim, p. 627, IEDM'2004
20. E. Bernard, T. Ernst, B. Guillaumot, VLSI'2008
21. O. Weber, p. 245, IEDM'2008
22. A. Asenov, ESSDERC'2008
23. K.E. Moselund, A. Ionescu, p. 191, IEDM'2007
24. C. Dupre, A. Hubert, S. Becu, IEDM'2008
25. T.Ernst, L. Duraffourg, C. Dupre, IEDM'2008
26. M.C. Lemme, IEEE Elec. Dev. Let. 28, 282, 2007
27. K. Bhuwalka., Jap. Journal of Appl. Phys., vol. 43, no. 7A, 4073, 2004
28. P.-F. Wang., Solid-State Elec., vol. 48, 2281, 2004
29. K. Boucart, A.M. Ionescu, IEEE Trans. on Elec. Dev., 1725, 2007
30. T. Krishnamohan, D. Kim, S. Raghunathan, K. Saraswat, IEDM'2008
31. A. Padilla, C.W. Yeung, C. Shin, C. Hu, T.-J. King Liu, IEDM'2008

Optimization of Blue/UV Sensors Using PIN Photodiodes in Thin-Film SOI Technology

O. Bulteel[a] and D. Flandre[a]

[a] Microelectronics Laboratory, Université catholique de Louvain, Louvain-la-Neuve 1348, Belgium

The design and fabrication of high efficiency UV detectors for biomedical or environmental applications is very challenging. The goal of this work is to demonstrate the ability of SOI technology to implement such optical sensors. In this paper, we propose a structure of PIN photodiode implemented in a silicon thin film of a SOI wafer with an additional anti-reflection coating in silicon nitride. The simulation and the characterization of this device fit very well together. We also propose improved structures for this photodiode in SOI and in silicon-on-nothing technologies. For all the structures we study the appropriate thicknesses of the anti-reflection coating that could maximize the optical responsivities of the devices. Finally, we present two applications of our photodiodes proving their efficiency in biomedical field, such as DNA concentration measurement over a large range of characterization.

Introduction

Optical detection at low wavelengths close to blue and ultraviolet (i.e. for λ<480 nm) is commonly used in biomedical and environmental fields. For example, DNA concentration measurement, bacteria or proteins detection require optical light whose emission wavelength is below λ=300 nm. On the other hand, UV and ozone rates are measured in upper wavelengths close to λ=400 nm. So, there is a real need for miniaturized sensors able to detect in this wavelength range. In addition, such applications require low levels of detection, due to the low power of the emitted optical signals to monitor. It's thus important to minimize photodiode dark currents and to achieve high optical responsivities in order to reach a detection level as small as possible. The optical responsivity of a photodevice can be defined as:

$$R = \frac{I_D - I_{Dark}}{P_{opt}} \qquad [1]$$

where I_D is the total current through the device, I_{Dark} is the dark current and P_{opt} is the incident optical power.

Whereas photodetectors in (Al)GaN-based technologies can target high optical responsivities below λ=300 nm thanks to their appropriate bandgap [1], silicon devices absorb light as a function of their thickness up to wavelengths close to 1200 nm. A solution to specifically and efficiently absorb low wavelengths with silicon-based technologies is to implement the sensors in thin silicon layers based on silicon-on-insulator (SOI) technology [2]. Thin-film structures are more and more used due to their insulating properties for integrated circuits. This provides the opportunity to implement a complete system-on-a-chip and is a way to fabricate a low-wavelength photodetector with low-cost of process. As this type of microsystem can be useful for complex manipulations in biomedical labs, there is an opportunity to replace existing large-sized and expensive apparatus with portable photosensors.

The photodiode: design and characterization

In this work, we extensively study the influence on light absorption, of the process parameters, and then discuss the optimization, of lateral PIN photodiodes designed and characterized in various SOI technologies. The basis of our whole structure is a standard SOI wafer. We can improve it with additional layers or by etching parts of layers. This section presents two main structures: an improved one with an anti-reflection coating (ARC) standard PIN photodiode and a SON (for silicon-on-nothing) PIN photodiode. Both of them can be customized with appropriate thicknesses as will be seen in this work.

Standard structure

A transversal view of one finger of a standard interdigitated photodiode is shown in figure 1.

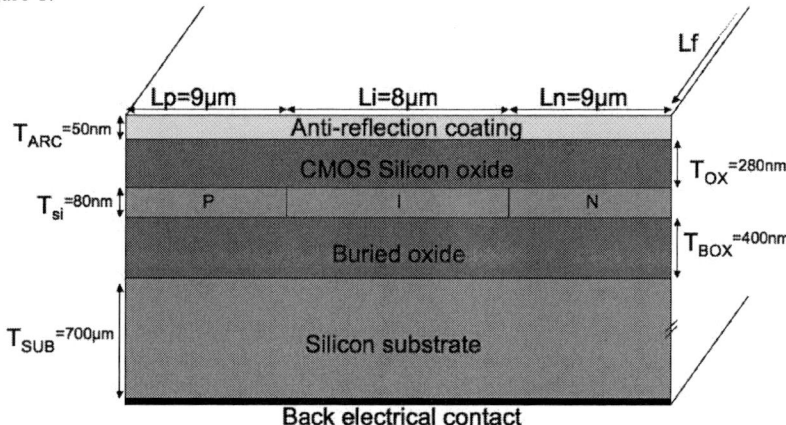

Figure 1: Transversal view of one finger of an interdigitated PIN lateral photodiode with typical dimensions of our UCL SOI technology.

Our standard SOI technology features an 80 nm thick silicon layer which allows absorption of light emissions below $\lambda=400$ nm. The thin Si film is separated from the silicon substrate by a 400 nm buried silicon oxide. A CMOS process silicon oxide of 280 nm, which consists in a gate oxide of 30 nm and an interdielectric oxide of 250 nm, covers the photodiode. In addition to this standard process, an ARC of 50 nm was deposited on the top of the wafer. This ARC can be either silicon nitride (Si_3N_4) or alumina (Al_2O_3) since their indexes of refraction are quite similar. Other dimensions are indicated in fig. 1. With a mathematical model implemented in Matlab, we simulated the spectral responsivity R. The design of the PIN diode is always a trade-off between a small intrinsic length that can increase photogenerated electron-hole pairs collection but also considerably increase the dark current and a larger intrinsic length that could reduce this dark current but increase electrons-holes recombination and thus decrease the photogenerated current through the device. Our choice was an intrinsic length of Li=8μm. The other lateral dimensions (i.e. anode and cathode lengths, respectively Ln and Lp) have to be the smallest possible to maximize the photosensible area of the device but their limit is fixed by the technology, which is in our basic case: Ln=Lp=9μm taking into account the contact and metal interconnects.

As the standard basis of SOI wafer (consisting in the substrate and the buried oxide) is imposed by the fabrication process, and the thin Si film and the CMOS process oxide thicknesses are also constant on the wafer, the only parametrical layer is the additional ARC, which replaces a standard SiO_2 passivation layer. As demonstrated in [3], we can minimize the reflected power by depositing two anti-reflecting coatings on top of a semiconductor layer. We first choose silicon nitride as the top anti-reflection coating, since its refraction index, close to 2 is commonly used, while the second ARC is the existing silicon oxide of 280 nm previously presented.

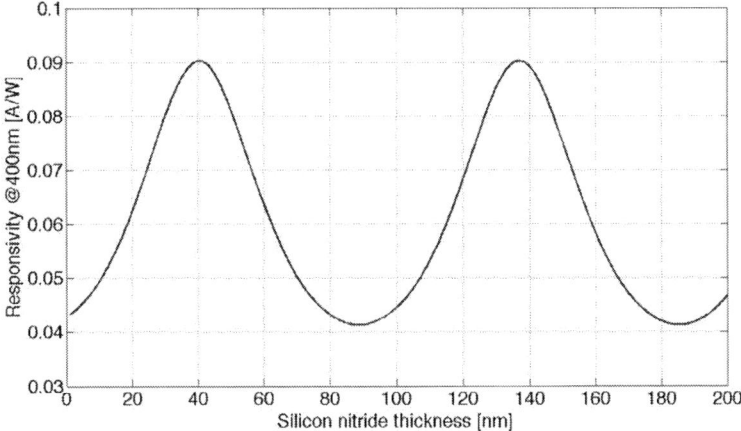

Figure 2: Simulation of the responsivity at λ=400 nm of PIN photodiodes with structure as in figure 1 as a function of the silicon nitride ARC.

The simulations of our model demonstrated a periodical responsivity of the total photodiode. To maximize responsivity at λ=400 nm, the adequate Si_3N_4 thickness is T_{ARC}= 40 nm + k*100 nm where k is 0 or a number close to 1 as can be seen in figure 2. Due to process issues, our mean nitride thickness was about 50 nm for the fabricated devices.

Their responsivity was measured by steps of 10 nm with optic fiber direct illumination, selected, with a monochromator, from a high power light source. The optical characterization of the fabricated devices shows that measurements correspond very well to the simulations of these photodiodes (using figure 1 dimensions) as shown by curves 1 and 2 in figure 3. The differences below λ=400 nm may be explained by process variations (e.g. slight thickness variations of the oxide and ARC over the total area of the photodiode).

We obtained high responsivity values in UV, close to R=0.1 A/W with a spectral cut-off wavelength of $λ_c$=450 nm and an attenuation of 90% over 50 nm. Measured dark current corresponds to the simulations with very low I_{Dark}=10 pA for a photodiode with a total area of 1 mm^2 and a reverse bias of Vd=-0.5 V.

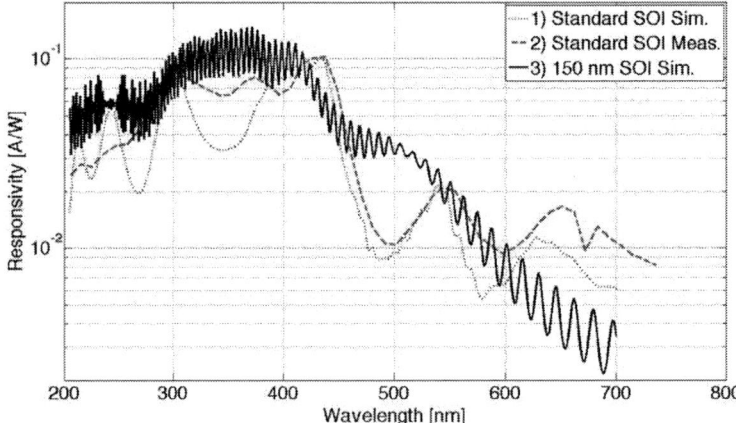

Figure 3: Simulations and measurements of spectral responsivity of the lateral PIN photodiodes in different technologies.

Improved SOI structures

Based on the same structure (i.e. using similar material layers superposed in the same order) but with lower thicknesses of the layers as well as lower lateral dimensions of the PIN diode, a new device has been investigated in a 150 nm fully-depleted SOI CMOS process [4]. Featuring a Si film thickness of T_{Si}=50 nm, our simulations (curve 3 in fig. 3) indicate higher responsivity values thanks to an increase of the sensitive area (the anode and cathode lengths are reduced to less than 1 μm) but the cut-off wavelength is lowered to λ_c=400 nm due to our reduction of the silicon film thickness and the attenuation of 90% is now achieved over 200 nm because of the thicker top oxide in that industrial process.

Another important improvement would be the reduction of the front silicon oxide thickness while keeping the 40 nm silicon nitride anti-reflection coating. It could reduce the oscillations in the bandwidth and considerably increase the responsivity values. But as we saw with the 150 nm CMOS process simulations, the best way to increase responsivity is to reduce the anode and cathode areas, as predicted by our simulations (curve 2 in fig. 5) where we can achieve a responsivity higher than R=0.2 [A/W] with T_{OX}= 6 nm and Ln=Lp=1 μm near λ=375 nm. The other dimensions (i.e. Li, T_{Si}, T_{BOX}, T_{SUB}) are as in Figure 1.

Silicon-on-Nothing structure

SOI technology can also be extended, for example, to SON technology [5] by etching a part of the buried silicon oxide. A sketch of a SON photodiode structure is shown in figure 4-a.

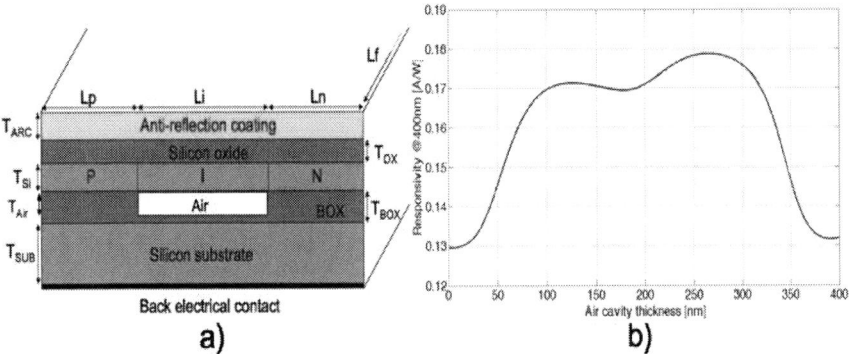

a)   b)

Figure 4: Structure of a silicon-on-nothing photodiode and its responsivity at λ=400 nm as a function of the air cavity thickness. At figure 4-a, T_{OX}= 6 nm, Ln=Lp=1 µm and the other dimensions (i.e. Li, T_{Si}, T_{BOX}, T_{SUB}) are as in Figure 1.

Figure 4-b shows the evolution of the responsivity of the SON structure at λ=400 nm with the air cavity thickness. We can observe a maximal value of R when T_{Air}≈250 nm. Simulations show that this structure could allow to reduce the spectral sensitive bandwidth, in order to be more selective around λ=375 nm while keeping high responsivity values (curve 3 in fig. 5). For this simulation, the dimensions are as indicated in the caption of figure 4.

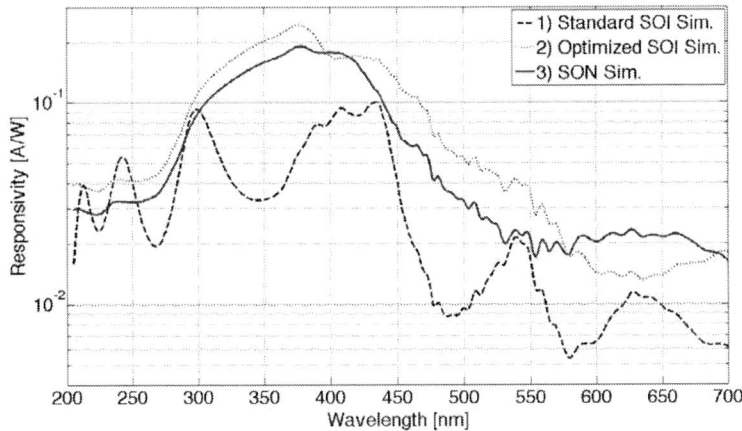

Figure 5: Simulations of lateral PIN photodiodes optimized in different technologies and their comparison to a standard SOI structure.

We can observe a slight lesser responsivity of the SON structure than the optimized SOI one at λ=375 nm but the visible blindness is accentuated for the SON structure. Moreover, the air cavity can also be used for special application (i.e. in introducing particles or gases in the cavity for a detection based on the refraction index variation).

Applications of the SOI PIN photodiodes

After designing these photodevices, we tested them in two different applications. The first one was a current-to-frequency converter implemented in our SOI CMOS process. The characterization of the fabricated integrated circuit showed a good linearity of the circuit output with the responsivity of the photodiodes [6].

For the second application, we used the photodiodes in a biomedical application: the measurement of DNA concentration. The purpose of this experiment was to measure DNA, in closed tube containers, at very low concentration levels. To achieve this measurement, we directly illuminate the photodiodes with a 280 nm ray of light and place the tube containers through the direct optical path, between the light source and the photodiode. With natural DNA absorption peak around this wavelength [7], we could expect a monotonic relation between the DNA concentration and the induced photocurrent (e.g. a lower photocurrent for larger DNA concentration, and a large photocurrent for the smallest DNA concentrations). It was verified by characterization over a large range of 5 orders of magnitude of DNA concentration (from 400 ng/μL to pg/μL), demonstrating higher efficiencies with respect to commonly used methods or apparatus in a biology and clinical lab [8].

Conclusion

In conclusion, SOI CMOS technology enables the implementation of high-performance sensors and analog signal processing circuits towards a complete system-on-a-chip. In this work, we designed and characterized photodiodes in SOI technologies, and proposed optimizations, to keep a low-cost SOI CMOS compatible process and target high responsivities and appropriate wavelength range for biomedical and environmental UV detection. The characterized responsivities fit very well with the simulations and other non standard structures have been studied. The simulated responsivities place them as very efficient UV sensors. Finally, we discussed two concrete applications of our photodiodes that demonstrate their efficiency.

References

1. M. Razeghi, *Proc. of the IEEE*, **90**, 1006 (2002).
2. A. Afzalian and D. Flandre, *IEEE Trans. on Electron Devices*, **52**, 1116 (2005).
3. B. Kumar, T. Baskara Pandian, E. Sreekiran and S. Narayanan, *Record of the Thirty-first IEEE Photovoltaic Specialists Conference/2005*, p. 1205, Bangalore (2005).
4. OKI semiconductor website : *http://www2.okisemi.com*
5. V. Kilchytska, M.T. Chung, B. Olbrechts, Ya. Vovk, J.-P. Raskin and D. Flandre, *Solid State Electronics*, **51**, 1238 (2007).
6. O. Bulteel, A. Afzalian and D. Flandre, in *Proc. of Taisa/2007*, p. 50, *Lyon* (2007).
7. A. Karczemska and A. Sokolowska, in *3rd International Conference on Novel Applications of Wide Bandgap Layers/2001*, p. 176, Zakopane (2001).
8. O. Bulteel, P. Dupuis, S. Jeumont, L.M. Irenge, J. Ambroise, B. Macq, J.-L. Gala and D. Flandre, in *Proceedings of eMBEC/2008*, p. 290, Antwerp (2008).

A Wide Range Temperature Sensor Using SOI Technology

R. L. Patterson[a], M. E. Elbuluk[b], and A. Hammoud[c]

[a] Richard Patterson, NASA GRC, MS 309-2, Cleveland, OH 44135, USA
[b] Malik E. Elbuluk, ECE Dept., University of Akron, Akron, OH 44325, USA
[c] Ahmad Hammoud, ASRC Aerospace Corp., MS 309-2, Cleveland, OH 4413, USA

> Silicon-on-insulator (SOI) technology is becoming widely used in
> integrated circuit chips for its advantages over the conventional silicon
> counterpart. The decrease in leakage current combined with lower
> power consumption allows electronics to operate in a broader
> temperature range. This paper describes the performance of an SOI-
> based temperature sensor under extreme temperatures and thermal
> cycling. The sensor comprised of a temperature-to-frequency
> relaxation oscillator circuit utilizing an SOI precision timer chip. The
> circuit was evaluated under extreme temperature exposure and thermal
> cycling between -190 °C and +210 °C. The results indicate that the
> sensor performed well over the entire test temperature range and it was
> able to re-start at extreme temperatures.

Introduction

Electronic systems in space exploration missions and in aerospace are expected to
encounter extreme temperatures and wide thermal swings. Such missions include
planetary surface exploration and deep space probes (hot and/or cold temperature), jet
engine distributed control architecture (hot), and the NASA James Webb Space
Telescope (cold). Electronics designed for such applications must, therefore, be able to
withstand exposure to extreme temperatures and to perform properly for the duration of
the missions.

Silicon-on-insulator (SOI) technology is becoming widely used in integrated circuit
chips for its advantages over the conventional silicon counterpart (1). The technology
offers low power consumption and high performance in many critical applications (2,3).
In addition, the low parasitic in SOI-based devices can expand CMOS technology in the
nanometer regime for high-performance applications (4). Although the SOI electronics
are geared for high temperature applications (5), they also offer reduced leakage currents
and good tolerance to radiation. NASA could utilize this technology in the design of
reliable standard as well as application-specific integrated circuits (ASIC) that are
essential in space exploration missions and aerospace where severe operational
conditions exist. Typically, SOI parts are designed for high temperature applications (>
200 °C) and improved radiation resistance. Very little information, however, exists about
their performance at very low temperature conditions.

An effort was undertaken to design and develop a temperature sensor for distributed
engine control architecture. Prompted by design considerations and ensuing benefits in
terms of weight and cost reduction and improved reliability, the sensor needs to be
located in the engine compartment, and it must be able to operate at temperatures

exceeding 150 °C. The output of the sensor, which consists of a stream of digitized pulses whose frequency is proportional to the sensed temperature, would be interfaced with a controller or a computer. The data acquisition system would then give a direct readout of the temperature through the use of a look-up table, a built-in algorithm, or a mathematical model.

Temperature Sensor Circuit

The temperature-to-frequency relaxation oscillator (6) circuit was constructed using a high temperature polyimide circuit board, Teflon wire interconnects, and high temperature lead-free solder. The circuit employed a high temperature SOI precision 555 timer. The timer is a ceramic-packaged chip and is specified for -30 °C to +225 °C operation (7). The other components of the circuit included a high temperature precision, thin film platinum RTD (Resistance Temperature Detector) as the temperature-sensing element, a solid tantalum input filter capacitor, and a couple of ceramic timing capacitors. A schematic of the circuit is shown in Figure 1a, and an example of a practical application is depicted in Figure 2b. The temperature sensor in this case is being utilized to monitor the temperature of a jet engine with the sensor circuit is physically placed inside the engine compartment where the temperature varies from about 120 °C to over 700 °C, depending on the location inside the engine. The power supply and the output processing device, on the other hand, are located externally and thus are not exposed to the high temperature harsh environment. This distributed engine control architecture results in minimizing the wiring harness interface, eliminates thermal management for the sensor electronics, simplifies maintenance, improves reliability, and reduces weight and cost. Figure 2 shows a photograph of the circuit board of this temperature sensor. The sensor circuit was evaluated at selected test temperatures over the range of -190 °C to +210 °C. The effect of thermal cycling was also investigated by subjecting it to a total of ten cycles within these temperature extremes using a rate of 10 °C/min and a dwell time of ten minutes. The sensor circuit was evaluated in terms of its frequency-to-temperature conversion and variation in the output signal duty cycle and rise time with test temperature.

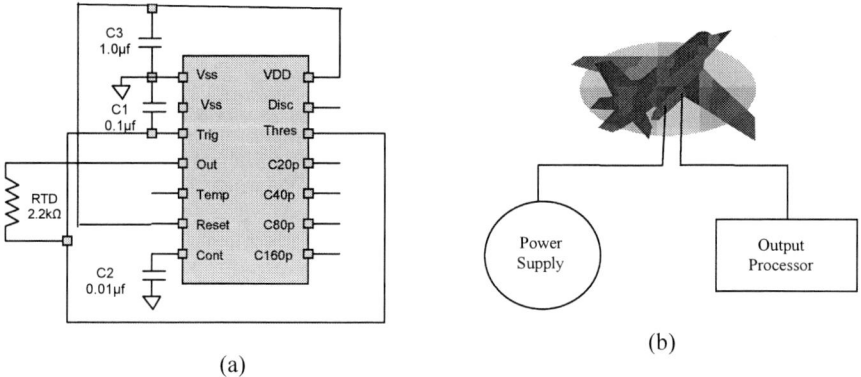

Figure 1. Sensor circuit schematic (a); practical application in jet engine environment (b).

Figure 2. Proto-board of the temperature sensor circuit.

Test Results and Discussions

The output response of the temperature-to-frequency conversion circuit, which comprised of a rectangular pulse train, is shown in Figure 3 at 25 °C. The signal at the threshold pin, which governs the charge/discharge cycle of the timing capacitor C1, is also depicted in this figure as a triangular waveform. Those obtained at the high temperature of +210 °C and at the cryogenic temperature of -190 °C are shown in Figures 4 and 5, respectively. It can be clearly seen that the relaxation oscillator circuit performed very well throughout the temperature range between +210 °C and -190 °C as the output signal exhibited no distortion and its pulse count, i.e. frequency, fluctuated in response to variation in the sensed temperature. For example, while the frequency of the output signal had a value of about 3.393 kHz at room temperature, it decreased to about 2.072 kHz at +210 °C and attained a magnitude of 15.048 kHz when the temperature approached -190 °C. This frequency response, which took on a hyperbolic trend with temperature, is depicted in Figure 6. Plotting the period of the output signal versus temperature reveals, as expected, a linear response as shown in Figure 7. No change was experienced by either the duty cycle or the rise time of the output signal throughout the test temperature range as shown in Figures 8 and 9, respectively.

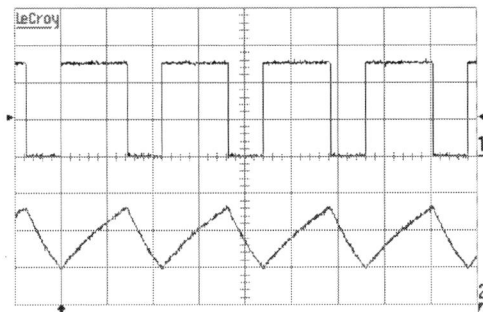

Figure 3. Output and threshold signal waveforms at 25 °C.

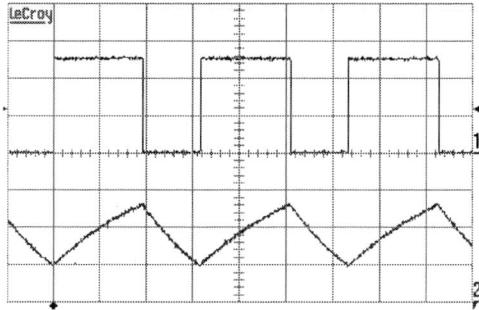

Figure 4. Output and threshold signal waveforms at +210 °C.

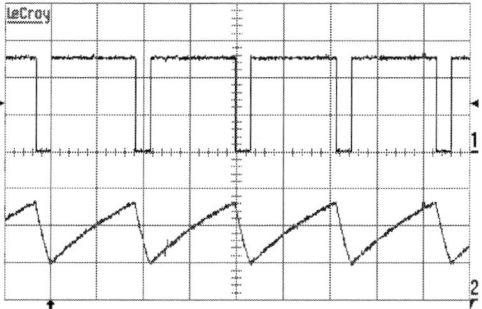

Figure 5. Output and threshold signal waveforms at -190 °C.

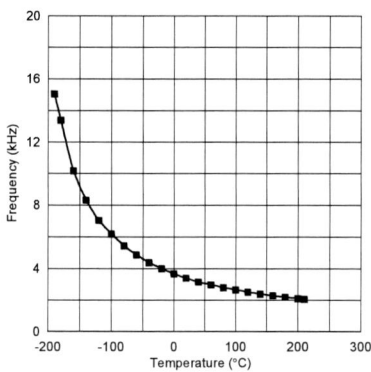

Figure 6. Frequency versus temperature.

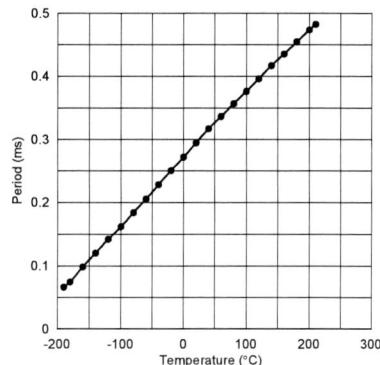

Figure 7. Period versus temperature.

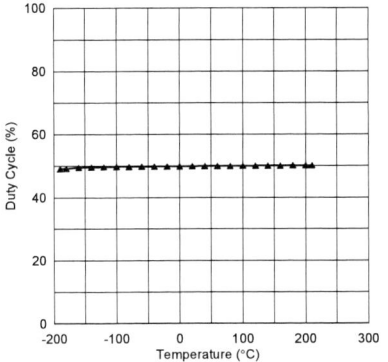
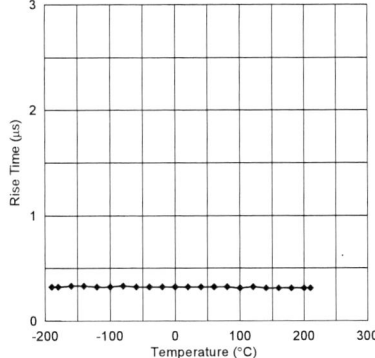

Figure 8. Duty cycle versus temperature. Figure 9. Rise time versus temperature.

The supply current of the circuit varied slightly with temperature as depicted in Figure 10. While the current hovered around 0.68 mA between +20 °C and +210 °C, it exhibited slight increase with decreasing temperature; reaching about 2.5 mA at the extreme cryogenic temperature of - 190 °C. This change in the circuit current with temperature is most likely attributed to the variation in the resistance value of the RTD element in the circuit as test temperature changed.

As was mentioned earlier, the performance of the relaxation oscillator circuit was also investigated after subjecting it to ten thermal cycles between +210 °C and -190 °C. Post-cycling measurements performed on the investigated parameters, i.e. frequency output, duty cycle, rise time, and supply current, revealed no major deviation from those obtained prior to cycling at any given test temperature. Therefore, this limited thermal cycling has had no impact on the operation of the oscillator circuit. In addition, the circuit demonstrated successful start-up operation while at the extreme temperatures, i.e. +210 °C and -190 °C.

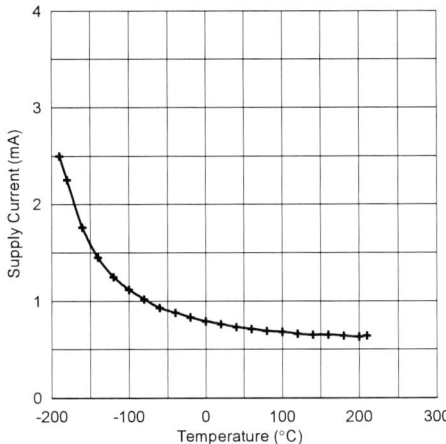

Figure. 10. Circuit supply current as a function of temperature.

Conclusions

A temperature-to-frequency relaxation oscillator circuit was constructed using COTS (Commercial-Off-The-Shelf) parts for application under extreme temperatures. The circuit employed a recently-developed high temperature silicon-on-insulator 555 timer, thin-film platinum RTD, and solid tantalum and ceramic capacitors. The circuit was designed mainly for hot jet engine environment but it was evaluated also for potential use under cryogenic conditions. Performance of the oscillator circuit was investigated in terms of its temperature-sensing response, output signal duty cycle and rise time, and supply current under a wide temperature range between -190 °C and +210 °C and after thermal cycling. The prototype circuit performed well throughout this temperature range in producing a frequency output in proportion to the sensed temperature, and no major changes were observed in its characteristics, i.e. duty cycle and rise time of the output signal, as a result of change in test temperature. In addition, all of the individual parts exhibited no physical or packaging damage due to the extreme temperature exposure or cycling. These preliminary results suggest that the circuit has good potential for use in extreme, both hot and cold, temperature environments.

Acknowledgments

This work was performed under the NASA Glenn Research Center, GESS-2 Contract # NNC06BA07B. Support was provided by the NASA GRC Fundamental Aeronautics / Subsonic Fixed Wing Program and by the NASA Electronic Parts and Packaging (NEPP) Program.

References

1. R. Patterson, A. Hammoud, and M. Elbuluk, *Silicon-On-Insulator Operational Amplifier for Extreme Temperatures*, NASA Report, http://www.nepp.nasa.gov.
2. J. L. Pelloie, *Using SOI to Achieve Low-Power Consumption in Digital*, Proceedings of 2005 IEEE International SOI Conference, p. 14, (2005).
3. W. Wu et al., *PSP-SOI: An Advanced Surface Potential Based Compact Model of Partially Depleted SOI MOSFETs for Circuit Simulations*, Solid-State Electronics, **53**, 18, (2009).
4. J. O. Plouchart, *SOI Nano-Technology for High-Performance System on-Chip*, Proceedings of 2003 IEEE International SOI Conference, p. 1, (2003).
5. M. Si Moussa et al., *An Investigation of Temperature Effects on CPW and MSL on SOI Substrate for RF Applications*, Proceedings of 2005 IEEE International SOI Conference, p. 70, (2005).
6. DeLiang Wang, *Relaxation Oscillators and Networks*, Wiley Encyclopedia of Electrical and Electronics Engineering, vol. 18, pp. 396-405, (1999).
7. Cissoid Company, *CHT 555 High Temperature Precision Timer*, Datasheet Doc. DS-07006 Rev. A, (March 28, 2007).

CHAPTER 8

DEVICE PROCESS TECHNOLOGY

Diamond-on-SOI Field Emission Device Patterning

X.C. LeQuan[a], B.K. Choi[a], W.P. Kang[a], J.L. Davidson[a]

[a] Department of EECS, Vanderbilt University, Nashville, Tennessee 37235, USA

Nanocrystalline diamond films deposited on SOI and electron beam patterning techniques are used to fabricate a field emission device with sub-micron cathode-anode separation. The subsequent aluminum evaporation and dry etch creates a mask for the electrode features, demonstrating that the nanodiamond (ND) roughness does not compromise the device delineation when proper EBL write parameters are used. A mixture of BCl_3 and Cl_2 were used to dry etch the aluminum, oxygen O_2 plasma was used to etch the nanodiamond, and SF_6 gas was used to etch the active layer of silicon followed by $KOH:DIH_2O$ wet etching.

Introduction

Field emission performance is highly dependent on cathode tip geometry and is designed to be sharp, with minimal cathode-anode distance for more efficient electron tunneling. The material of choice is CVD diamond due to an inherent wide band gap (5.5 eV) that promises operational capability at high temperature and radiation intensities. Other attractive properties of CVD diamond include high thermal conductivity and low electron affinity (2), characteristics beneficial for an efficient emitter material. Coupling ND with silicon-on-insulator (SOI) technology gives many advantages, namely lower power dissipation, high radiation tolerance, lower parasitic capacitance, and manufacturing compatibility (3). A cathode device in the lateral configuration allows a higher extent of lithographic control versus the vertical configuration. Planar lateral field emitters had been difficult to pattern, moreover the lack of well accepted etching techniques for patterning makes the fabrication process less straightforward. The motivation for this research is to develop a reference process flow for achieving a sub-micron emission gap device. Our earlier generations of the nanodiamond lateral device were process limited (4); but with EBL as the key tool in our new technique, we are able to delineate sub-micron cathode-anode spacing.

We explore combining the proven optical lithography method with EBL as a viable avenue for building the sub-micron cathode emission gap. Optical lithography is relatively fast, while EBL enables finer geometry patterning, a feat that is further complicated by the due roughness of the CVD diamond. In our work we created a dosage array to help find the optimal dose-to-clear parameter combination for a given nanodiamond film thickness of 0.65 μm. Subsequent material etching further introduces factors affecting the final gap distance however our systematic approach enabled us to achieve a cathode-anode distance that is consistently less than 1μm.

Experimental

The nanodiamond films used for this work were created using microwave plasma enhanced CVD (MPECVD) using CH_4, H_2, and N_2 as the reactant feed gas. We incorporate nitrogen because it induces conductive behavior by way of defect states formed on the film surface and grain boundaries. The substrate is pretreated by mechanical polishing with detonation diamond powder and ultrasonicated with the same class of powder dispersed in acetone. The film is grown at 800°C, 800 Watts, 20 Torr, and gas flows were set at 15 sccm nitrogen, 15 sccm methane, and 135 sccm hydrogen. High magnification SEM imaging seen in Fig. 1 a) and b) show that the 212 Ω-cm film has a cluster formation resembling cauliflower, similar to the diamond substrates used for the previous generation of lateral devices (4). The SOI substrate used consisted of a 2μm active Si layer on 4μm oxide all on a 525 μm Si bulk layer, as seen in Fig. 1 c). The 4 μm buried oxide will serve as the dielectric isolation for the nanodiamond electrodes and adjacent devices on the same substrate. This SiO_2 layer is protected from the aluminum and diamond plasma etching by the active Si layer. Fig. 1 d) shows a cross section of the device substrate prepped for EBL, consisting of a 0.65 μm ND topped with a 350 nm Al and 300nm PMMA layers, respectively.

Fig.1. a) 60, 000 × SEM of nanodiamond substrate, b) 100, 000 × SEM of nanodiamond substrate showing clustering of crystals, c) 1, 000 × low magnification of device cross section, d) cross sections of PMMA-Aluminum-Nanodiamond-SOI where the inset shows a higher magnification image of PMMA-Aluminum-Diamond thicknesses.

The nanodiamond cathode is comprised of teeth or finger-like structures having a high aspect ratio with tapering down to sharp tips. The nanodiamond anode is of straight edge shape for rectification. The critical device area is contained in a 100μm × 20μm area. The lateral devices in this manuscript were designed to have a 100 nm emission gap, as shown by the inset of Fig. 2. However obtaining an actual emission gap that would be the same as the mask design proves to be a challenge because the subtractive process used to create the device involves striking a delicate balance between the dose-to-clear parameters and the corresponding emission gap width; a higher dosage would certainly clear the features however at the expense of a larger emission gap width. Any over etch of masking layers will directly result in the widening of the anode-cathode gaps. Fig. 2 also shows a 4 × 4 array mapping that enables us to simultaneously study the effect of changing the exposure dosage factors in one batch while keeping all other write parameters constant. The base dosage was 205 $\mu C/cm^2$ at 20 kV. Fig. 4 shows the actual results of varying the exposure dosage on the emission gap. A smaller dosage resulted in a smaller emission gap but also less clearance below the critical device area.

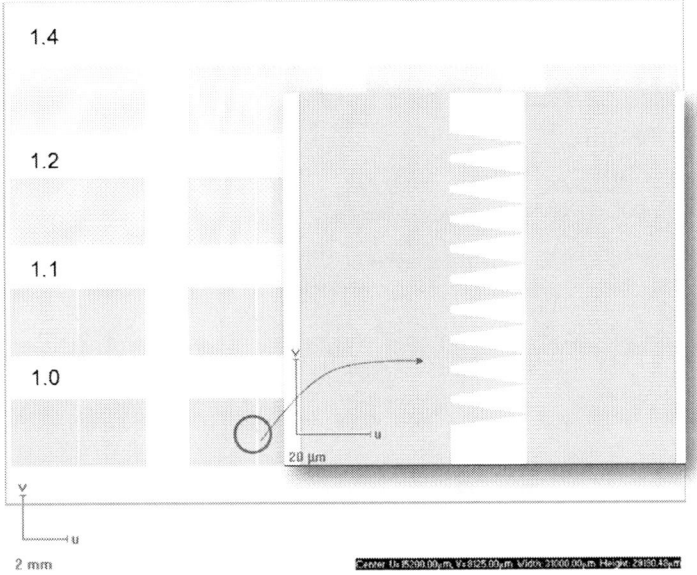

Fig. 2. Matrix array illustrating variation in dose factors within the same batch of devices. Inset shows mask design for a 10 cathode tip and anode pattern.

The fabrication steps are outlined in Fig. 3. The AZ5214E photoresist was spin deposited onto CVD nanodiamond films at 4000 rpm for 45 seconds, then baked at 120 °C, producing a thin layer for optical patterning. The larger patterns of the device are then patterned over on top of the photoresist. Following, an Al thickness of 350 nm was deposited by electron beam evaporation and lift-off followed using warm acetone. Lift off removes the unwanted materials surrounding the device area. PMMA 950 A6 was then spin coat deposited at 4000 rpm for 45 seconds and soft baked at 200°C for 2 minutes. Using an e-Line Raith EBL tool, the direct patterning of the critical cathode structure was performed. The exposed areas from EBL were developed using an MIBK:IPA mixture for 1 minute and rinsed with IPA for 30 seconds. This is the final patterning step prior to etch processing of aluminum, diamond, and silicon.

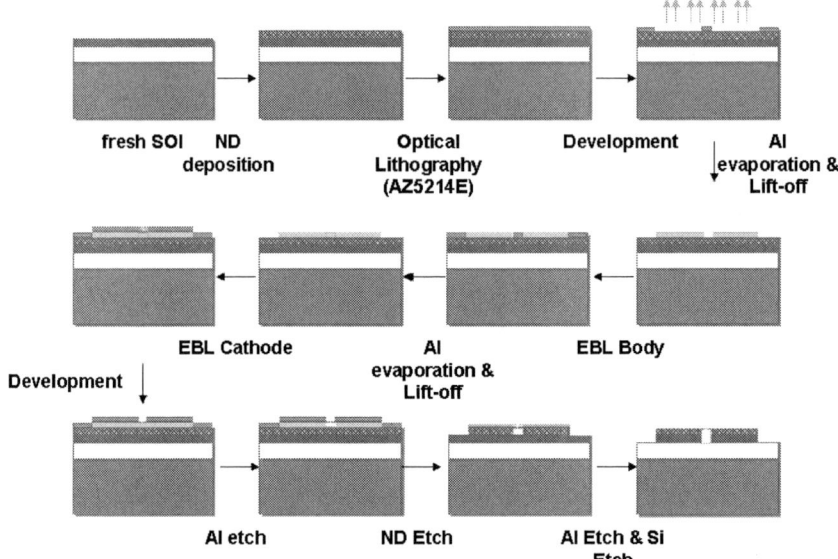

Fig. 3. Process flow for fabricating the sub-micron gap lateral configuration field emission device. Start with fresh SOI wafer (2 μm active Si, 4 μm buried oxide, 525 μm Si bulk), deposit CVD nanodiamond, then perform feature alignment of optical and electron beam lithography patterning followed by metal mask, diamond, and silicon etch for device isolation.

Results

Dry etch of the aluminum mask provides the benefit of one-directional etching, however it hardens remaining PMMA by inducing cross-linking, resulting in the formation of particles visible in the lower plane below the cathode tips in Fig. 4. These particles do not interfere with subsequent etching steps. The optimal Al etch recipe should etch the metal mask faster than the PMMA, which must remain as a mask for areas outside of the critical device area, as illustrated in the aforementioned process flow seen in Fig. 3. Rainey et al. (5) reported that a gas flow rate of $BCl_3:Cl_2$ of 40:3 gave the best sidewalls, however when we used this recipe, the PMMA etched completely while the aluminum remained fully intact. The main etchant of aluminum is Cl_2 (6); the purpose of BCl_3 is to etch the aluminum oxide that forms during the process. The optimal Al etch would bore through the metal layer and expose the diamond since any trace of Al would inhibit diamond etch and compromise further processing. The PlasmaTherm SLR aluminum mask etch conditions were BCl_3 and Cl_2 gases at a ratio of 20:16 sccm under 20 mTorr at 200W. The STS AOE anisotropic RIE etch conditions for selectively removing the diamond film below the aluminum mask were 10 mTorr, 600W/100W Coil/Platen, and 30 sccm of O_2 for 11 minutes. A second aluminum etch step is performed to reveal the diamond as the top layer followed by using a fluorine-based silicon RIE etch condition was 13 Torr, 200W, and 30 sccm of SF_6 to reveal the SiO_2.

One of the critical issues is preserving a pattern true to the mask design and from both Fig. 4 and Fig. 5 we can see that the nanodiamond grain structures do not pose problems in terms of preferential etch along the grain boundaries. The fabrication technique can be further improved by optimizing parameters to where there would be less partial cross-linking in the PMMA layer (particles in layer below critical features). In order to further improve the fabrication technique, an additional polymer based resist layer can be introduced. Other e-beam work based on CVD diamond as the substrate refer to using flowable oxide (FOx-12) on top of PMMA as an additional mask for finer feature delineation (7). Although this work targeted sub-micron hole array features for the fabrication of diamond photonic crystals, the method may be a possible extension to our developed process.

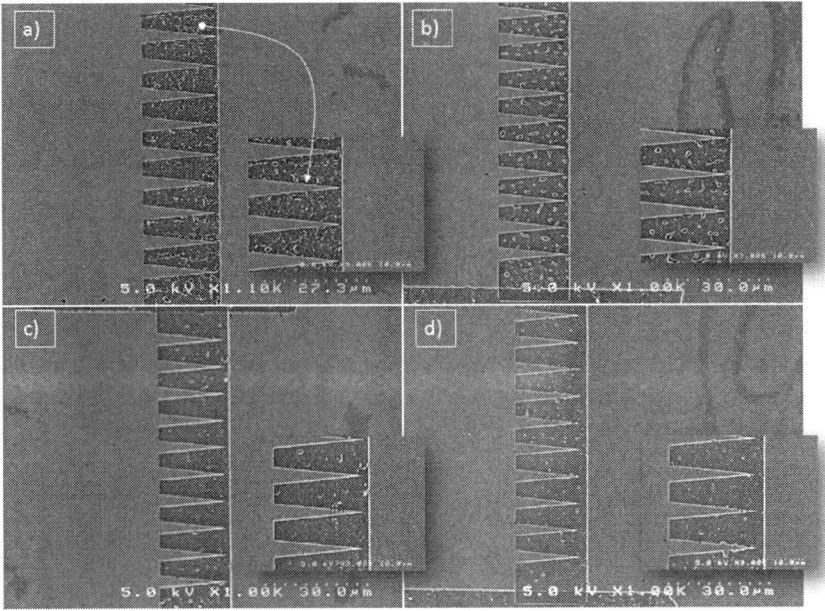

Fig. 4. Device delineation results after nanodiamond and aluminum etch. The lowest dosage resulted in the smallest sub-micron emission gap in a).

The silicon beneath the diamond is first removed using dry etch, however silicon crystal residue will remain underneath the critical feature since dry etch is directional. A silicon wet etch step using KOH and DI H_2O is performed to clear out the silicon crystals and one of the final results are show in Fig. 5. The tilted angle view shows the columnar structure of the diamond substrate. The silicon layer is undercut, however the remaining silicon support elevates the diamond from the insulator and substrate, ensuring emission will be from the tips only.

Fig. 5. Tilted angle view of Fig 4. c) after silicon wet etch using KOH: DIH_2O.

Conclusions

A patterning method combining both optical lithography and EBL towards achieving a sub-micron emission gap delineation has been developed and described. Optical lithography was used to pattern the larger features of the field emission device. Aluminum was evaporated and lift off was performed to remove unwanted features. After PMMA deposition followed by e-beam exposure and development, Al dry etch of critical device area using the PlasmaTherm SLR equipment series. Following, advanced oxide etch of the nanodiamond to yield the the cathode-anode structures was performed. After the final Al etch step was performed the silicon was etched both in fluorine based plasma and in $KOH:DIH_2O$ solution to yield device isolation and elevation. Fig. 5 shows the results for the device from Fig. 4 c) which has slighter larger than $1\mu m$ gap, indicating that the same processing applied to the sub-micron gap device in Fig. 4 a) will result in an isolated device with a sub-micron emission gap width.

References

1. V. Melanović, *IEEE Transactions on Electron Devices*, **48**, No. 1, (2001).
2. R.J. Nemanich, P.K Baumann, J. Van der Weide, *Proceedings of the Applied Diamond Conference*, **1**, 17 (1995).
3. V. Jaju, "Silicon-on-Insulator Technology", *EE530 Advances in MOSFETs*, 1-3 (2004).
4. K. Subramanian, W.P. Kang, J.L. Davidson, B.K. Choi, *Diamond & Related Materials* **17** 1808–1811 (2008).
5. A. Rainey, *Aluminum Etching using Reactive Ion Etching (RIE)*, 7-32 (2003).
6. T. Tsukada, *Japanese Journal of Applied Physics*, **30**, No. 11A, 2956-2964 (1991).
7. J. Baldwin, M. Zalalutdinov, T. Feygelson, J. E. Butler, B. H. Houston, *J. Vac. Sci. Technol.* B 24, **1**, 51-53 (Jan/Feb 2006).

ECS Transactions, 19 (4) 195-200 (2009)
10.1149/1.3117409 ©The Electrochemical Society

H-shaped Vertical Polycrystalline Silicon Thin Film Transistor on Insulator

H. D. Toure, T. Gaillard, N. Coulon, O. Bonnaud

Groupe Microélectronique, Institut d'Électronique et de Télécommunications de Rennes,
UMR CNRS 6164 Université de Rennes 1, Campus Beaulieu,
35042 Rennes Cedex, France

We introduce a new design of vertical fully depleted type of the polycrystalline silicon (polysilicon) on insulator thin film transistor (TFT). This new structure fabricated based on polysilicon is characterized by vertical channels, an H-shaped geometry and also a technology similar to classical silicon on insulator (SOI) one. In practice, the design corresponding to a FinFET one is close to multigate SOI MOSFET and thus requires less area than planar design. It allows increasing easily the channel width simultaneously to better control and reduction of channel length, leading to an increase of the average driven current and device densities. This paper describes the low temperature fabrication process of this new structure that is compatible with SOI or glass substrates. A first run involving simple plastics masks allowed electrical characterizations of the transistors, mainly output and transfer characteristics that are discussed, confirming the potentiality of such a structure.

Introduction

In a world abundantly marked by ever growing applications of electronic and microelectronic technologies, innovative devices are more and more needed to enlarge the fields of applications, that is included in the trend called "more than Moore" evolution. Among them, polycrystalline silicon (polysilicon) thin film transistor technology appears as one of the challenging devices thanks to is low temperature process that is compatible with glass and even flexible substrates available to flat panel displays (1), but also to above IC (Integrated Circuits) device capability. These technologies allow a hybrid assembling combining VLSI devices to low temperature process devices. Some challenge remains on this type of technology, mainly the frequency response that is strongly linked to the parasitic capacitance, series resistance and equivalent current densities driven by the transistors. The development of high speed and low-power components since the propagation delay is roughly proportional to the square of the channel length is a general goal in the field of microelectronics devices. For deposited silicon film, the main problem is the quality of the material, with regard to the mobility of the free carriers. The amorphous silicon easily deposited for a lot of applications (mainly large area electronics and flat panel displays, has however a strong drawback, the very low mobility of the free carriers, lower than 1 cm^2/Vs for the electrons. A first improvement was got thanks to the control of the crystallization of this film. The most convenient and the cheaper technique compatible with glass substrates is base on solid-phase crystallization while a laser crystallization may lead to better results but for a much higher complexity of the global process (2). In previous works (3), we developed a relatively low temperature process (< 600°C) involving a thermal annealing of the film just at the end of the silicon film deposition. This process demonstrated the possibility to reach a field effect mobility in the range of 140 cm^2/Vs for the electrons, to get a good reproducibility and a high reliability thanks to the low content of hydrogen in the silicon film at the end of the process. However, these TFTs had a rather large area due to the glass substrate and photolithography constraints. Indeed, the usual channel length is several micrometers and thus the electrical performances of the transistors are mainly limited by the geometry. An improvement of these electrical performances is thus governed by the size of the device and by the total driven current, that both needs a short channel length and a large channel width for the same silicon or glass substrate area. Similarly to ultra large scale integration, one way consists to move to a three dimensional geometry, in our case a quasi-vertical structure similarly to FinFET technologies, for which two channels are vertically oriented with very low thickness and very low height of the fin.

For our new structures, the vertical aspect design is performed thanks to a key technological step that consists to define the length of the channel by a deposition of the channel film. Because, the control of the thickness is much better than a photolithography patterning, the channel length can be adjusted by the layer deposition that leads to a much better miniaturization in comparison with its lateral geometry counterparts (4).

However, similarly to SOI technologies, some phenomena that were negligible in the lateral technology are becoming of importance in this new approach, such as short channel effects, punch-through effects, drain induced barrier lowering (DIBL), or off-stage leakage, and are limiting the scalability and deteriorate the performances (5). Thus, to maintain or improve these performances, several designs and technologies are proposed to minimize or solve these problems. (6-8). Since the first fabrication of the fully depleted double gate SOI MOSFET, a lot of technological improvements have been performed to overpass the limits of classical technologies. Among them, polysilicon TFTs exhibit the beneficial characteristics as silicon on insulators (9-10).

Vertical thin film transistor principle

As preliminary mentioned, in order to improve the performances of such transistors mainly to reduce the size with the same driven current and to scale down the associated circuits, a new design was proposed with a quasi-vertical double channel having a high width thanks to an H-shaped design and involving low temperature processes compatible with low temperature substrates or above IC circuits. Figure 1 shows a schematic cross-section of the generic thin film transistor. Source, channel, and drain silicon films are stacked and deposited on an oxide. This oxide can be thermally grown from silicon, or deposited. This configuration is thus very close to a SOI one. The silicon film of the three regions is deposited by Low Pressure Chemical Vapor Deposition (LPCVD) technique during the same run at a temperature of 550°C. At this temperature, the silicon material is mainly amorphous. Source and drain regions are in-situ doped during the LPCVD step. The stacked films are afterwards crystallized by a thermal annealing at a temperature of 600°C, that corresponds to solid-phase crystallization.

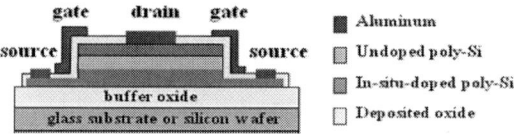

Figure 1. Schematic cross-section of the vertical thin film transistor on insulator. On the principle, this device is close to a FinFET. The main difference comes from the nature of the semiconducting films for source, channel and drain, in this case, polysilicon one.

On the principle, the channels are on the edges symmetrically positioned and the gate oxide and gate contact films play the role of lateral walls. This device is thus in the family of the multi-gate channel devices and therefore beneficiates of the potential performances of these structures (7). In fact, the recent studies on SOI multi-gate devices over the last decade show that technology is a good reliable alternative to meet the future technological requirements allowing to push the limits of silicon integration beyond the limits of classical technologies (11). We move on the same direction with this new structure.

The effective structure was defined by involving SILVACO Virtual Wafer Fab and CADENCE tools. Even if some physical parameters should be different (for example the interface state densities), globally the electrical behavior is not so far of the SOI technology, the mechanisms governing the conduction being more or less the same. On the base of previous works (3), electrical and physical parameters were deduced to initiate an electrical simulating of the generic transistor. With the help of CADENCE tool, it was then possible to design a four-mask process with different widths and lengths, and different shapes, more especially like a comb (12) or like an H as shown in figure 2.

Figure 2. H-shaped thin film transistor design. The lateral walls correspond to a large width of the equivalent transistor. The thickness of the non-intentionally doped silicon defines the channel length. The source is deposited on an oxide. Source and drain are in-situ doped.

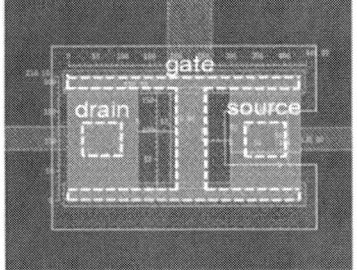

Figure 3. Layout of the H-shaped thin film transistor design. To be compatible with the plastics masks, the size of the patterns are rather large.

In order to go faster, to get cheaper, but also to check and validate the process, the first set of masks was fabricated on plastics support that had relaxed the design rules and consequently had significantly increased the total area of the structure. Indeed, the minimum size of the patterns is about 5µm, and thus the total common surface of the stacked film is close to 105 µm^2, that had some consequences on the final electrical characteristics, mainly in term of leakage current between drain and source. From these size and geometry, the CADENCE tool allows predicting approximately a behavior of the final H-shape structure that was fabricated. Figure 3 shows the layout of the H-shaped structure.

Vertical thin film transistor technology

After cleaning glass substrates or oxidized silicon oxide, a thick buffer oxide is deposited by atmospheric pressure chemical vapor deposition (APCVD). This 500nm thick oxide plays the role of a diffusion barrier in order to avoid a potential contamination by the substrate of the active layers. It is also a mean to always insure whatever the substrate material the same conditions of growth of the silicon films. The active silicon layers are successively deposited by low pressure chemical vapor deposition (LPCVD) at 550°C; let us notice that at this temperature, using silane as precursor gas, the films are mainly amorphous. Drain and source are 600 nm and 360 nm thick, respectively, and in-situ doped by phosphorus. After heavily doped drain layer deposition, a specific shielding of the furnace for two hours is proceeded in order to minimize the doping remanence of the furnace and thus to get a more abrupt doping profile between drain and channel regions. The 1 μm thick silicon channel is non-intentionally doped. After source layer deposition, the multi-layer film is crystallized by solid phase crystallization (SPC) for 72 h at 600°C.

A first photolithography defines the edges of the global structure (geometry of the bottom polysilicon film) corresponding to the external source area. The second photolithography is the most important in this process, because this step defines the geometry, the size, the multi-gate aspect, and the channel width. Figure 2 shows a perspective schematic representation of the structure after this step. H-shaped geometry allows a strong increase of the effective width, corresponding to the length of the lateral vertical walls. After these two photolithography steps, the substrates are cleaned by a conventional RCA, and then a 100 nm thick gate oxide is deposited by RF sputtering and further densified during 2 hours at 200°C under nitrogen flow in order to improve its insulating properties. After contact opening using a third mask, a 500 nm-thick aluminum film is deposited by thermal evaporation. The fourth mask is used to define the contact pads and lines.

In practice, the second photolithography involves a reactive ion etching step (R.I.E) that governs the quality of the channel surface. The edges must be covered by the gate oxide and the gate contact. First, the silicon stacked layer must be etched with a good surface quality allowing a continuous oxide film. In fact, depending of the physical conditions of the R.I.E step, some overhanging was observed that prevents the deposition of aluminum by thermal evaporation all along the channel; a mask effect occurs due to the overhanging (see figure 4). Pressure and gas flow were thus optimized to get a plane surface of the edge but it was tilted and thus not vertical.

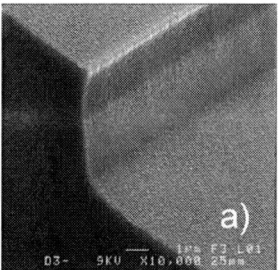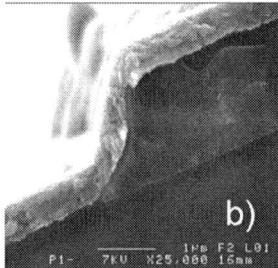

Figure 4. Optimization of the edge etching. a). Vertical etching but with an overhanging on the top that strongly affect the quality of the vertical gate contact. b) Tilted edge with gate oxide and gate aluminum contact presenting a good aspect ratio.

Figure 5. Top view of the structure at the end of the fabrication process.

Figure 5 shows a top view of the fabricated H-shaped transistor that will be electrically characterized. This devices is fabricated a silicon substrate on which an APCVD oxide is deposited. The size of the different regions is rather large due to the choice of plastics masks.

Electrical characterization

The thin film transistors were electrically characterized. The output characteristics shown in figure 6 confirm the transistor effect with typical variations of the drain-source current in function of the gate voltage and of the drain-source bias. The threshold voltage is close to 2V that is a good result for this type of transistor made of polycrystalline semiconductor. The saturation regimes for the low gate biases indicate a low Early effect and no kink effect is detected.

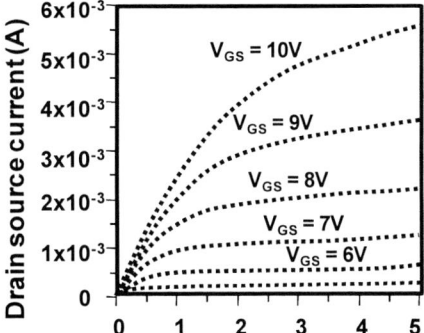

Figure 6. Output characteristics of the transistor. The displayed shapes of the curves confirm the MOSFET effect. For the chosen gate oxide thickness, the gate voltage (V_{GS}) may reach 10 volts.

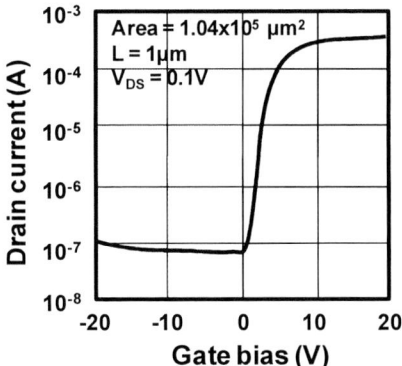

Figure 7. Transfer characteristics of the H-shaped structure. The sub-threshold slope is good for such a transistor. The I_{on}/I_{off} ratio is relatively low due to a high drain-source leakage current.

The transfer characteristics shown in figure 7 give evidence of the good quality of the gate contact. The subthreshold slope is much higher than in the case of usual integrated SOI technologies, due to the high defect density in the polycrystalline semiconductor and at the gate oxide interfaces, but is considered as rather well for polysilicon-based transistors. The reverse current is large, but it is well explained by considering the area of the common coverage of source and drain; indeed, the common area is in this case higher than 105 μm^2, than induces a global leakage current even if the current density is rather low, that decreases significantly the I_{on}/I_{off} ratio. A new layout of the masks with a much better definition of the pattern should decrease significantly this effect.

Conclusion

A fabrication process of thin film transistor on insulator with quasi-vertical channels was developed and optimized. Even if the first run involved a simple design, the electrical characterizations confirmed the potentialities of such a structure. The electrical behavior is close to the usual fully depleted CMOS on SOI. However, the quality of the semiconductor affects the electrical parameters, mainly the subthreshold slope and the I_{on}/I_{off} ratio. A new design involving chromium masks with a much better definition should lead to improvement of electrical properties.

References

1. J.-T. Lin, K.-D. Hunag, S.-F. Hu, Solid-State Electronics, 51, 1056 (2007).
2. Y. Helen, R. Dassow, K. Mourgues, O. Bonnaud, T. Mohammed-Brahim, J. Kholer, J. Werner, Proc. IEDM'99, Dec. 1999, pp.12.3.1.-12.3.4
3. K. Mourgues, F. Raoult, L. Pichon, T. Mohammed-Brahim, D. Briand, O. Bonnaud, Mat. Res. Soc. Symp. Proc., **471**, 155 (1997).
4. J. Moers, Applied Physics A, **87**, 531 (2007).
5. M. Heuser, M. Baus, B. Hadam, O. Winkler, B. Spangenberg, R. Granzner, M. Lemme, H. Kur, Microelectronic Engineering, **61-62**, 613 (2002).
6. S.K. Jayanarayanan, S. Dey, J.P. Donnelly, S.K. Banerjee, Solid-State Electronics, **50**, 897 (2006).
7. K. Akarvardar, S. Cristoloveanu, M. Bawedin, P. Gentil, B.J. Blalock, D. Flandre, Solid-State Electronics, **51**, 278 (2007).
8. P.E. Thomson, G. Jernigan, J. Schulze, I. Eisele, T. Suligoj, Materials Science in Semiconductor Processing, **8**, 51(2005).
9. H. Wang, M. Chan, S. Jagar, V.M.C Poon, M. Qin, IEEE Transactions on Electron Devices, **47**, 1580 (2000).
10. J.P. Colinge, Microelectronic Engineering, **84**, 2071 (2007).
11. C.-W. Lee, D. Lederer, A. Afzalian, R. Yan, N. Dehdashti, W. Xiong, J.-P. Colinge, Solid-State Electronics, **52**, 1815 (2008).
12. H.D. Toure, T. Gaillard, N. Coulon, O. Bonnaud, ECST, Volume **16**, Issue 9, 165 (2008).

ECS Transactions, 19 (4) 201-207 (2009)
10.1149/1.3117410 ©The Electrochemical Society

Issues Associated To Rare Earth Silicide Integration In Ultra Thin FD SOI Schottky Barrier nMOSFETs

G. Larrieu,[a,*] D.A. Yarekha [a], E. Dubois [a], D. Deresmes [a], N. Breil [a,c], N. Reckinger [b], X. Tang [b] and A. Halimaoui [c]

[a] Institute for Electronics Microelectronics and Nanotechnologie - Avenue Poincaré - Cité Scientifique - BP 69 - 59652 Villeneuve d'Ascq Cedex - France
[b] Universite Catholique Louvain - place du Levant 3, B-1348 Louvain La Neuve, Belgium
[c] STMicroelectronics, 850 rue Jean Monnet, 38926 Crolles Cedex France

* guilhem.larrieu@isen.iemn.univ-lille1.fr

The paper focuses on specific issues associated to rare earth silicide integration on UTB-SOI substrate with a particular attention to erbium and ytterbium silicides. Due to the limited Si source, defects generation on SOI is prevented compared to bulk substrate. Reaction of RE with dielectric materials limits the temperature of silicidation. It is shown that RE S/D MOSFETs are still limited in current-drive by the Schottky barrier height.

Introduction

As CMOS technology is entering in the decananometre era, the contact resistance associated to the silicide/silicon interface is identified as one of the biggest challenge to solve in order to preserve current drive capabilities. In that context, source/drain (S/D) engineering takes an increasing importance in the development of leading edge CMOS generations because of the increasing impact of S/D series resistances on transistor performance (1). In order to further pursue down-scaling of MOSFETs in the sub-32 nm range of gate lengths, novel devices that hierarchically combine alternative materials as well new substrate concepts such as silicon on insulator (SOI) substrate or strained substrate have been proposed. Considering that the aforementioned innovations are expected to contribute to a higher current drive at shallower junction depth and reduced silicide thickness, extremely severe constraints are placed on the junction and contact technologies. Indeed, the formation of S/D region on ultra thin body (UTB) SOI is a critical issue because of the huge difficulties to form ultra thin highly doped junction on such UTB-SOI film. One promising approach is related to the S/D architecture to replace ohmic contacts on highly doped junctions by metallic junctions (2). From that standpoint, the Schottky Barrier (SB) MOSFETs architecture provides abrupt junctions, reduced capacitive coupling with the advantage of process simplification and reduced thermal budget. For SB pMOSFETs, the implementation of PtSi contacts on UTB-SOI has been demonstrated and characterized (2)-(4). For the n-type counterpart, only few demonstrations (2)(5)-(7) exist but no specific study on integration has been disclosed so far. For the fabrication of n-type SB MOSFETs, rare-earth (RE) silicides present a great interest because they provide the lowest barrier to electrons on n-type silicon. Er silicide ($ErSi_{2-x}$) is a very interesting compound which presents a low SBH to electrons (Φ_{Bn}) of 0.28 eV and a smooth interfacial morphology. Alternatively, among the RE metals, Yb features the lowest work function (2.59 eV) (8). Therefore, Yb silicide ($YbSi_{2-x}$) is

201

potentially a very promising material to obtain a low SBH that promotes electron injection. This paper focuses on specific issues associated to rare earth silicide integration on UTB-SOI substrate with a particular attention to erbium and ytterbium silicides.

Sample Preparation

RE silicides are usually formed by *ex situ* rapid thermal annealing (RTA). Because of the extremely high sensitivity of the RE metals to oxygen, it is necessary to use a protective cap layer (9) to limit or suppress any oxidation reaction at the silicide/silicon interface However, the implementation of an additional cap layer makes the process of devices fabrication more complicated and implies several restrictions on the choice of the barrier material: i) no intermixing between the cap and the silicided metals, and ii) the sacrificial cap has to be removed selectively to the other materials such a metallic gate, silicide contacts, dielectrics An alternative solution to prevent oxidation during the silicidation is to form the silicide under ultra-high vacuum condition (UHV).

The rare earth silicide films were grown on (10^{15} cm^{-3}) p-type (100) Si substrates and on ultra thin SOI substrates with an active lowly doped p-Si layer of 12 nm. First, the wafers were cleaned in sulfuric peroxide mixture (SPM, 1:1) during 5 min, rinsed in de-ionized water (DI H$_2$O) and blown dried with the nitrogen (N$_2$). The wafers were subsequently dipped into 1% hydrofluoric acid (HF) for 30 s to remove the native oxide, rinsed, and dried as previously. Immediately after the cleaning procedure, the samples were loaded into the introduction chamber of the evaporator. A 15 nm thick layer of Er or Yb was deposited by e-beam under UHV (~ 5 x 10^{-9} mbar) with a low deposition rate (1 Å/s). Before each deposition, the metal target was cleaned by evaporation of several nanometers of metal, the shutter being closed. Finally, samples were transferred to the annealing chamber (5 x 10^{-9} mbar), without breaking the vacuum and annealed for 1 hour at different temperatures. After cooling down to room temperature, the samples were unloaded and the unreacted RE overlayer was removed by SPM (1:1).

Defects Formation

It is known that the formation of rare earth silicides proceeds through a nucleation controlled reaction where Si atoms constitute the diffusing specie. One major drawback of RE compounds is the extreme reactivity with oxygen. Another property of RE silicides grown on bulk Si substrate is the formation of structural defects (10) which lead to an inhomogeneous and rough film. A lot of papers describe the mechanism of pinhole formation (see for example (10)). The major parts of them explain this phenomenon by the Si over consumption from the substrate. During the RE deposition, an amorphous interlayer is unavoidably formed due to a large driving force for the interdiffusion of RE metals and Si of the substrate. After higher temperature annealing, epitaxial RESi$_{2-x}$ nuclei are formed at the interface between the *a*-interlayer and Si substrate. These islands act as a diffusion barrier for the further interdiffusion of RE and Si and result in the formation of pinholes. Silicon atoms then continue to diffuse from the substrate around the RESi islands and to form amorphous RE-Si regions. As a result, the growth of a-interlayer can slow down and the RESi$_{2-x}$ regions grow to form a continuous thin film. Therefore, to prevent the pinhole growth, Si can be deposited on RE and being consumed for the silicidation instead of Si from the substrate. But this approach is not compatible with the implementation in a self aligned process.

SEM and AFM characterizations have been conducted on Yb- and Er-based silicides on bulk and UTB SOI. Fig. 1 shows a comparison between a 15 nm thick ErSix layer obtained on Si bulk (top) and its 12 nm counterpart on SOI (bottom). Starting from a bulk substrate, rectangular shaped penetrating pits with [110]/[1-10] oriented sides is due on inhomogeneous nucleation and lateral Si atoms out diffusion. As presented below, the origin of the pinholes come from localized islands of epitaxial $RESi_{2-x}$ nuclei which are oriented with the (100) substrate. From AFM analysis, the root mean square (RMS) roughness is around 19 nm, relevant of a perturbed topology. In contrast, it is shown that this effect does not take place on UTB-SOI where the surface is perfectly flat with a RMS roughness around 0.4 nm. The contrast in terms of topology between the two silicides obtained in the same condition is striking. It is demonstrated for the first time that the limited reservoir of silicon of the UTB SOI substrate prevents the formation of defects associated to rare earth silicide. The same observations and conclusions have been done with the growth of ytterbium silicide on SOI.

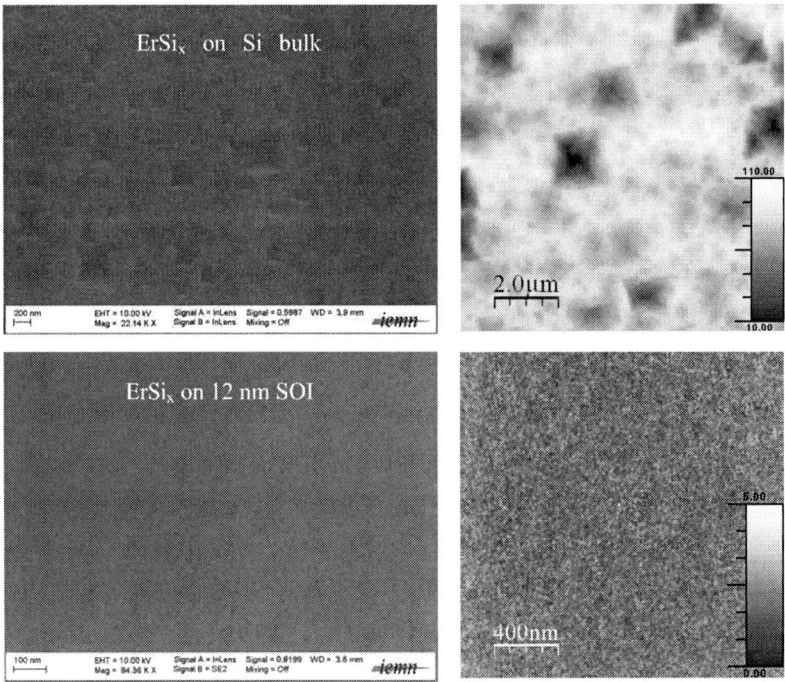

Figure 1: Comparison of structural defect formation on $ErSi_x$ layer obtained at 500°C on 12 nm SOI and Si bulk substrate.

Ytterbium and Erbium silicides were also formed with similar conditions on SOI wafers with integrated oxide (densified HSQ) lines in order to simulation the gate integration with oxide spacers. After the annealing, RE in excess were removed with $H_2SO_4+HNO_3$ (2:1) solution. Fig.2 shows the cross sectional SEM view of a) $ErSi_x$ and b) $YbSi_x$ samples after the unreacted metals stripping. It is observed that the entire silicon layer was consumed in both case and that homogeneous silicide films are obtained. It is also observed that the integrity of the channel under the gate is preserved, indicating that silicon is not laterally pumped by the silicidation reaction as it can be the case with the silicidation on bulk substrate.

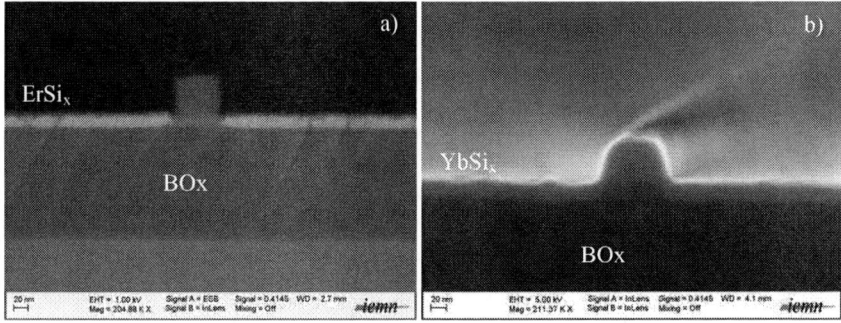

Figure 2: SEM cross sections of a) $ErSi_x$ and b) $YbSi_x$ contacts with a narrow gate on 12 nm SOI substrate.

Device Integration

Lastly, integration of RE silicide on SOI MOSFET with active film of 8 nm has been conducted. These devices feature a HK dielectric with TiN / polysilicon gate (Equivalent Oxide Thickness ~ 1.7nm). The devices were isolated using a mesa topology down to the SiO_2 BOX. The devices are used as a test vehicle to integrate and to compare rare earth silicides obtained under various conditions. It is worth noting that no back-end metal is deposited to eliminate the potential consequences that would arise from a reaction of the back-end metal with the RE silicide.

Fig. 3 shows the comparison of the static I_d-V_g and I_d-V_d characteristics of long channel (1μm) Er S/D MOSFETs obtained at a) 500°C and b) 600°C. The presence of a high Schottky barrier is perceptible on the transfer characteristics, in particular, current flattening on the I_d-V_g characteristics at the transition between the subthreshold and the weak accumulation regime and the non saturated the I_d-V_d characteristics. Then, device at 500°C exhibits a moderate current drive of 14μA/μm at V_g=V_d=2V. From the material study, the Er silicide formed at higher temperature offers a SBH slightly lower. This conclusion is validated at the device level, where the current is less limited by the contact resistance and the channel resistance tends to dominate the total resistance. The transfer characteristic of Er silicide contacts at 600°C seems to be less affected by the SBH leading to a max current drive 4 time higher (I_{on}=65μA/μm @V_g=V_d=2V). This point shows that the modulation of the SBH is a real leverage to enhance the performance of metallic S/D MOSFETs. Higher temperature can not be implemented at the device level because the reaction of erbium with oxide isolation becomes so severe that short cuts between gate and S/D regions prevent the proper operation of the transistor. Lastly, the same approach has been applied with ytterbium silicide on MOSFET devices. An Yb layer of 12 nm have been deposited and annealed under UHV for 1h at 480°C/1h. The static characteristics are relevant of a high Schottky barrier (~ 0.28eV) at the S/D junction. From the material study, the SBH of YbSi remains constant with the temperature of silicidation.

In both case (Yb and Er), for long channel device, the current is limited by the high value of the Schottky barrier height (i.e. 0.3 eV range). This is all the more true for shorter gate length device where the current limitation induced by the contact is more severe. In order to be attractive for shorter gate length devices, RE silicide should be coupled to a low temperature dopant segregation technique (11) that can substantially decrease the specific contact resistance for n-type devices.

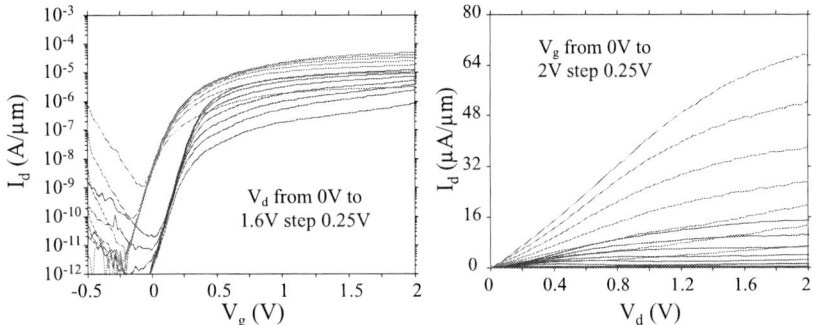

Figure 3: Comparison of static I_d-V_g and I_d-V_d characteristics of nMOSFET (L_g=1µm) with Er S/D contacts obtained at 500°C and 600°C on 8 nm SOI substrate

Lastly, the reaction of the RE with dielectric layers (SiN spacer, SiO_2 from MESA) has been characterized as function of the temperature. Before the RE layer stripping step, the presence of a residual film on the SiO_2 isolation field is observed. Below an activation temperature of 550°C for Yb and 620°C for Er, this film is not conductive. At higher temperature, this film is characterized by a larger density of grains that ultimately form a continuous electrically conductive film. Fig. 4 shows a SEM view of a device region including the gate, the isolation oxide and the silicide after the selective etch step of Yb silicidation performed at a) 480°C and b) 520°C. The reaction with the buried oxide is clearly higher at 520°C. A higher temperature leads to high leakage currents between the S/D silicide and the gate resulting in no working transistor.

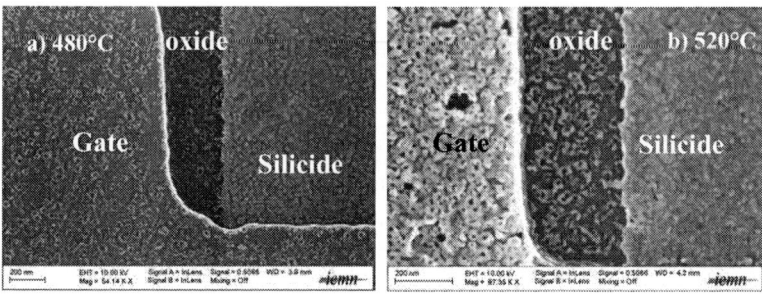

Figure 4: SEM view of a part of a device including the gate, the isolation oxide and the silicide after the selective etch step of Yb silicidation performed at a) 480°C and b) 520°C.

Conclusion

Integration of RE silicide on UTB SOI substrate has been studied. It was demonstrated for the first time that $ErSi_x$ and $YbSi_x$ formed on SOI with thin Si body (< 17 nm) are pinholes free. No diffusion of Si was observed from the transistor channel during the RE silicide formation. Rare earth metals were also selectively removed from the oxide and SiN spacers up to the temperature of 550°C for Yb and 620°C for Er. RE S/D MOSFETs are severely current-limited by a large Schottky barrier height but a moderate SBH reduction (~ 40 meV) leads to a drastic improvement of device performance.

Acknowledgments

This work was supported by the European Commission through the METAMOS project (016677) and NANOSIL project (216171).

REFERENCES

1. W.J. Taylor, E. Verret, C. Capasso, J.-Y. Nguyen; Le Boi La; E. Luckowski, A. Martinez, C. Happ, J. Schaeffer, M. Raymond, P. Tobin, *Ext. Abs. of 4th Inter. Workshop on Junction Technol.*, Shangai (2004) 107.
2. J. Kedzierski, P. Xuan, E.H. Anderson, J. Bokor, T.J. King, C. Hu, IEDM Tech. Dig., 57 (2000).
3. A. Itoh, M. Saitoh, M. Asada, Jpn.J.Appl.Phys., 39, 4757 (2000).
4. G. Larrieu and E. Dubois, Trans . Elect. Dev., 52, 2720 (2005).
5. W. Saitoh, A. Itoh, S. Yamagami, M. Asada, Jpn.J.Appl.Phys., 38, 6226 (1999).
6. H. Sato, H. Sato, T. Iguchi, M. Asada, Jpn.J.Appl.Phys., 43, 6038 (2004).
7. M. Jang, K. Kang, S. Lee, K. Park, Appl. Phys. Lett. 84, 741 (2004).
8. M. P. Lepselter and S. M. Sze, Proc. IEEE 56, p. 1400, 1968.
9. Y.-L. Jiang, Q. Xie, C. Detavernier, R. L. Van Meirhaeghe, G.-P. Ru, X.-P. Qu, B.-Z. Li, A. Huang, P. K. Chu J. Vac. Sci. Tech. A 25, p. 285, 2007.
10. G. Shen J., C. Chen, C. H. Lou, S. L. Cheng, and L. J. Chen., J. Appl. Phys. 84, 3630, (1998).

11. G. Larrieu, E. Dubois, R. Valentin, N. Breil, F. Danneville, G. Dambrine, J.P. Raskin, J.C. Pesant, IEDM Tech. Dig., 145 (2007).

CHAPTER 9

SIMULATIONS FOR NANO-SCALE SOI DEVICES I

ECS Transactions, 19 (4) 211-220 (2009)
10.1149/1.3117411 ©The Electrochemical Society

Device Physics and Simulation Techniques for Nanoscale SOI-MOSFETs

H. Tsuchiya, Y. Yamada, S. Souma, and M. Ogawa

Department of Electrical and Electronics Engineering, Kobe University,
1-1, Rokko-dai, Nada-ku, Kobe 657-8501, JAPAN

The silicon-based planar CMOS technology is expected to face
fundamental limits in the near future, and therefore, new types of
nanoscale devices are being investigated aggressively. To
understand the device physics and assess their performance limits,
new types of simulation techniques considering quantum
mechanical phenomena, carrier's ballistic transport and atomistic
effects are also needed. In this paper, we present our recent
approaches toward the quantum and atomistic transport modeling,
and make a performance projection of emerging nano-MOS
transistors employing new device structures and new channel
materials.

Introduction

The remarkable progress of integrated circuit technology has been primarily based on
downsizing of MOSFETs that has continued to the present day. Due to extensive efforts,
a nanoscale MOSFET with gate length shorter than 10nm has been fabricated in research
and development (1). In such an ultimately-scaled device, carrier transport essentially
changes from that in the conventional devices, because carriers suffer only a few
scattering events inside the channel, and then quantum mechanical and atomistic effects
begin to appear on the device characteristics. On the other hand, to meet all the
requirements for suppressing leakage current, minimizing short channel effects and
maintaining high drive current, the application of technology boosters such as high
mobility channel materials and multi-gate architectures is highly expected (2). To make a
performance projection of the most-advanced devices, a state-of-the-art simulation
technique based on the quantum mechanical transport theory is required.

Quantum transport theoretical models for nanoscale devices are summarized as
shown in Fig. 1. The non-equilibrium Green's function method is the most fundamental
theoretical model, because it is generally applicable to incoherent quantum transport
problems involving scattering processes (3,4,5). It is important to note that the self-
energy Σ^R becomes a function of $G^<$ in the case of incoherent transport, and we need to
solve the two Dyson equations shown in Fig. 1 self-consistently, as well as the Poisson's
equation. Therefore, practical solution of NEGF model needs large computational efforts.
The Wigner function is defined as Fourier transform of $-iG^<$ with respect to relative
coordinates (6), and its quantum transport equation can be derived both from the
Schrödinger equation (7) and the non-equilibrium Green's function formalism including
scattering terms (8). By adopting the same scattering integrals used in the semi-classical
Boltzmann equation, the Wigner function model provides an efficient quantum transport
model considering the physics-based scattering mechanisms (9). The quantum-corrected
semi-classical model, which has been derived by introducing an approximation taking
only $O(\hbar^2)$ correction terms in the Wigner function model (10,11), provides a practically
useful approach since we can perform device analysis including quantum effects with

211

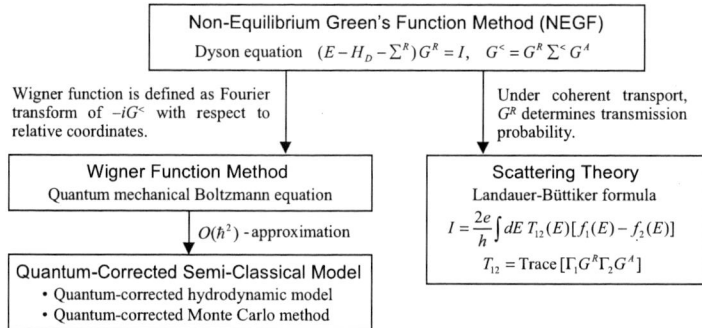

Figure 1. Systematic diagram of non-equilibrium quantum transport models.

computational resources developed based on the semi-classical approaches such as drift-diffusion and MC methods.

On the other hand, under full ballistic (coherent) transport, the scattering theory is applicable. Current through nanostructures can be calculated by using the Landauer-Büttiker formula (12,13) with transmission probability $T_{12}(E)$ estimated from the retarded Green's function G^R. In the following sections, we will present practical applications of those quantum approaches to nanoscale devices.

Quasi-Ballistic Transport in Si-MOSFETs

Since the frequency of carrier scattering reduces in ultrashort channel MOSFETs, carrier transport becomes ballistic and a drive current enhancement is expected. In reality, it is known to be difficult to achieve a full ballistic transport in MOSFETs with a 10nm order of channel lengths (14), so the investigation on the role of scattering is still important. In this section, we study the influence of phonon scattering on the carrier transport based on the quantum-corrected MC method (15).

Fig. 2 shows the characteristic aspects of elastic and inelastic phonon scattering at on-state. Carriers are injected into the channel with kinetic energy of the order of kT at a potential bottleneck point and they suffer scattering events in the channel. The total kinetic energy of a carrier is conserved before and after elastic scattering, while it is significantly altered by inelastic scattering via phonon absorption and emission processes. As is well known, the g-LO inelastic phonon with $\hbar\omega_{op} = 61\text{meV}$ is dominant in silicon. As a consequence, the inelastic phonon emission is suppressed at the beginning of the channel because the carrier energy is initially too small. In addition, phonon absorption is rare at ordinary temperatures, so elastic scattering (such as of acoustic phonons) becomes dominant at the beginning of the channel, and occasionally returns carriers back to the source electrode. If a carrier survives this region, it is subsequently exposed to frequent optical phonon emission and immediately loses its energy in multiples of kT (mostly of $\hbar\omega_{op} = 61\text{meV}$). Since the carrier then has little chance of returning to the source and is eventually absorbed into the drain, the current reduction from its ballistic value becomes only due to the backscattering at the beginning of the channel. From this point of view, Natori has suggested that inelastic emission processes suppress the backscattering and then improve the ballisticity at the on-state (16). We examined this Natori's prediction by using the quantum-corrected MC simulation for a double-gate MOSFET shown in Fig. 3.

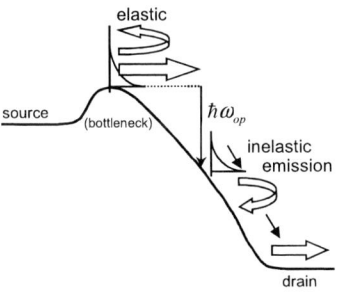

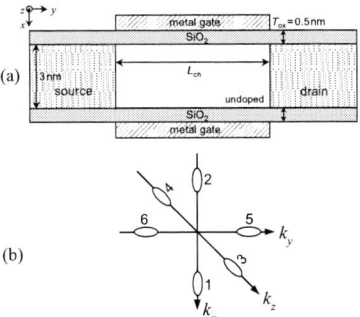

Figure 2. Characteristic aspects of elastic and inelastic phonon scattering on carrier transport.

Figure 3. (a) Double-gate MOSFET device model and (b) conduction band valleys of Si.

First, Fig. 4 shows the influences of elastic and inelastic phonon scatterings on the average electron velocity profiles computed for (a) L_{ch}=30nm and (b) 10nm. The lower figures summarize the source-end velocities v_s defined at the bottleneck point. It is found that elastic and inelastic absorption processes decrease v_s for both channel lengths. On the other hand, inelastic emission processes have the opposite effect on v_s for the longer channel device. That is, the inelastic emission process increases v_s for L_{ch}=10nm, in agreement with Natori's prediction, while it decreases v_s for L_{ch}=30nm.

To further probe the dependence on the inelastic emission processes, we examined variations in the potential profiles as shown in Fig. 5, where (a) L_{ch}=30nm and (b) 10nm. Interestingly, the bottleneck barrier for L_{ch}=30nm is found to broaden due to inelastic phonon emission, which means that the accumulated charges caused by the inelastic phonon emission influences the carrier transport via the modulation of source-end potential in a long-channel device, and then decrease the drain current. On the other hand, the source-end potential profile hardly changes due to scattering for the shorter channel

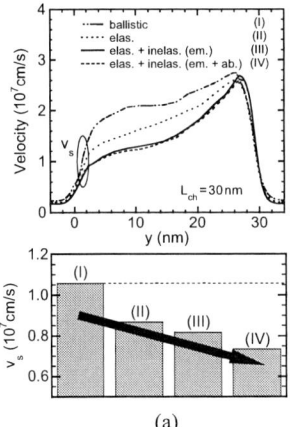

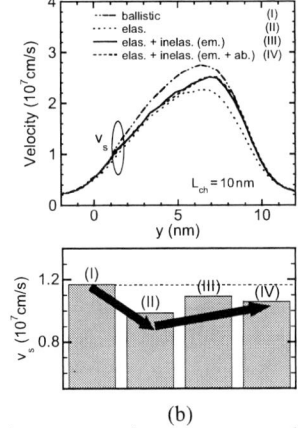

(a)

(b)

Figure 4. Influences of elastic and inelastic phonon scatterings on average electron velocities computed for (a) L_{ch}=30nm and (b) 10nm. V_G=0.5V and V_D=0.6V.

ECS Transactions, 19 (4) 211-220 (2009)

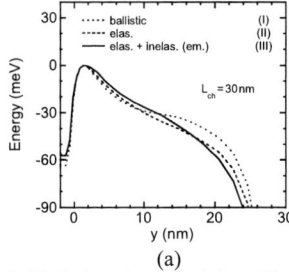
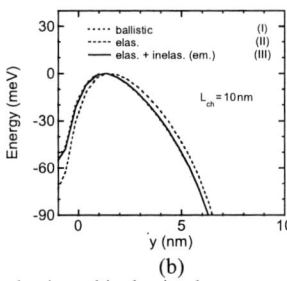

(a)   (b)

Figure 5. Variations in potential profiles due to elastic and inelastic phonon scatterings.

device as shown in Fig. 5(b), which implies that the charge accumulation has a negligible influence on the potential profile in ultrashort channel MOSFETs. As a result, the suppression effect of backscattering due to inelastic phonon emission contributes to improve the ballisticity in the 10nm device. From further investigation, we have shown that the ballisticity can be enhanced by reducing the bottleneck barrier width even in a longer channel device (source-end potential engineering) (15).

High Mobility Channel MOSFETs

To achieve a continuous enhancement in drive current of MOSFETs, a variety of high mobility channel materials have been intensively studied. To precisely estimate the device performance of high mobility channel MOSFETs, a device simulation considering bandstructure, scattering and quantum mechanical effects is indispensable. In this section, we make a performance projection of III-V semiconductors, Ge and strained-Si channel n-MOSFETs based on the quantum-corrected MC method (17).

Fig. 6 shows the device model used in the simulation, where ultrathin-body (UTB) structure is employed. This is to minimize the effects of inversion layer capacitance, which becomes significant for III-V semiconductors because of their smaller confinement effective mass (17,18). The surface orientation is (100), but the (111) orientation is also considered for Ge because the smaller transport effective mass is expected in this surface orientation (19). Here, we should pay attention to the physical limits of solid solubility of donors in III-V semiconductors. Namely, activated donor concentrations larger than $2 \times 10^{19} \text{cm}^{-3}$ cannot be obtained in III-V semiconductors (20), so we assumed the donor

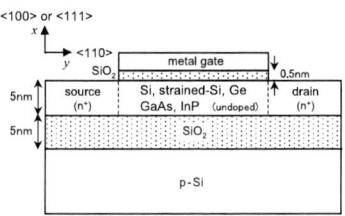

Figure 6. Device model with ultrathin-body (UTB) structure. The channel region is assumed to be undoped.

TABLE I. Valley information and physical parameters of each material.

		Si	Ge	GaAs	InP
mass (Γ)		–	0.037	0.067	0.082
mass (X)	m_t (m_0)	0.19	0.29	–	–
	m_l (m_0)	0.98	1.35	–	–
mass (L)	m_t (m_0)	–	0.082	0.127	0.153
	m_l (m_0)	–	1.59	1.538	1.878
nonparabolicity α		0.5 (X)	0.65 (L, Γ, X)	0.61 (Γ) 0.46 (L)	0.61 (Γ) 0.49 (L)
$\Delta E_{\Gamma L}$ (eV)		–	–	0.33	1.492
$\Delta E_{LF} / \Delta E_{LX}$ (eV)		–	0.14 / 0.18	–	–
permittivity ε_r		11.9	16.0	12.9	12.6
E_f of S/D (eV)		0.064	0.138	0.278	0.326

214

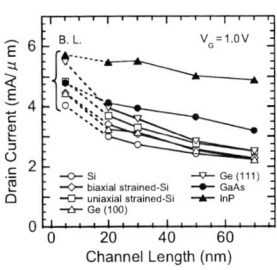

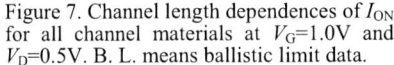

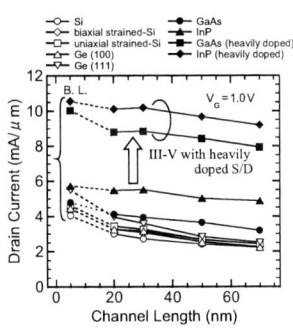

Figure 7. Channel length dependences of I_{ON} for all channel materials at $V_G=1.0$V and $V_D=0.5$V. B. L. means ballistic limit data.

Figure 8. Drive current enhancement due to heavily doped S/D in III-V MOSFETs.

concentration to be 2×10^{19}cm^{-3} in the source and drain (S/D) regions for GaAs and InP, while a higher donor concentration of 1×10^{20}cm^{-3} was used for Si and Ge.

Table 1 shows the physical parameters of each material. As pointed out by S. Takagi (18), the energy differences between the lowest and higher valleys as $\Delta E_{\Gamma L}$ and ΔE_{LX} are important physical parameters, because an electron transfer to the higher valleys with heavier transport effective mass, degrades the device performance significantly. Here, note that the Fermi energies in S/D electrodes of Ge and GaAs are comparable with the energy differences from the higher valleys, and thus certain amount of carriers already occupy the higher valleys in S/D electrodes, which can affect the injection velocity at ballistic transport (21). For strained-Si devices, biaxially and uniaxially 1% tensile strains are applied to both the channel and S/D regions. The strain effects are included in terms of energy splitting between the 2-fold and 4-fold valleys and effective mass reduction due to the <110>-oriented uniaxial tensile strain (22). The scattering processes considered are impurity and phonon scatterings, while roughness and electron-electron scatterings are ignored in this study, so as to evaluate the intrinsic device performance of each material.

Fig. 7 shows the channel length dependences of I_{ON} computed for all channel materials. The ballistic limit (B. L.) data were computed by ignoring all scattering processes in the channel. It is found that the drive current enhancement due to the ballistic transport is more effective in group IV materials than in group III-V ones, because the carrier transport in group III-V materials are already quasi-ballistic in the present range of channel length. The above results mean that III-V materials lose their advantage as the channel length is scaled down to the ballistic regime. As a result, Ge (111) and <110>-oriented uniaxially-strained Si channels indicate the equivalent device performance with those of III-V materials at the ballistic limit. On the other hand, if the S/D parasitic resistance can be reduced down to the same level of group IV materials with heavily doped S/D electrode, the drive current of III-V MOSFETs drastically increases as shown in Fig. 8. Therefore, lower resistive S/D technology such as metal S/D technique might be necessary for III-V MOSFETs to show their own real abilities (17).

Si Nanowire Transistors

Si-nanowire transistors (SNWTs) have drawn much attention as extremely downscaled MOSFETs for the future Si-VLSIs, because they have superior electrostatic gate control of the channel potential owing to gate-all-around (GAA) configuration, and higher current capability due to bandstructure modulation (22,23) and ballistic transport. While the significant progress in fabrication technology and in understanding of electrostatics and carrier transport in SNWTs has been achieved rapidly (24), huge challenges remain to be addressed to make this new device architecture reach the level of manufacturing. In this section, we study the physical properties of carrier transport in SNWTs based on a three-dimensional (3D) quantum mechanical device simulation.

Recently, a number of computational studies on electrical properties of SNWTs have been reported based on the semi-classical Boltzmann transport equation where quantum mechanical transport along the channel direction is disregarded. Therefore, they are only applicable for decananometer scaled SNWTs with channel lengths longer than 10nm (25,26,27). However, to assess their performance limits in sub-10nm regime, it is important to consider quantum mechanical phenomena along the channel direction, such as source-drain tunneling. To this end, we have developed a 3D quantum simulator for SNWTs based on the multisubband Wigner transport equation, coupled with Schrödinger-Poisson algorithm (28). The Wigner transport equation is a fully quantum mechanical equation, and it can handle scattering effects in the same manner as in the semi-classical Boltzmann equation (9).

The SNWT device model used in this study is shown in Fig. 9(a). The donor concentration in the S/D regions is 10^{20}cm^{-3} and the channel region is undoped. The conduction band valleys of Si is shown in Fig. 9(b), where the channel direction is taken as <100> and the confinement directions <010> and <001> for simplicity. In the present approach, the 3D Wigner transport equation is decoupled into the 1D Wigner transport equation along the source-drain (x) direction and the 2D Schrödinger equation in the transverse (y, z) cross-section. Then, Wigner function is defined for each quantized subband n for the three-pairs of the conduction band valleys, $v = $ 1-1', 2-2' and 3-3'. This so-called mode-space approximation is considered to be valid for the uniform nanowire configuration used in this study. The quantized subbands are computed by solving the 2D Schrödinger equation for each y-z cross-section, and by using the obtained subband profiles the 1D Wigner transport equation is solved along the S/D direction by introducing a relaxation time approximation in the scattering term. Further, to analyze the gate control in SNWTs, the 3D Poisson equation is solved self-consistently with the 2D Schrödinger equation and the 1D Wigner transport equation mentioned above.

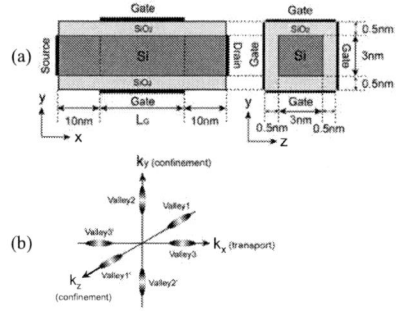

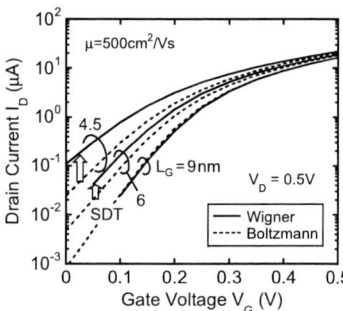

Figure 9. (a) GAA-SNWT device model and (b) conduction band valleys of Si.

Figure 10. I_D - V_G characteristics for L_G = 9, 6 and 4.5nm, computed by using the Wigner and Boltzmann approaches.

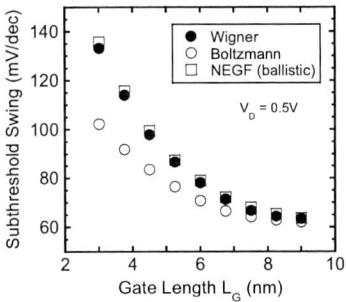
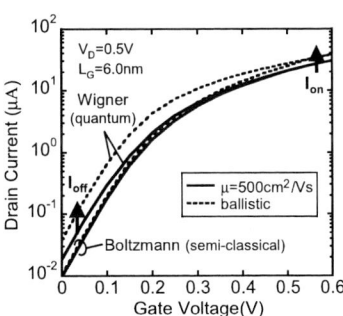

Figure 11. Channel length dependences of subthreshold swings computed by using the Wigner, Boltzmann and ballistic NEGF approaches.

Figure 12. Influence of ballistic transport on I_D - V_G characteristics computed for $L_G = 6$nm.

Fig. 10 shows the computed I_D - V_G characteristics in sub-10nm regime, where each curve is compared with that from the semi-classical Boltzmann approach (the 1D Boltzmann transport equation is solved along the S/D direction, instead of the Wigner transport equation). The relaxation time due to scattering is given by using the mobility of 500cm^2/(V·s) and a relation of $\tau_v = m_v^* \mu / e$ for each conduction band valley. It should be noted that the subthreshold current in the Wigner approach becomes larger than that in the Boltzmann approach for $L_G \leq 6$nm, which was confirmed due to source-drain tunneling (SDT) (28). On the other hand, the on-currents at higher V_G conditions are almost the same for both approaches because thermal injection is a dominant carrier injection process at the on-state. Furthermore, the subthreshold swings (SS) are calculated to compare with those from ballistic NEGF method (29,30) as shown in Fig. 11. It is found that the Wigner and ballistic NEGF results coincide closely for a wide range of gate length shorter than 10nm. Since we have verified that SS values are barely affected by carrier scattering in the subthreshold regime (28), the comparison with ballistic NEGF approach in Fig. 11 is valid. Consequently, the present quantum simulator based on the Wigner approach provides reliable simulations. Furthermore, we have investigated the influence of ballistic transport on I_D - V_G characteristics as shown in Fig. 12, where the solid and dashed lines represent the results with and without scattering, respectively. Looking at the Wigner results, not only the on-current, but also the off-current increase due to the ballistic transport. This off-current increase is due to the enhanced S/D tunnelling under ballistic transport, and causes nonnegligible lowering of the threshold voltage. Since it could counteract the on-current enhancement due to ballistic transport, we need to make a continued investigation in more detail.

Zigzag-Edged Graphene Nano-Ribbon Devices

Recent successful fabrication of single layer graphite (graphene) has opened up the possibility to realize a new type of nanoscale devices based on graphene (31). While the graphene itself shows various interesting properties such as the quantum Hall effect in room temperature, even more interesting properties appear if the graphene is patterned into ribbon like geometry called the graphene nano ribbon (GNR). For instance, it is known that the zigzag-edged GNR (Z-GNR) with hydrogen termination shows partly-flat

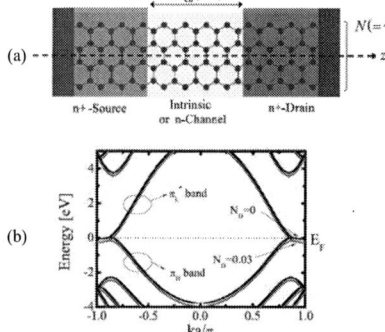

(a)

(b)

Figure 13. (a) Schematic top view of Z-GNR device. The left and right hatched regions represent source and drain electrodes, respectively. (b) Energy bandstructures computed for undoped (N_D=0) and electron doped (N_D=0.03 electrons per atom) Z-GNR with the width of N=4.

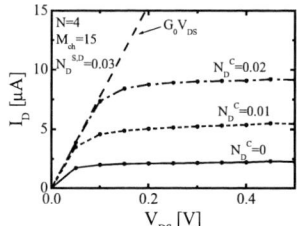

Figure 14. I_D -V_{DS} characteristics for various doping levels N_D^C in the channel.

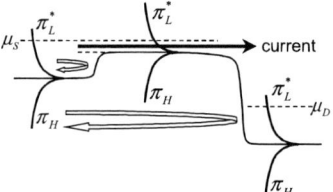

Figure 15. Schematic diagram of current flow in the Z-GNR device.

electronic band structures near the Fermi energy, and the corresponding wavefunction is localized near the edges of the GNR. Previous studies on electron transport in the Z-GNR have mainly focused only on the zero bias conductance as a function of the Fermi energy (32). There is also an interesting numerical study on the armchair-edged GNR (A-GNR) MOSFET as an alternative digital switch used in the integrated circuits (33). The A-GNR is known to show a finite bandgap when the ribbon width satisfies a certain condition, which is due to the spatial confinement of the ribbons in their transverse direction. Motivated by such previous studies, we investigate numerically the electron transport in the Z-GNR in the presence of finite bias voltage, taking into account the effect of doping in the electrode and channel regions.

Fig. 13(a) shows the schematic top view of a Z-GNR device with hydrogen termination. Here, the central channel region has the width of N (number of zigzag lines along the transverse direction) and length M_{ch} (number of unit cells within the channel region), and is sandwiched in between the left (source) and right (drain) Z-GNR electrodes. The source and drain Z-GNR electrodes are assumed to be doped heavily so as to induce the additional N_D^S and N_D^D electrons per atom, respectively, while the central Z-GNR channel is intrinsic or doped only weakly with doping level N_D^C per atom. Recent advances in the experimental technique have made it possible to realize such local doping by means of the local gating method (34). To perform transport simulations in such an atomistic device, we employ the density functional tight-binding (DFTB) method along with NEGF method, where device Hamiltonian H_D in the Dyson equation is expressed by a LCAO Hamiltonian matrix expanded using non-orthogonal atomic orbital basis set, and the current is calculated by using the Landauer-Büttiker formula (35).

Fig. 13(b) shows the bandstructures of intrinsic and doped Z-GNRs. In the intrinsic case, the flat-bands induced by the edge states are seen at around the Fermi energy (E=0), while in the electron doped case the energy of the flat-bands is shifted lower to acquire more electrons. Hereafter, the highest π (valence) band and the lowest π^* (conduction) band are referred to as the π_H and the π_L^* bands, respectively.

Having understood how the bandstructure changes with doping, we computed the I_D-V_{DS} characteristics as shown in Fig. 14 for the width $N=4$ (7.1Å) and the channel length $M_{ch}=15$ (36.9Å), where results for three different channel doping levels are compared. It is interesting to note that the I_D-V_{DS} curves of the Z-GNR show clear saturation behavior in spite of metallic bandstructure as shown in Fig. 13(b), with larger saturation current for larger channel doping. It is also found that the conductance in the linear current regime is estimated to be the quantized conductance of $2e^2/h$, which means the presence of a single conducting channel via the π_L^* band. The origin of such peculiar current behaviors can be cast into the fact that π_L^* wavefunction is symmetric and π_H wavefunction is antisymmetric with respect to the z-axis (see Fig. 13(a)), respectively, in the case of the Z-GNR with $N=4$. Then, the transfer of electrons between the π_H and π_L^* bands is prohibited due to the spatial orthogonality between them. Therefore, the intra-band current from the π_L^* band in the source to the π_L^* band in the drain is greatly reduced by the presence of the π_H band in the channel as schematically shown in Fig. 15. In other words, the π_H band in the channel acts as a "barrier" for π_L^* electrons in the source and drain. Moreover, the inter-band current from the π_H band in the source to the π_L^* band in the drain is exactly zero due to the spatial orthogonality as shown in Fig. 15. Accordingly, the current flows only through the energy region between the Fermi energy and the "barrier" of π_H band in the channel, so the "barrier" height lowering by the additional channel doping leads to the appearance of the linear current increase with the quantized conductance and then the larger saturation current is observed. As mentioned above, the current control mechanism due to the "barrier" height modulation in the present Z-GNR device is the same as that in Si-MOSFETs, so the Z-GNR is a potential material for FET applications, as well as the A-GNR.

Conclusions

The integrated circuit devices seem to continue the downsizing toward 10nm order channel lengths (2). While we pursue a rigorous quantum simulation of such nanoscale devices, the physics related to the nanoscale transistor operation needs to be well articulated so that simple compact models, including quasi-ballistic transport and realistic bandstructures, can be developed for circuit design.

Acknowledgments

This work was partially supported by a NEDO/MIRAI project and the Semiconductor Technology Academic Research Center (STARC) of Japan. The authors would like to thank Honorary Professor T. Miyoshi of Kobe University for his continuous support and encouragement.

References

1. H. Wakabayashi, S. Yamagami, N. Ikezawa, A. Ogura, M. Narihiro, K. Arai, Y. Ochiai, K. Takeuchi, T. Yamamoto, and T. Mogami, *IEDM Tech. Dig.*, 989(2003).
2. ITRS. [Online] Available: http://www.itrs.net/
3. L. V. Keldysh, *Sov. Phys. JETP*, **20**, 1018 (1965).

4. L. P. Kadanoff and G. Baym, *Quantum Statistical Mechanics*, Benjamin, New York, 1962.
5. C. Caroli, R. Combescot, P. Nozieres, and D. Saint-James, *J. Phys. C: Solid State Phys.*, **4**, 916 (1971).
6. E. Wigner, *Phys. Rev.*, **40**, 749 (1932).
7. W. R. Frensley, *Phys. Rev.* **B36**, 1570 (1987).
8. H. Tsuchiya and T. Miyoshi, *J. Appl. Phys.*, **83**, 2574 (1998).
9. D. Querlioz, J. S.-Martin, K. Huet, A. Bournel, V. A.-Fortuna, C. Chassat, S. G.-Retailleau, and P. Dollfus, *IEEE Trans. Electron Devices*, **54**, 2232 (2007).
10. M. G. Ancona and G. L. Iafrate, *Phys. Rev.* **B39**, 9536 (1989).
11. H. Tsuchiya and T. Miyoshi, *IEICE Trans. Electron.*, **E82-C**, 880 (1999).
12. R. Landauer, *IBM J. Res. Develop.*, **1**, 223 (1957).
13. M. Buttiker, *IBM J. Res. Develop.*, **32**, 317 (1988).
14. H. Tsuchiya, K. Fujii, T. Mori, and T. Miyoshi, *IEEE Trans. Electron Devices*, **53**, 2965 (2006).
15. H. Tsuchiya and S. Takagi, *IEEE Trans. Electron Devices*, **55**, 2397 (2008).
16. K. Natori, *Applied Surface Science*, **254**, 6194 (2008).
17. T. Mori, Y. Azuma, H. Tsuchiya, and T. Miyoshi, *IEEE Trans. Nanotechnology*, **7**, 237 (2008).
18. S. Takagi and S. Sugahara, *Ext. Abst. of SSDM*, Yokohama, 1056 (2006).
19. S. Takagi, *Ext. Abst. of SSDM*, Tokyo, 10 (2004).
20. M. V. Fischetti and S. E. Laux, *IEEE Trans. Electron Devices*, **38**, 650 (1991).
21. Y. Azuma, T. Mori, and H. Tsuchiya, *Phys. Stat. Sol. (c)*, **5**, 3153 (2008).
22. T. Maegawa, T. Yamauchi, T. Hara, H. Tsuchiya, and M. Ogawa, *IEEE Trans. Electron Devices*, to be published in April issue, 2009.
23. N. Neophytou, A. Paul, M. Lundstrom, and G. Klimeck, *IEEE Trans. Electron Devices*, **55**, 1286 (2008).
24. Special issue on Nanowire Transistors: Modeling, Device Design, and Technology, *IEEE Trans. Electron Devices*, **55**, 2008.
25. S. Scaldaferri, G. Curatola, and G. Iannaccone, *IEEE Trans. Electron Devices*, **54**, 2901 (2007).
26. S. Jin, T. W.-Tang, and M. V. Fischetti, *IEEE Trans. Electron Devices*, **55**, 727 (2008).
27. M. Lenzi, P. Palestri, E. Gnani, S. Reggiani, A. Gnudi, D. Esseni, L. Selmi, and G. Baccarani, *IEEE Trans. Electron Devices*, **55**, 2086 (2008).
28. Y. Yamada and H. Tsuchiya, *Ext. Abst. of SISPAD*, Hakone, 281 (2008).
29. S. Datta, *Electronic Transport in Mesoscopic Systems*, Cambridge, U. K.: Cambridge University Press, 1995.
30. H. Fitriawan, M. Ogawa, S. Satofumi, and T. Miyoshi, *IEICE Trans. Electron.*, **E91-C**, 105 (2008).
31. K. S. Novoselov, A. K. Geim, S. V. Morozov, D. Jiang, Y. Zhang, S. V. Dubonos, I. V. Grigorieva, A. A. Firsov, *Science*, **306**, 666 (2004).
32. K. Wakabayashi, *Phys. Rev.* **B64**, 125428 (2001).
33. G. Liang, N. Neophytou, D. E. Nikonov, and M. S. Lundstrom, *IEEE Trans. Electron Devices*, **54**, 677 (2007).
34. B. Ozyilmarz, P. J.-Herrero, D. Efetov, D. A. Abanin, L. S. Levitov, and P. Kim, *Phys. Rev. Lett.*, **99**, 166804 (2007).
35. S. Souma, M. Ogawa, T. Yamamoto, and K. Watanabe, *J. Comp. Electron.*, **7**, 390 (2008).

Analytical Model for Conduction Band Nonparabolicity to Transport Analysis of Nano-Scale SOI MOSFET

Y. Omura

ORDIST, Department of Electronics, Kansai University, Yamate-cho, Suita, Osaka 564-8680 Japan

This paper reconsiders the mathematical formulation for the conventional nonparabolic band model, and discusses how the nonparabolicity of the conduction band impacts the effective masses of electrons that are confined within the barriers. To support this consideration, this paper proposes an analytical expression for the conduction-band effective mass of electrons including the term of band nonparabolicity.

Introduction

This paper reconsiders the mathematical formulation for the nonparabolic band model, and discusses how the nonparabolicity of the conduction band impacts the effective masses of electrons. Basically, the conventional simplified model for band nonparabolicity does not include the external potential effect as a perturbation. Accordingly, this paper examines whether this perturbation can be implemented in the conventional model for convenience. In the following discussion, we focus on the low-dimensionality electron system with insulator barrier confinement, so this study is also applicable to future device analyses. Here we discuss the impact of the nonparabolic conduction band in Si on the effective mass, and propose an analytical expression for the effective mass of electrons including band nonparabolicity.

Theoretical Basis of Modeling Non-Parabolic Band Structure

In theoretical formulations, band nonparabolicity appears as a consequence of using the $k \cdot p$ perturbation method when some interaction terms are introduced to Schroedinger's equation for the free electron system. The expression is often simplified approximately as [1]

$$E_{n-par}\left(1+\alpha E_{n-par}\right)= E_{par} =\frac{\hbar^2 k^2}{2m^*}, \qquad [1]$$

where α is the non-parabolic band factor, E_{n-par} is the energy of carriers in the non-parabolic band, and E_{par} is the carrier energy expression in the form of the parabolic band scheme. When an external field effect is considered, we must examine whether the following formulation is theoretically valid or not.

$$E_{n-par}\left(1+\alpha E_{n-par}\right)= E_{par} + \beta E_{ext}, \qquad [2]$$

where E_{ext} is the perturbation energy corresponding to the 1st order perturbation of the external electric field and β is the perturbation factor.

In order to examine the validity of Eq. (2), E_{n-par} is changed into Taylor's series, and it is replaced with the following operator representation.

$$\hat{H}_{n-par} = \frac{1}{2\alpha}\left\{\left(1+2\alpha(\hat{H}_{par}+\beta\hat{H}_{ext})+\sum_{n=2}^{\infty}c_n(\hat{H}_{par}+\beta\hat{H}_{ext})^n\right)-1\right\} = \hat{H}_{par}+\hat{H}'$$ [3]

$$\hat{H}' = \beta\hat{H}_{ext} + \frac{1}{2\alpha}\sum_{n=2}^{\infty}c_n(\hat{H}_{par}+\beta\hat{H}_{ext})^n .$$ [4]

We can extract the eigen energy of H_{n-par} after mathematical algebra from eq. (3) when $H' < H_{par}$. As a result, we have the following equation for the energy of 2-D electrons [2].

$$E_{n-par} = \frac{\sqrt{1+4\alpha\big((E_n(k_n)-E_C+\beta E_{ext})+E_{free}(k_{\parallel})\big)}-1}{2\alpha}.$$ [5]

Effective mass of electrons in non-parabolic bands

From eq. (5), we can calculate the electron effective mass. Using the result given, we can derive the effective mass tensor component ($m^*_{n-par,ij}$) for the three-dimensional system as

$$\frac{1}{m_{n-par,ij}} = \frac{1}{\hbar^2}\frac{\partial^2 E_{n-par,n}(k)}{\partial k_j \partial k_i}$$

$$= \big(1+4\alpha\big((E_j(k_j)-E_C+\beta E_{ext})+E_{free}(k_{\parallel})\big)\big)^{-\frac{1}{2}}\frac{1}{m_{par,ij}}$$

$$-2\alpha\big(1+4\alpha\big((E_j(k_j)-E_C+\beta E_{ext})+E_{free}(k_{\parallel})\big)\big)^{-\frac{3}{2}}v^c_{par,i}v^c_{par,j},$$ [6]

where 'j' means the confinement direction and $v^c_{par,j}$ is the "*effective group velocity* " along the direction labeled 'j'[3]. When label 'i' indicates the transport direction, we have

$$v^P_{par,i}(k_i) = \frac{1}{\hbar}\frac{\partial(E_{par}(k)+\beta E_{ext})}{\partial k_i}.$$ [7]

When label 'j' indicates the confinement direction, we have

$$v^c_{par,j} = \frac{1}{\hbar}\frac{\partial(E_j(k_j)+\beta E_{ext})}{\partial k_j}$$

$$= \hbar k_j / m^*_{jj} + \frac{1}{\hbar}\frac{\partial\beta E_{ext}}{\partial k_j},$$ [8]

where k_j is the discrete wavenumber and m_{jj}^* is the effective mass along the confinement direction. The 2nd term of the right-hand side is the perturbation term imposed by the external field. When the semiconductor layer thickness is of the order of nanometers, the contribution of the 2nd term of eq. (8) is quite small. Thus the effective mass tensor ($m^*_{n-par,ij}$) value of low-dimensionality electron systems can be calculated around the subband bottom [4]; the important point is the fact that the effective mass of electrons is larger by the factor of $(1+4\alpha(E_j(0)-E_C+\beta E_{ext}))^{1/2}$ than that estimated assuming a parabolic band.

Calculation results of effective mass of low-dimensionality electrons

Assuming a thin Si layer, we calculated effective mass values (m_l/m_0 and m_t/m_0) for (001) surface orientations, where it is assumed that 2-fold and 4-fold X-band electrons are confined along the <001> axis.

Fig. 1 shows the effective mass of electrons occupying the ground state as a function of Si layer thickness (t_S), where the nonparabolicity factor α is assumed to be 0.5 eV^{-1} and, for comparison, 1st principle calculation results [5] are also shown in Fig. 1; the impact of band non-parabolicity on the effective mass of conduction band electrons appears when $t_S < 5$ nm [5]. The present model successfully reproduces the 1st principle calculation results.

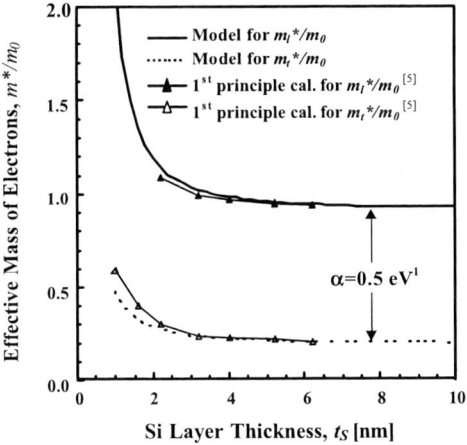

Figure 1. Calculated effective mass values of electrons occupying the ground state as a function of semiconductor layer thickness. It is assumed that $\alpha = 0.5$ eV^{-1} in order to calculate the non-parabolic band effect.

On the other hand, Fig. 2(a) shows the effective mass of electrons occupying the 1st-excited states as a function of Si layer thickness (t_S), the nonparabolicity factor α was assumed to be 0.5 eV^{-1}. For comparison, 1st principle calculation results [5] are also shown in Fig. 2(a). The present model does not reproduce the 1st principle calculation results; it is suggested that the conventional value of α is not appropriate. We, therefore, varied the value of α until the calculated curves fitted the 1st principle calculation results. Fig. 2(b) shows the calculation results of the effective mass value of conduction band electrons occupying the 1st-excited states, including nonparabolicity effect, where it is assumed that $\alpha = 0.1$ eV^{-1} for m_l/m_0 and $\alpha = 0.05$ eV^{-1} for m_t/m_0. Discarding the higher-order terms of Taylor's series in eq. (3) for the derivation of eq. (5) is the primary reason why the value of α is smaller than the conventional value. It is seen that the present model for effective mass successfully reproduces the theoretical simulation results

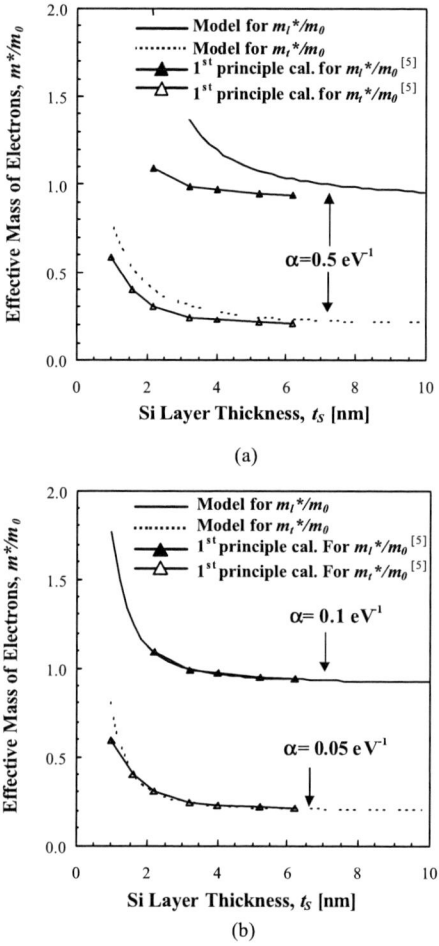

(a)

(b)

Figure 2. Calculated effective mass values of electrons occupying the f^t-excited states as a function of semiconductor layer width.

(a) Effective mass dependence on Si layer thickness. It is assumed that $m_l/m_0 = 0.916$ and $m_t/m_0 = 0.19$ for the parabolic band parameters, and that $\alpha = 0.5$ eV^{-1} in order to calculate the nonparabolic band effect.

(b) Effective mass dependence on Si layer thickness. It is assumed that $m_l/m_0 = 0.916$ and $m_t/m_0 = 0.19$ for the parabolic band parameters, and that $\alpha = 0.1$ eV^{-1} for m_l/m_0 and $\alpha = 0.05$ eV^{-1} for m_t/m_0 in calculating the nonparabolic band effect.

Semiconductor layer thickness dependence of quantum energy levels

Following the calculations of the effective mass for the nonparabolic band, we can evaluate the impact of band nonparabolicity on subband energy levels of the two-dimensional Si quantum well. Calculation results for Si show that the nonparabolicity of conduction bands significantly reduces the energy when t_S<5 nm (not shown here).

From the viewpoint of device applications, it should be noted that lowering the ground-state energy level brings out desirable effects with regard to the retention time of non-volatile nano-crystalline dot memory when t_S< 5 nm. In contrast, it can be concluded that, for t_S> 5 nm, the conventional consideration, which assumes a parabolic band, remains valid. From the above discussion, it is expected that the impact of band nonparabolicity is much stronger in Ge than in Si.

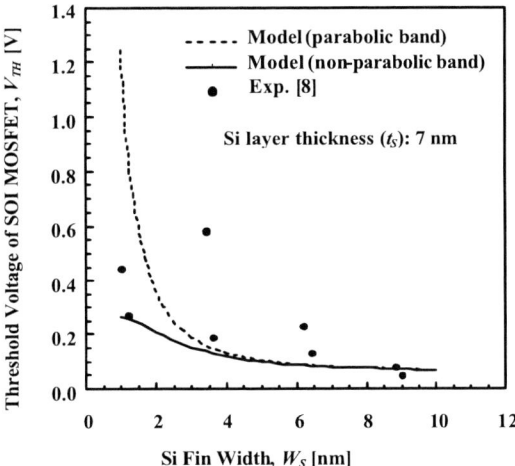

Figure 3. V_{th} dependence on fin width of Si SOI MOSFET device 2-dimensionally confined with (001) and (011) surfaces [8]. It is assumed that the gate oxide layer is 3-nm thick, the body silicon layer is 7-nm thick, and α=0.5 eV^{-1}. It is also assumed that m^*/m_0= 0.916 for (001) Si surface and m^*/m_0= 0.314 for (011) Si surface.

Applying Effective Mass Model to Express Threshold Voltage of Nano-Scale MOSFET's

The effective mass model is applicable to transport analysis of nano-scale SOI MOSFET. Since the threshold voltage (V_{th}) of ultra-thin body SOI MOSFET's is effectively ruled by the lowest energy level [6], it is easily anticipated that the impact of band nonparabolicity on V_{th} is very significant. Recently, the author clarified that the rise in threshold-voltage stems from not only the quantum-mechanical mechanism [6], but also the semi-classical mechanism [7]. When it is assumed that the threshold-voltage rising stems from the quantum confinement [4, 6], V_{th} for ultra-thin body SOI MOSFET can be expressed as

$$V_{th} = V_{FB} + \phi_{sth} + Q_B / C_{ox},$$ [9]

$$\phi_{sth} = \phi_{F,QM} + \left(\frac{E_G}{2q}\right) + \frac{E_{n,ij} - E_C}{q},$$ [10]

$$E_{n,ij} - E_C = \frac{h^2}{8}\left(\frac{i^2}{m^*_{nz,i}t_S^2} + \frac{j^2}{m^*_{nx,j}w_S^2}\right),$$ [11]

where h is Planck's constant, t_s means the fin height, w_s means the fin width, $E_{n,ij}$ is the quantum level energy and $m^*_{nx,j}$ (or $m^*_{nz,i}$) is the effective mass of electrons occupying the corresponding subband.

Calculated threshold voltage dependence on fin width of a SOI FinFET two-dimensionally confined with (001) and (011) surfaces is shown in Fig. 3, where the top surface of the Si fin body is (001) and the side surface is (011). Physical parameters assumed in the calculations are the same ones as those appearing in [8]. In this calculation of threshold voltage, it is assumed that electrons of the Si body can be represented as quasi-two-dimensional system because the value of t_S is larger than the minimal value of w_S. The conventional parabolic band model overestimates the quantum confinement effect in the range defined by sub-5-nm t_S. This suggests that the nonparabolicity should be taken account when estimating the threshold voltage values of nano-scale SOI MOSFET's [9, 10].

Acknowledgment

The author wishes to express his thanks to Dr. Shingo Sato (presently, Renesas Technology, Japan) for his assistance in performing the calculations.

References

1. B. K. Ridley, "*Quantum Processes in Semiconductors*," 2nd Ed. (Clarendon, Oxford, 1988).
2. Y. Omura, *J. Appl. Phys.*, **105**, 014310 (2009).
3. L. F. Register, E. Rosenbaum, and Kelvin Yang, *Appl. Phys. Lett.* **74**, 457 (1999).
4. R. Granzner, F. Schweirz, and V. M. Polyakov, *IEEE Trans. Electron Devices* **54**, 2562 (2007).
5. K. Nehari, N. Cavassilas, J. L. Autran, M. Bscond, D. Munteanu, and M. Lannoo, *Solid State Electron.*, **50**, 716 (2006).
6. Y. Omura, S. Horiguchi, M. Tabe, and K. Kishi, *IEEE Electron Device Lett.*, **14**, 569 (1993).
7. Y. Tahara and Y. Omura, *Jpn. J. Appl. Phys.*, **45**, 3074 (2006).
8. H. Majima, H. Ishikuro, and T. Hiramoto, *IEEE Int. Electron Devices Meet.* (Washington, D. C., Dec., 1999) p. 379.
9. Y. Omura and D. Kyokane, Abstr., the 2nd IEEE Nanotech. Mat. and Dev. Conf. (NMDC), (Kyoto, Oct., 2008) pp. 64.
10. Y. Omura, *IEEE Trans. Electron Devices*, to be published.

CHAPTER 10

SIMULATIONS FOR NANO-SCALE SOI
DEVICES II

ECS Transactions, 19 (4) 229-234 (2009)
10.1149/1.3117413 ©The Electrochemical Society

Three-Dimensional NEGF Simulations of Constriction Tunnel Barrier Silicon Nanowire MUGFETs

Aryan Afzalian, Chi-Woo Lee, Ran Yan, Nima Dehdashti Akhavan, Cindy Colinge and Jean-Pierre Colinge

Tyndall National Institute, Prospect Row, Cork, Ireland

In this paper, we investigate the effect of cross-section variations in nanowire MUGFETs through 3D quantum simulations based on Non-Equilibrium Green's Function (NEGF) formalism. We show that the effect of cross-section variations in a nanowire results in energy barriers in the conduction band that can be interpreted as a local increase of the bandgap. In narrow nanowires with a cross section of a few nanometers, this can strongly influence the current and the characteristics of the device. A small constriction resulting in a barrier of the order of a 0.1eV is shown to be an effective way to create a tunnel barrier that can be used to improve the on/off current ratio in ultra-scaled transistors.

INTRODUCTION

In a constant effort to increase performance, the dimensions of MOSFETs are continuously being shrunk. In ultra-scaled devices with cross-section dimensions and channel length of a few nanometers, quantum effects are playing a crucial role on device characteristics [1]. To fully benefit from scaling at that level, new device architectures and new simulations tools are needed to tame and exploit these new effects as well as possible. Among the new simulations methods developed for that purpose, the NEGF (Non-Equilibrium Green's Function) method [2-5] has gained popularity and shown a real potential for modeling the quantum effects at the scale of a few nanometers.

Here we use a 3D quantum simulator based on a coupled-mode-space version of the NEGF [5] to explore the effect of local constriction on the performance of ultra-scaled SOI nanowires. We show that constriction results in local energy barrier. We give a physical explanation to this phenomenon and show that there is a window of constriction size that allows an improvement of the on-to-off current ratio.

CONSTRICTION-INDUCED, TUNNEL-BARRIER NANOWIRE

Using our simulator we have investigated the effect of a local cross-section variation on the transport properties of nanowire MOSFETs. Such a cross-section variation can be introduced intentionally or can result from non-uniformities in the photolithography process used to make the devices. Here we show the effect of two constrictions at the source-channel and drain-channel interface as well as the resulting energy subband profile (Fig. 1). By forming constrictions in the nanowire, energy barriers are created locally [6]. Simulations show that the wave function and the energy levels variations are sharp enough to be considered in equilibrium with the local structure dimensions. Therefore, the width of the barrier can be tuned by varying the length of the constriction,

229

L_C, in the x (transport) direction, since the barrier has the same length as the physical discontinuity, while the height of the barrier can be adjusted by modifying the diameter of the cross-section, as shown in Fig. 2. The energy levels and the wave functions in the discontinuity are in equilibrium with the dimension of this discontinuity, *i.e.* they take the same value they would have in a nanowire with a cross-section equal to that of the discontinuity. The resulting barrier results in an increase of the bandgap or/and the energy of the first subband. This can be predicted as a first approximation by using an analytical model of the energy level of a constant potential well:

$$\Delta E_1(t_{si}, W_{si}, \Delta t_{si}, \Delta W_{si}) = E_1(t_{si} - \Delta t_{si}, W_{si} - \Delta W_{si}) - E_1(t_{si}, W_{si}) \qquad [1]$$

where W_{si} and t_{si} are the nanowire width and thickness, respectively, while ΔW_{si} and Δt_{si} are the variation of width and thickness in the constriction, respectively. Using the effective mass approximation (parabolic E-k dispersion relationship), E_1 is given by:

$$E_1^{parab}(t_{si}, W_{si}) = \frac{\hbar^2}{2}\left(\left(\frac{\pi}{m_y^* t_{si}}\right)^2 + \left(\frac{\pi}{m_z^* W_{si}}\right)^2\right) \qquad [2]$$

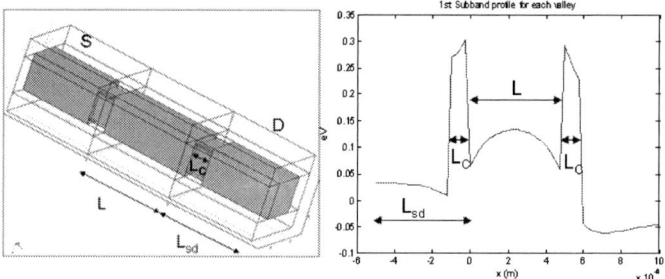

Figure 1. Schematic view of the simulated gate-all-around (GAA) nanowire with the 2 constrictions of the semiconductor cross-section at the Source extension (S)-body and body-Drain extension (D) boundaries (Left). 1st subband *vs.* x in a 2nm × 2nm [100] GAA) nanowire with the 2 constrictions of the semiconductor cross-section. $\Delta t_{si} = \Delta W_{si} = 0.5$ nm (Right). $L_{sd} = L = 5$nm. $L_C = 1$nm. The energy barrier resulting from the constriction can clearly be observed. $V_g = V_d = 0.1$V.

When the diameter of the cross-section is smaller than 3nm, however, the E-k dispersion relationship is no longer parabolic and the deviation of the effective mass in the transport direction from the bulk value needs to be taken into account in order to obtain accurate results [7]. Fig. 2.Left shows a comparison of the simulated barrier height and the barrier that is predicted analytically by such a model assuming a corrected effective mass in both the simulations and the analytical model [7]. The smaller the cross-section, the larger the bandgap mismatch is for a given diameter reduction. As can be observed in Fig. 2.Left, this means that relatively large barrier heights can be obtained in narrow nanowires with relatively small cross-section reduction.

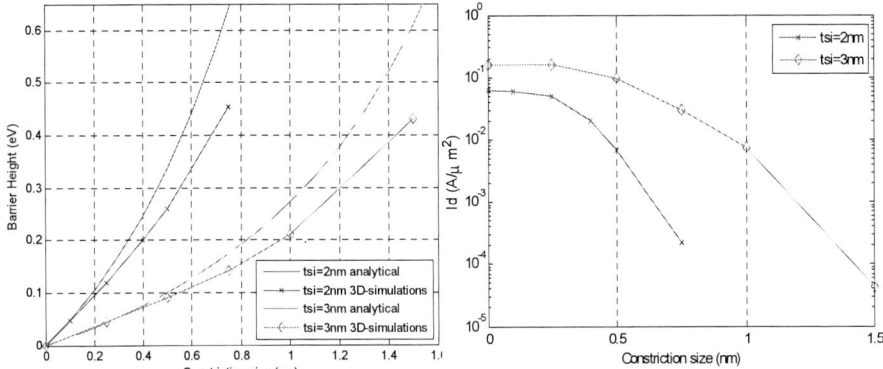

Figure 2. Barrier Height vs. diameter reduction of the constriction (Δt_{si}). The diameter of the cross-section in the constriction is t_{si}-Δt_{si}, while its area is equal to $(t_{si}-\Delta t_{si})^2$ (Left). Simulated current vs. Δt_{si} near threshold. $V_g=V_d=0.1V$ (Right) for a reference diameter (t_{si}) of 2 and 3 nm in a square [100] nanowire.

Due to the resulting barrier, electrons from the source have to tunnel through the constriction to reach the channel. This locally reduces the electron density in the constriction and reduces the current. If the resulting tunnel barrier (TB) height is low, *i.e.* lower than the top of the channel barrier, TBC, the drain current reduction is low as it is mostly limited by the channel barrier. In the opposite case, the drain current becomes quite sensitive to the tunnel barrier height and reduces drastically with it (Fig. 2 Right). This means that the current will be more sensitive to diameter change in narrower cross-sections (since the slope of the energy barrier *vs.* diameter variation is higher) and in strong inversion regime (where the channel barrier is the lowest and situated around the Fermi level, which is less than 0.1 eV above E_c for usual doping values).

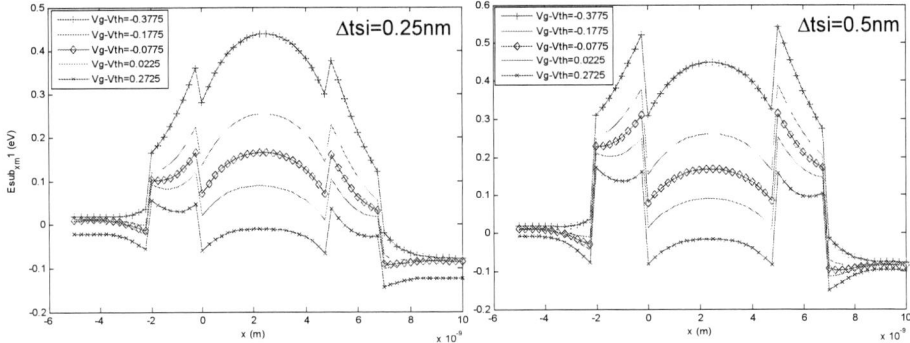

Figure 3. Evolution with V_g-V_{th} of the 1st subband *vs.* x profile in the 2nm × 2nm [100] GAA nanowire with the 2 constrictions of the semiconductor cross-section. $\Delta t_{si}=\Delta W_{si}=0.25$ nm (Left). $\Delta t_{si}=\Delta W_{si}=0.5$ nm (Right). $L_{sd}=L=5$nm. $V_d=0.1V$. Constriction width, $L_C=2$nm.

In Fig. 3, the energy of the first subband is plotted along the channel for different values of V_g–V_{th} for a [100] 2×2 nm^2 nanowire with a channel length of 5nm and two constrictions offering a diameter reduction of 0.25nm and 0.5nm and a barrier width, L_C, of 2nm . The corresponding I_d-V_g curves are shown and compared to that of the same nanowire without constriction (Δt_{si}=0nm) on Fig. 4. For the nanowire with a 0.25nm constriction, TB is lower than TCB up to about V_g –V_{th}=0V, which explains why the current at 0V is the same than that in the transistor without constriction.

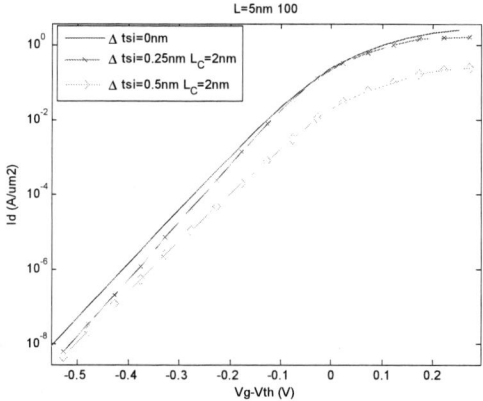

Figure 4. Id-Vg curves of the 2nm x 2nm [100] GAA nanowire without and with 2 constrictions of the semiconductor cross-section with Δt_{si}=ΔW_{si}=0.25 nm and Δt_{si}=ΔW_{si}=0.5 nm. L_{sd}=L=5nm. Vd=0.1V.

However, for values of V_g–V_{th} below -0.1V, the current of the 0.25nm constriction nanowire is lower than in the standard nanowire despite the fact that TB is lower than TCB. Our explanation, confirmed by the current spectral distribution simulation results is the following: The off-current is composed of two components: a source-to-channel tunneling current that takes place for energies below the source-to-channel barrier, and a diffusion current for electrons with energies above the barrier [4].

Below -0.1V, the channel barrier in the classical nanowire is high enough so that the subthreshold current is dominated by the current that tunnel under the channel barrier. The TBs resulting from the 0.25nm constrictions are high enough to reduce significantly the tunnel current under the channel barrier. Above -0.1V, the subtheshold current of the classical nanowire is dominated by the diffusion current. Therefore, as the TBs in the 0.25nm constriction nanowire are below TCB up to V_g–V_{th} =0V, both transistors have the same currents. Above V_g–V_{th} =0V, the TBs pass above the channel barrier but only slightly, reducing the on-current but by less than a factor 2, while the Off-current can be improved by nearly a factor 4. This allows an improved on to off current ratio in the transistor with the 0.25nm constriction compared to a classical nanowire.

Increasing the height of barrier too much so that the TBs height is always above the TBC in the applied voltage range as in the case of the 0.5nm constriction nanowire, does not reduce much further the off-current, as the diffusion current reduction that it allows in not that significant, while resulting in a drastic reduction of the on current and, therefore, of the overall on/off current ratio.

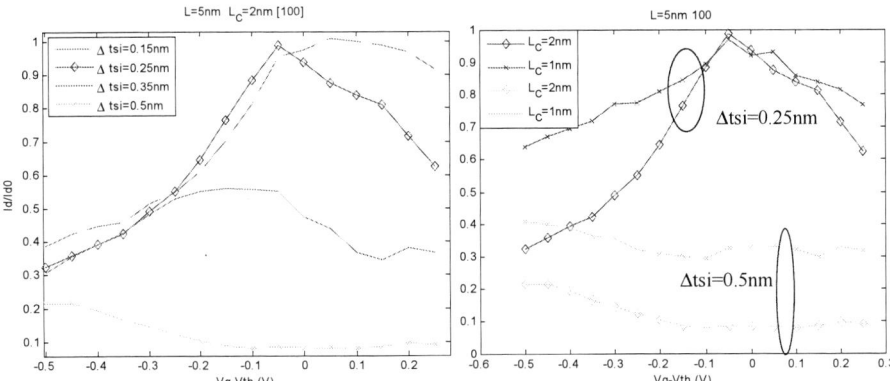

Figure 5: Ratio of the drain current with and without constriction *vs.* V_g-V_{th} for different diameter reduction length, Δt_{si}. L_C=2nm (Left). Influence of the constriction length, L_C, on the ratio of the drain current with and without constriction vs. V_g-V_{th} (Right).

The impact of the constriction on the current characteristic can be more conveniently compared on a graph of the ratio of the drain current with and without constriction *vs.* V_g-V_{th}. In Fig. 5.Left we further investigate the optimal diameter reduction length, Δt_{si}, and compare the case already studied in Fig. 4 for a [100] 2×2 nm^2 nanowire with a channel length of 5nm, a barrier width, L_C, of 2nm and two constrictions offering a diameter reduction of 0.25nm and 0.5nm with two more diameter reductions of 0.15 and 0.35nm.

The optimal barrier height is in fact located around the Fermi level E_F (around 0.065eV above E_C in our case), *i.e.* for diameter reduction Δt_{si} of approximately 0.15 to 0.25nm (Fig. 2). Indeed, the transistor is switched on (*i.e.* threshold is reached) when the TBC is around the Fermi level (around mean in a few k_BT) where there are plenty of electrons to drive the current. It is why the transistors with a tunnel barrier height close to E_F (*i.e.* for the 0.15 and 0.25nm constriction cases) have about the same current at threshold than the standard transistor. It also allows that above threshold, when the TBC passes below E_F, the TBs do not completely shadow the electron-rich energy band around the Fermi level and therefore ensure a sufficient on-current. This is not the case anymore when increasing the height of the barrier too much above E_F as in the case of the 0.35 and 0.5nm constriction nanowires. Reducing too much the height of the barrier below E_F on the other hand results in a reduction of the filtering of the off current. However, again, a barrier height around E_F seems already sufficient to filter a big part of the tunnel current under the channel barrier.

In Fig. 5.Right we now study the impact of the constriction length and compare the case already studied in Fig. 4 for a [100] 2×2 nm^2 nanowire with a channel length of 5nm and two constrictions offering a diameter reduction of 0.25nm and 0.5nm but for two different constriction lengths: L_C=1 and 2 nm. For both length of the barrier, we can observe the same behavior than explained before with the reduction of the cross-section diameter, *i.e.* an enhanced (resp. reduced) on-to-off current ratio for a diameter reduction of 0.25 (resp. 0.5) nm when compared to a nanowire without constriction. Due to the relatively low effective mass in the *x*-direction of the 4-fold degenerated unprimed valleys that dominate the current, the 1nm length constriction barriers are not sufficient to suppress completely the tunnel current and a 2nm length barrier can result in a significant improvement of the off-current. In the case of the 0.25nm constriction, this reduction of the tunnel current does not penalize the current at threshold at all as explained above and the subthreshold slope is significantly increased with L_C.

CONCLUSIONS

We have shown here that the effect of cross-section change in a nanowire results in energy barriers in the conduction bands that can be related to the increase of the bandgap or the level of the subbands as the cross-section is reduced. This can be predicted quite precisely by using an analytical model of the energy level of a constant potential well. In narrow nanowires of a few nanometers cross-section this can strongly influence the current and the characteristics of the devices. A small constriction resulting in a barrier of the order of a 0.1eV is shown to be an effective solution to the problem of the channel barrier tunneling current and a way to improve the on-to-off current ratio in ultra scaled transistors.

Acknowledgments

This material is based upon works supported by Science Foundation Ireland under Grant 05/IN/I888.

References

1. J.P. Colinge (Ed.), FinFETs and Other Multi-Gate Transistors, Springer, 2007.
2. S. Datta, "Nanoscale device modeling: the Green's function method". *Superlattices and Microstructures*, Vol. 28 (4), pp. 253-278 (2000).
3. J. Wang, E. Polizzi and M. Lundstrom, "A three-dimensionnal quantum simulation of silicon nanowire transistors with the effective-mass approximation", *Journal of Applied Physics*, Vol. 96 (4), pp. 2192-2203 (2004).
4. R. Venugopal, M.Paulsson, S. Goasguen, S. Datta and M. Lundstrom, "A simple quantum mechanical treatment of scattering in nanoscale transistors", *Journal of Applied Physics, 93, 5613 (2003)*.
5. A. Afzalian, D. Lederer, C.W. Lee, R. Yan, W. Xiong, C.R. Cleavelin, J.P. Colinge, 'MultiGate SOI MOSFETs: Accumulation-Mode vs. Enhancement-Mode', *IEEE 2008 Silicon Nanoelectronics Workshop*, P1-6, Honolulu, Jun. 2008.
6. X. Tang, X. Baie, J.P. Colinge, P. Loumaye, C. Renaux, and V. Bayot, "Influence of gate geometry on SOI single-hole transistor device characteristics", *Microelectronics Reliability*, Vol.41, pp. 1841-1846, 2001.
7. J. Wang, A. Rahman, A. Ghosh, G. Klimeck, and M. Lundstrom, "On the Validity of the parabolic effective mass Approximation for the I-V Calculation of Silicon Nanowire Transistors", *IEEE Trans. Electron Dev.*, Vol,52, pp.1589-1595, 2005.

Simulation of Hole Mobility in DGSOI Transistors

L. Donetti, F. Gamiz, F.G.Ruiz, N. Rodriguez, and A. Godoy

Departamento de Electrónica, Universidad de Granada, 18071 Granada, SPAIN

In this paper we study carrier mobility in double gate SOI p-channel devices. A self-consistent solver for the six-band $k \cdot p$ and Poisson equations is employed to compute the wavefunctions and dispersion relations of hole subbands. Hole mobility is evaluated employing the Kubo-Greenwood formalism. Both phonon and surface roughness scattering mechanisms are taken into account. In the case of surface roughness scattering, an extension of a model previously developed for n-channel double gate devices has been developed. To do so, new assumptions were necessary to maintain reasonable simulation times as $k \cdot p$ approach is considered. The results show that volume inversion has a lower impact in DGSOI p-channel devices than the one shown for DGSOI n-channel devices.

Introduction

It is well known that Silicon On Insulator (SOI) technology is a very promising candidate for extending the life of traditional silicon technology as we approach the end of the ITRS roadmap (1, 2). Among its several advantages over bulk silicon devices, some of the most relevant ones consist in the immunity to short channel effects, in the reduction in junction capacitance and in a lower sensitivity to process variability. SOI technology also offers the possibility of non-conventional device concepts such as the use of multiple gates. In the case of double gate (DG) SOI devices, besides the improved electrostatic control due to the use of a second gate, an important advantage is given by volume inversion. This phenomenon occurs when the inversion layer is not formed only at the top and bottom of the silicon slab, but throughout the entire film (3). As a consequence, the number of inversion carriers can be enhanced and, at the same time, the mobility can be larger due to the reduced influence of the scattering associated with oxide and interface charges. For these reasons, DGSOI devices have been the focus of intense studies in recent years. However, simulations have mainly considered n-channel devices, because of the intrinsic difficulties deriving from hole bandstructure, which do not allow a simple analytic description. While the energy dispersion of electrons can be obtained with a good accuracy with the effective mass approximation, using ellipsoidal valleys and a relatively simple approach to take into account non-parabolicity, more complex approaches have to be employed in the case of holes. Indeed, a sufficiently accurate description of holes bandstructure can be obtained using the $k \cdot p$ model. However, computing the hole properties in a confining potential using the $k \cdot p$ approach is a very demanding task from a computational point of view, especially if the potential profile is self-consistently computed at the same time, because very large eigenvalue problems have to be solved for a large number of points in the plane of the parallel momentum $k_{\parallel}$. To obtain the hole subband dispersion relations and wavefunctions we employ a fully self-consistent solver for the $k \cdot p$ and Poisson equations, in which a reduced computational time is obtained thanks to a specially designed mesh in the $k_\parallel$ plane (4, 5). This procedure has been used successfully to compute the valence band properties for p-channel MOSFETs with different substrate and channel orientations. We have performed the transport simulations based on these results, taking into account phonon and surface roughness (SR) scattering. A good agreement with experimental data was found(5).

In the case of DGSOI devices, however, SR scattering has to be addressed in a proper special way because of the presence of the two Si-SiO₂ interfaces close to each other. Indeed, not only the carrier scattering is produced by the roughness of both interfaces, but also the perturbation potential corresponding to each of them is strongly affected by the presence of the other one. Taking into account these issues, a suitable model was developed for n-channel devices (6, 7). Following the procedure outlined in that work, we now develop a SR scattering model for p-channel DGSOI devices, taking into account the different steps needed in this case to compute the perturbation potential and the matrix elements, because of the characteristic features of the valence band. In next section we derive the SR scattering model; in the following one we compute hole mobility employing Kubo-Greenwood formula taking into account phonon and SR scattering. We then discuss the simulation results and the dependence of hole mobility on silicon thickness.

Surface roughness model

The surface roughness is modeled by considering an abrupt barrier at the silicon-oxide interfaces whose position $\Delta(r)$ in the z direction (the confinement direction) depends on $r=(x,y)$ (6,7). The two interfaces of the SOI device are considered separately and independently, and the perturbation potential relative to the roughness of each of them is computed as the difference between the self-consistent potential $V(z)$ obtained for an ideal device with $\Delta=0$, and the self-consistent potential $V_{\Delta(r)}(z)$ computed for a device where the channel-oxide interface is given by $\Delta(r)$:

$$H_{SR}(r,z) = -e\, \Delta V_{SR}(r,z) = -e\, [V_{\Delta(r)}(z) - V(z)] \qquad [1]$$

In the calculation of $V_{\Delta(r)}(z)$, to avoid the lengthy solution of the $k \cdot p$ equation in all $k_\parallel$ plane, the dispersion relations of hole subbands of the unperturbed device are assumed, so that the potential is obtained by solving the $k \cdot p$ equation only at $k_\parallel=0$, self-consistently with the Poisson equation. We verified that in a large range of silicon thicknesses and applied biases this approximation does not produce a significant error, as can be seen in Figure 1(a) for a device with a silicon channel thickness of 10 nm.

Then, the perturbation potential $\Delta V_{SR}(r,z)$ is assumed to be proportional to the interface perturbation $\Delta(r)$ (6, 7). This means that only one self-consistent calculation need to be performed using $\Delta(r) = \Delta_m$, where Δ_m is the rms value of $\Delta(r)$ and:

$$\Delta V_{SR}(r,z) = \Delta(r)\, [V_{\Delta m}(z) - V(z)] / \Delta_m \qquad [2]$$

This assumption is also verified in all the considered cases and an example is shown in Figure 1(b). Taking advantage of the linear dependence of the perturbation potential on $\Delta(r)$ (equation 2) the matrix elements of the perturbation potential $H_{SR}(r,z)$ can be computed as:

$$\left|M_{ij}(\boldsymbol{q})\right|^2 = e^2 \left| \int \frac{V_{\Delta_m}(z) - V(z)}{\Delta_m} \psi_j^+(z)\psi_i(z)dz \right|^2 \left|\Delta(\boldsymbol{q})\right|^2 \tag{3}$$

where ψ_i is the six-component hole envelope function in the i-th subband, and $\Delta(\boldsymbol{q})$ is the Fourier transform of $\Delta(\boldsymbol{r})$ for which an exponential model is assumed (8). Dielectric screening due to mobile charge redistribution is then taken into account following the procedure reported in Ref. (9). Scattering rates are finally computed by integrating the screened matrix element over all possible final states (4, 5):

$$\frac{1}{\tau_{ij}(\boldsymbol{k})} = \frac{2\pi}{\hbar} \int \frac{d\boldsymbol{k}'}{2\pi^2} M_{ij}^{(s)}(\boldsymbol{k}' - \boldsymbol{k}) \delta\left[E_i(\boldsymbol{k}) - E_j(\boldsymbol{k}')\right] \tag{4}$$

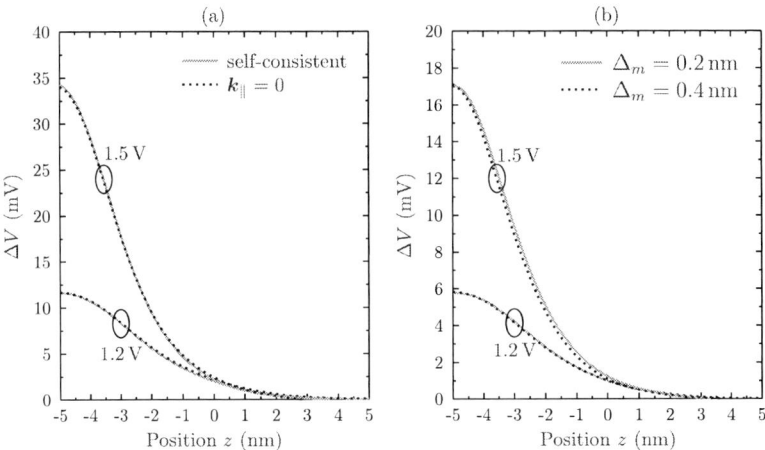

Figure 1. (a) Perturbation potential computed by solving the $\boldsymbol{k \cdot p}$ equation only for $\boldsymbol{k}_\parallel=0$ and using a fully self-consistent solution for a device with T_{Si}=10 nm, Δ=0.4 nm. (b) Perturbation potential computed according to equation 2 with $\Delta(\boldsymbol{r})$=0.2 nm and Δ_m=0.2 or 0.4 nm.

Simulation results

Mobility calculations are performed by employing the Kubo-Greenwood formula, obtained by a linearization of the Boltzmann equation (9). Potential profile, energy dispersion of hole subbands and wavefunctions are computed by self-consistently solving the $\boldsymbol{k \cdot p}$ and Poisson equations. The momentum relaxation times are computed using

$$\frac{1}{\tau_{ii}(\boldsymbol{k})} = \frac{2\pi}{\hbar} \int \frac{d\boldsymbol{k}'}{2\pi^2} M_{ii}^{(s)}(\boldsymbol{k}' - \boldsymbol{k}) \left(1 - \frac{k_x'}{k_x}\right) \delta\left[E_i(\boldsymbol{k}) - E_j(\boldsymbol{k}')\right] \tag{5}$$

where the x axis is taken along channel direction, and only intra-subband scattering is considered. The integral over the final momentum in equation 5 is performed as reported in Ref. (5).

We computed the hole mobility for different SOI double gate devices with silicon thickness ranging from 4 to 20 nm. Firstly we obtained the phonon-limited mobility, by considering optical and acoustic phonons as scattering mechanisms, employing the parameters used in a previous paper to fit the universal mobility curves (5). Then we added the surface roughness scattering, considering the two Si/SiO$_2$ interfaces to be independent and characterized by the same parameters: the rms height $\Delta_m = \Delta_1 = \Delta_2$ and the correlation length $\Lambda = \Lambda_1 = \Lambda_2$.

Figure 2 shows hole mobility computed for the devices with T_{Si} = 4, 6, 10, 20 nm with and without taking into account the surface roughness; in the former case we used Λ = 1.5 nm and Δ_m = 0.2 or 0.4 nm. For small values of the silicon layer thickness, hole mobility shows a weak dependence on the effective field; indeed, if only phonon scattering is considered, for a device with a 4 nm Si layer mobility is almost constant. This can be explained by the fact that, in this case, the wavefunction confinement (which is normally responsible for the increased phonon scattering rate at high values of the effective field) is essentially due to the geometry (the barriers at the two very near Si-SiO$_2$ interfaces) and it is only weekly influenced by the applied bias. Moreover, the slight decrease of the mobility in each subband is almost entirely compensated by a redistribution of the carriers between subbands: the occupation of the first subband slightly decreases while the population of the next subband (which has a larger mobility) increases.

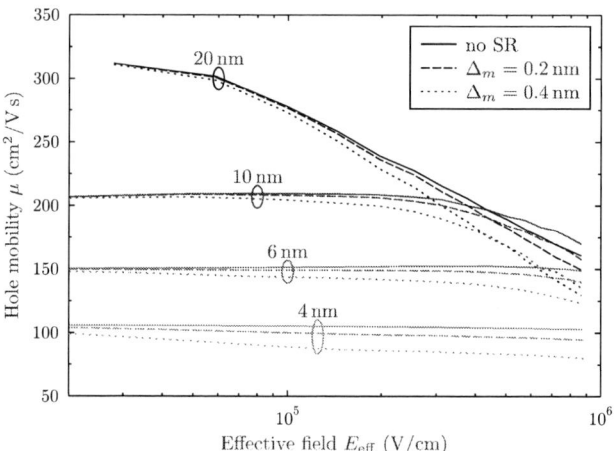

Figure 2. Simulated hole mobility as a function of effective field, for different silicon thicknesses. The mobility is computed considering only phonon scattering, and with surface roughness scattering using Δ_m = 0.2 or 0.4 nm and Λ = 1.5 nm.

It is interesting to analyze the dependence of hole mobility on the silicon layer thickness when phonon and SR scattering are considered. Figure 3 shows the simulation results: two different behaviors can be observed depending on the effective field value. For small E_{eff}, mobility is a monotonic function of T_{Si}, showing a strong mobility degradation at very small silicon layer thicknesses. On the contrary, for large values of effective field, mobility grows until it reaches a maximum for T_{Si}, in the range of 10-12 nm, then slightly decreases and tends to the mobility value obtained for bulk devices. The maximum value of the hole mobility curve is slightly higher than the value obtained for bulk devices. This means that for a small range of silicon thicknesses, hole mobility in a DG-MOSFET is higher than in a bulk MOSFET. This is a consequence of volume inversion effect. However, it is interesting to highlight here the difference with the electron mobility behavior. In that case, the mobility increase with regard to the bulk value is much more pronounced (10).

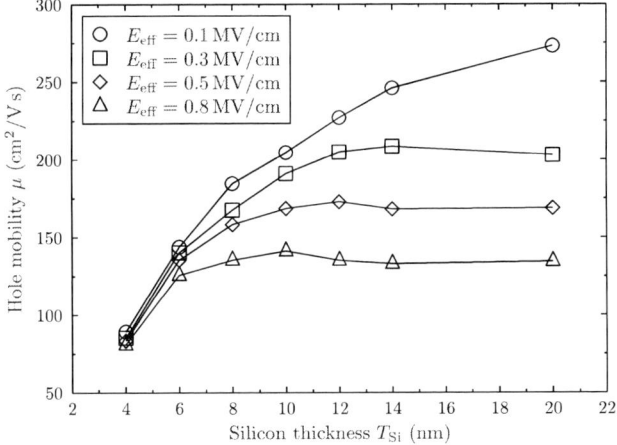

Figure 3. Simulated hole mobility as a function of silicon layer thickness at different values of effective field; $\varDelta_m = 0.4$ nm and $\varLambda = 1.5$ nm

Conclusions

We have calculated the hole mobility in DGSOI transistors by using the Kubo-Greenwood approach. Quantum effects and valence band properties are taken into account by using a self-consistent solver for the six-band $k \cdot p$ and Poisson equations. Phonon and surface roughness scattering were taken into account. The dependence of the mobility with effective field and with silicon thickness has been studied. Our results show that the influence of volume inversion in the case of hole mobility in DGSOI transistors is much less acute than in n-channel DGSOI transistors, i.e., there is a region of silicon thicknesses where hole mobility is higher than in bulk devices, but this increase is much smaller than the one observed in the case of n-channel DGSOI transistors.

Acknowledgments

This work has been carried out within the framework of projects TIC-2006-1899 (Junta de Andalucia) and NANOSIL Network of Excellence (FP7-NOE-216171) funded by EU. Support by Ramón y Cajal program of Spanish Ministry is also acknowledged.

References

1. http://public.itrs.net/.
2. G. K. Celler and S.Cristoloveanu., *J. Appl.Phys.*, **93**, 4955 (2003).
3. F Balestra, S. Cristoloveanu, M. Benachir, J. Brini, and T. Elewa, *IEEE Electron Device Letters*, **8**, 410 (1987).
4. L. Donetti, F. Gamiz, A.Godoy, and N. Rodriguez., *Proceedings of the 38th European Solid-State Device Research Conference*, 254 (2008).
5. L. Donetti, F. Gamiz, and N. Rodriguez, *Semicond. Sci. Technol.* **24**, 035016 (2009).
6. F. Gamiz, J. B. Roldan, P.Cartujo-Cassinello, J. A. Lopez-Villanueva, and P. Cartujo, *J. Appl. Phys.*, **89**, 1764 (2001).
7. F. Gamiz, J. B. Roldan, J. A. Lopez-Villanueva, P.Cartujo-Cassinello, and J. E: Carceller, *J. Appl. Phys.*, **86**, 6854 (2001).
8. S. M. Goodnick, D. K. Ferry, C. W. Wilmsen, Z. Liliental, D. Fathy, and O. L. Krivanek, *Phys, Rev. B*, **32**, 8171 (1985).
9. M. V. Fischetti, Z. Ren, P. M. Solomon, M. Yang, and K. Rim, *J. Appl. Phys.*, **94**, 1079 (2003).
10. F. Gamiz and M. V. Fischetti, *J. Appl. Phys.*, **89**, 5478 (2001).

CHAPTER 11

CIRCUIT TECHNOLOGY I

242

ECS Transactions, 19 (4) 243-256 (2009)
10.1149/1.3117415 ©The Electrochemical Society

Floating-Body SOI Memory: Concepts, Physics and Challenges
M. Bawedin[a], S. Cristoloveanu[b], D. Flandre[c], F. Udrea[a]

[a] CAPE, University of Cambridge, Cambridge CB30FA, U.K.
[b] IMEP, INPG-Minatec, Grenoble 38016, France
[c] DICE, Université Catholique de Louvain, Louvain-la-Neuve 1348, Belgium

We present an overview of the single-transistor memory cells (1T-DRAMs), which are based on floating-body effects in SOI MOSFETs. The typical device architectures, principles of operation and key mechanisms for programming are described. The various approaches (Z-RAM, MSDRAM, etc) are compared in terms of performance and potential for aggressive scaling.

Introduction

The scaling of conventional DRAMs is facing the imminent problem of further miniaturization of the storage capacitor. An attractive solution, offered by the SOI technology, is to avoid the use of a bulky capacitor. The floating body of SOI transistors can indeed be used for charge storage. Several types of single-transistor SOI DRAM (1T-DRAM) have been proposed and will be reviewed in this paper. In all variants, bit '1' reflects the presence of excess majority carriers in the body which increase the potential and the drain current (Fig. 1). Conversely, bit '0' is characterized by a lower current due to the removal of majority carriers from the body.

All 1T-DRAMs take advantage of the floating-body effects in SOI transistors. These memory cells can be classified in three groups according to their 1-state programming methods. During the 1-state programming, the high current level I_1 originates from the excess of majority carriers (holes) charge generated into the body. This additional charge can be achieved by (1) impact ionization, (2) bipolar junction transistor (BJT) effect and (3) band-to-band (B2B) tunneling generation. The 0-state (low current level I_0) programming, or body hole charge removing, is usually performed by forward biasing the drain- or source-body junction. The programming methods based on the threshold voltage shift (due to the body hole charge variation) have produced poor cell signal sense margin, i.e. low difference in current $|I_1 - I_0|$. Indeed, due to the lack of efficiency for removing holes during the 0-state programming, the current level I_0 remains of the same order of magnitude as I_1. To overcome this problem, dynamic coupling between front and back gates in fully depleted SOI or double-gate MOSFETs can be used.

The paper is organized as follows. Sections 2 and 3 introduce the methods for programming and reading the '1' and '0' states. The physics principles of the various types of 1T-DRAM, their performance in terms of programming speed, bias and power consumption, and the technological solutions for further optimization and scaling are critically reviewed in section 4. Finally, the best candidates able to survive aggressive miniaturization and to meet the demand for higher speed and lower power consumption will be presented. We focus on n-channel MOSFETs, i.e. with P-type body, but the same considerations can be applied to p-channel 1T-DRAM.

243

Programming 1-State

During the 1-state programming, the amount of majority carriers (holes) in the body is enlarged. This induces a dynamic decrease of the threshold voltage ($V_{TH} \rightarrow V_{TH} - \Delta V_{TH1}$) and an increase in drain current I_1. During the 1-state (or 0-state) programming, the threshold voltage shift ΔV_{TH1} (or ΔV_{TH0}) results from the combined effects of the hole charge and the corresponding body potential variations.

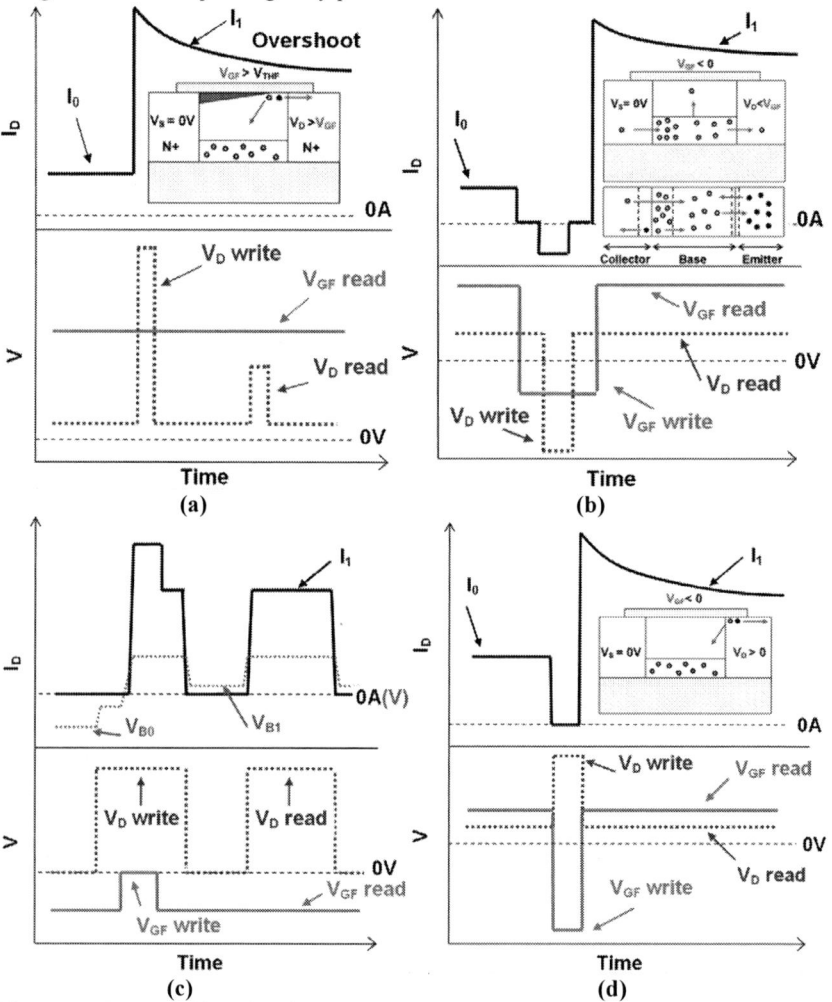

Figure 1. External bias signal sequences and schematics for the write/read "1"-state programming by (a) impact ionization, (b) bipolar junction transistor (BJT) effect 1st method, (c) BJT effect 2nd method and (d) band-to-band tunnelling (B2B) generation. Plotted versus time are the front-gate voltage V_{GF}, drain voltage V_D, body potential V_B and drain current I_D.

Impact Ionization (II)

The most commonly used method for 1-state programming consists in generating holes inside the body by impact ionization (Fig.1.a) [1-9]. A relatively high positive drain voltage V_D is applied while the front interface is in inversion mode ($V_D > V_{GF} > V_{TH}$). The holes generated at the pinch-off region, close to the drain, move into the body (Fig.1.a., inset). During the programming, the front-gate voltage V_{GF} remains unchanged (same as in hold and read conditions), while the drain voltage V_D is shifted from a low value (~ mV) to a higher one (~V). Since the body/drain junction is reverse biased, when the holes are filling the body, the floating body potential increases above the level it had before the programming stage. For 1-state reading (or hold), V_D is switched back to its low level. As the excess hole charge remains inside the body, the potential shift ΔV_B achieved during the programming period is preserved, inducing a "dynamic" V_{TH} lowering, hence an increase in drain current. During the reading/holding period, as the body potential is higher than its steady-state value, the stored holes are gradually evacuated through the body-to-source/drain junctions. This leads to a drain current overshoot or return to equilibrium with time (Fig.1.a). Notice that the reading is non-destructive since performed at a low drain voltage (~mV).

It is clear that the high drain voltage for 1-state programming can cause a premature degradation of the gate oxide, through hot carrier injection, affecting the memory cell retention time and shifting the threshold voltage. Besides, for the coming DRAM generations, in order to achieve a faster programming, the impact ionization rate will have to be enhanced. This unfortunately means higher V_D and hence higher power consumption.

Bipolar Junction Transistor (BJT)

The second method of 1-state programming takes advantage of the intrinsic BJT effect which can be activated in the floating body of SOI MOSFETs. In that case, the source (N+), body (P) and drain (N+) act as the emitter (or collector), base and collector (or emitter) of the BJT (Fig.1.b and 1.c). To turn on the BJT effect, a hole current has to be generated in the base. Since the base is floating, it results in an increase of the body potential during the programming. As for the previous method, this produces a threshold voltage lowering, hence a current increase. Two BJT programming methods are reviewed, their difference residing in the way the excess holes are created.

(a) Historically, the first BJT programming method [1,2] was introduced at the same time as the impact ionization method. It consists in applying a negative V_D pulse embedded into a negative gate V_{GF} pulse (Fig.1.b). The programming starts by switching V_{GF} to a negative value which has to be large enough (~V) to enable and keep an accumulation layer at the front interface. Then, V_D is also pushed to a high negative value. At this stage, the body/drain and body/source junctions are forward and reverse biased, respectively. If V_D is large enough, the energy of electrons flowing from the drain to the source enables the generation of electron/hole pairs by impact ionization at the body/source junction (Fig.1.b, inset). Before the reading (or holding) stage, the drain voltage is switched back to its reading low value (~ mV) prior to the gate, in order to secure the stored hole charge in the body.

The main advantage of this method is the power consumption reduction due to the low effective current flowing between the drain and the source during the programming. Indeed, on the one hand, the BJT gain β is low ($I_S = I_D \beta/(1+\beta)$) and, on the other hand, the base hole current I_h resulting from the impact ionization is small ($I_S = \beta I_h$).

(b) In the second programming method [10-12], the bipolar effect is initiated with different bias conditions (Fig.1.c). To enable the BJT, a V_{GF} pulse is embedded into a high positive drain voltage pulse. Notice that during the reading or holding, V_{GF} is kept at a negative value lower than the front flat-band voltage V_{FBF} in order to preserve the hole charge stored previously. At the programming onset, V_D is increased while V_{GF} is held below V_{FBF}: holes are generated by impact ionization in the body through the reverse biased body/drain junction. Nevertheless, as explained above, the resulting hole current I_h is small and the BJT effect is not significant. Consequently, the next step consists in amplifying the hole generation. In order to do so, V_{GF} is pushed to a higher value (> V_{FBF}) to lift up the body potential V_B by dynamic gate coupling. As the body potential is increased, the BJT turns on increasing the collector current I_D. Hence, the impact ionization and the resulting hole base current I_h are drastically enhanced. Prior to reading (or holding), V_{GF} is switched back to its negative reading value while V_D is maintained high. Since I_h is now large enough, the body potential (and current I_D) becomes "pinned" to its high level. The negative V_{GF} (<V_{FBF}) stores the generated holes at the front interface. To perform the reading, a high positive V_D pulse is applied. As a result, the body potential V_B increases by dynamic coupling and turns on the BJT and drain current.

The fact that the back gate is always grounded ($V_{GB} = 0V$) stands as the main advantage of the second BJT method. It can be applied to a variety of advanced SOI technologies [12] such as partially depleted (PD), fully depleted (FD), double-gate, FinFET, or Tri-Gate (Fig.2.a). Compared to the previous programming techniques, other improvements are the fast read and write operations (~ 2ns) [10]. It is worth noting that the impact ionization through the drain junction is not the only mechanism generating holes during the programming and reading. Indeed, the net potential drop at the gate edges can reach values up to 4V [11]. Therefore, at the gate-to-drain (source) overlap region, holes can also be generated by band-to-band tunneling or impact ionization. This aspect is beneficial for improving the programming speed and stabilizing the 1-state drain current, but can deteriorate the retention time (see section 4).

Band-to-Band (B2B) Tunneling

The holes generated by B2B arise from the gate-to-drain (source) overlap region (Fig.1.d, inset) [5,13-16]. In order to produce the holes, the drain/oxide interface is in strong inversion; the band bending and the local electric field must be large enough to allow holes (electrons) tunneling from the conduction (valence) band into the valence (conduction) band. The holes are collected by the body and the electrons flow toward the drain contact. During the programming, a negative V_{GF} pulse ($\leq V_{FBF}$) is applied while V_D remains positive (Fig.1.d). At the programming onset, if a high negative gate bias is used ($V_{GF} \ll V_{FBF}$), the body potential decreases by dynamic gate coupling and becomes negative (Fig.2.b). As the holes cannot be supplied and accumulated instantly, the front interface remains initially depleted. While the holes are gradually filling the body, the potential increases until it reaches equilibrium (~ 0V) when the accumulation layer is completed. Compared with the other programming methods, the holes are not able to escape through the source because the body potential is negative. Therefore, the storage and the body potential variation are more efficient. Obviously, if the programming time is too short, the body potential increase and the programming efficiency are reduced. For reading, V_{GF} is increased and the holes stored at the front interface are pushed down toward the bottom interface.

If the net potential drop $|V_G - V_D|$ on the gate is increased, the electric field and hence the B2B generation are enhanced. As a result, the programming time can be readily shorted by increasing either $|V_G|$ or V_D. To reach a competitive programming speed compared with the II method, the total signal swing $|V_G - V_D|$ has to be increased by about 20% [14] but it remains similar to the BJT method. Moreover, the B2B method features a striking advantage which is low power consumption. Indeed, the B2B current is much lower than the drain current arising from impact ionization or BJT. Finally, the reliability of the memory cell, programmed with B2B (or BJT), is expected to be superior to the impact ionization.

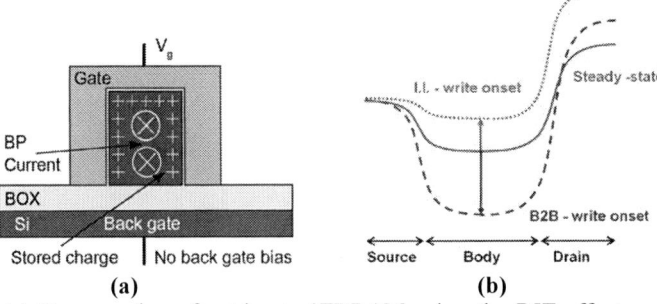

(a)   (b)

Figure 2. (a) Cross-section of a tri-gate 1TDRAM using the BJT effect programming technique [12] and (b) comparison between the impact ionization and the B2B tunnelling injection methods in terms of body potential variations during the "1"-state writing [14].

Programming 0-State

To program the 0-state, a deficit of holes has to be created in the body. During the programming, the holes are removed and the resulting body potential decreases. As a result, the effective threshold voltage is enhanced ($V_{TH} \rightarrow V_{TH}+\Delta V_{TH0}$) and the drain current lowered. It is obvious that the 0-state programming technique has to be compatible with that of the 1-state especially for embedded applications where the memory cell array configuration must be taken into account. Basically, two methods are available for removing the holes from the body.

Forward Biased Junction

In order to expel the holes (Fig.3.a), the front-gate voltage V_{GF} is kept at the reading voltage ($\geq V_{THF}$) [1,2,4] (or pushed to a negative value [5,7,9]) while a negative voltage ($\sim V$) is applied on the drain contact. The body/drain and body/source junctions become forward and reverse biased, respectively. Consequently, the hole current injected from source inside the body is much lower than the one extracted through the drain junction: the holes are readily extracted from the body.

This technique is reliable and fast since the time response of the forward body/drain junction to the bias switch is nearly instantaneous. Nevertheless, the resulting body potential variation is not really efficient and the 0-state current level usually keeps the same order of magnitude as the 1-state. This augments the power consumption during the reading and alters the retention time.

Capacitive Coupling

Let's assume that V_{GF} is lower than the front flat-band voltage V_{FBF} and a hole accumulation layer stands at the front interface. A fast removal can be performed by applying a front-gate pulse higher than V_{FBF} (Fig.3.b) [10-12]. When the gate voltage increases, the body potential follows by capacitive coupling. As the potential increases, the front interface becomes depleted and the body/drain and/or body/source junctions become "forward biased". Therefore, the holes are evacuated through the junctions. For reading, the gate is switched back to a negative value and the body potential decreases by capacitive coupling to a negative value. Since the holes cannot be accumulated instantly, the body potential is pinned to its negative value. The return to equilibrium during the reading or holding depends on hole charging mainly via the junction leakage current I_{LEAK}. Leakage is detrimental because it makes the 0-state drain current increase, altering the retention time.

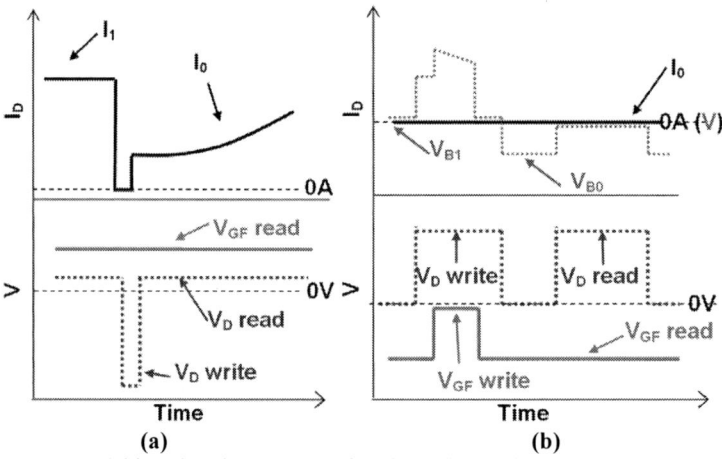

Figure 3. External bias signal sequences for the write/read "0"-state operations by (a) "forward bias junction" and (b) capacitive coupling.

In Fig.3.b, the capacitive coupling method is applied to produce a body potential V_B decrease which prevents the BJT activation during the reading. During the 0-state programming, the drain voltage keeps a high positive value for compatibility with the whole BJT method. This means that a positive source voltage has to be applied in order to avoid that the BJT turns on and charges the body. On the other hand, the source voltage cannot be too high in order to allow the holes to leave the body.

In FD SOI or double-gate 1T-DRAM, the capacitive coupling method can also be used to cut-off an inversion channel standing at one interface by applying a negative voltage pulse on the opposite gate [15, 16]. This method, currently used in the MSDRAM (see last section below), allows achieving very fast programming time and ultra low 0-state drain current level ($1nA/\mu m$). Consequently, the drain current sense ratio (I_1/I_0) and the retention time are greatly improved.

The main difficulty resides in combining all these 0 and 1-state programming methods to achieve simultaneously the best compatibility with a specific technology (PD SOI, FD SOI, double-gate, FinFET…) and high performances, i.e. low programming voltage, high retention time and high signal sense margin (maximum ΔV_{TH}). The retention time is defined by the minimum current difference ($I_1 - I_0$) detectable by the sensing circuit. All together, the performance improvement implies a reduction in power consumption.

When a memory cell is embedded into an array, the entire bias configuration and the disturb conditions have to be taken into account. During the programming of a cell standing in a word line, the word (row) and bit (column) line biases are also applied to other cells belonging to the same row and column. Moreover, the number of signal lines has to be limited in order to reduce the complexity of the control circuits.

Performance Improvement and Scaling

The partially depleted (PD) SOI 1T-DRAM (Fig.4.a) allows reducing the memory cell size compared to usual bulk-Si DRAMs. The next question concerns the scalability and reliability of the 1T-DRAM as well as its compatibility with more advanced SOI technologies (e.g. FD, FinFET).

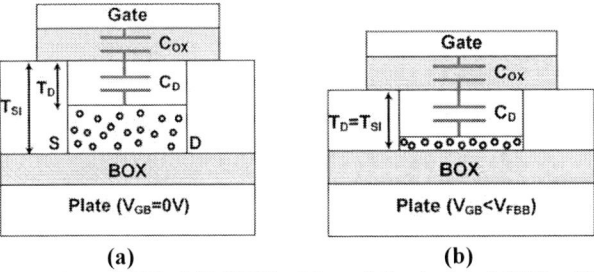

(a) (b)

Figure 4. Cross-section of SOI nMOSFETs: (a) partially depleted (PD) with a hole quasi-neutral zone and (b) fully-depleted (FD) with a back accumulation layer. T_D and T_{SI} are the depletion layer and silicon film thicknesses, respectively. C_{OX} and C_D are the front-gate oxide and depletion layer capacitances and V_{GB} and V_{FBB} are the back-gate and back flat-band voltages.

As the effective channel length L_{CH} decreases, the improvement of the retention time and the current signal margin ($|I_1-I_0|$ or ΔV_{TH}) remains the critical target and also the 1T-DRAM weak spot. While L_{CH} shrinks, the hole storage volume is reduced and the maximum electric field is enhanced. This is beneficial for improving the 1-state performances, i.e. to reduce the programming voltage and/or time. However, the 0-state retention time can be noticeably altered by parasitic hole generation. On the other hand, short-channel effects (SCE) can reduce further the storage volume and the threshold variation ΔV_{TH} during the programming. If the current signal margin can be increased, the retention time is improved because it takes longer for the memory cell to recover its steady state and reach the minimum acceptable $|I_1-I_0|$ difference.

In PD devices, the channel doping can be increased in order to suppress the SCE and enlarge ΔV_{TH} (= $\Delta V_{TH0}+\Delta V_{TH1}$) (Fig. 5.a) [4,7]. Hence, the junction depletion effect is reduced while a higher current sense margin is achieved. It corresponds to increased effective storage volume and current difference $|I_1-I_0|$, leading to longer retention time. Unfortunately, a too high body doping augments the junction leakage current I_{LEAK} during the reading and induces random dopant fluctuation (RDF) effects [4,17]. While I_{LEAK} degrades the retention time, the additional RDF effect produces random changes in ΔV_{TH} and fluctuations in read current (Fig.5.b). Since the read current also varies with time and only its statistical distribution is known, the comparison with a reference cell current to identify the 0 and 1-states is problematic and requires special sensing circuit design [18,19]. Notice that, when the impact ionization programming technique is used, if the body doping decreases, the maximum electric field at the body/drain junction is lowered. Then, higher V_D and/or longer programming time are required to achieve the same performance. This means that it becomes difficult for a PD 1T-DRAM to be scaled aggressively [3, 13].

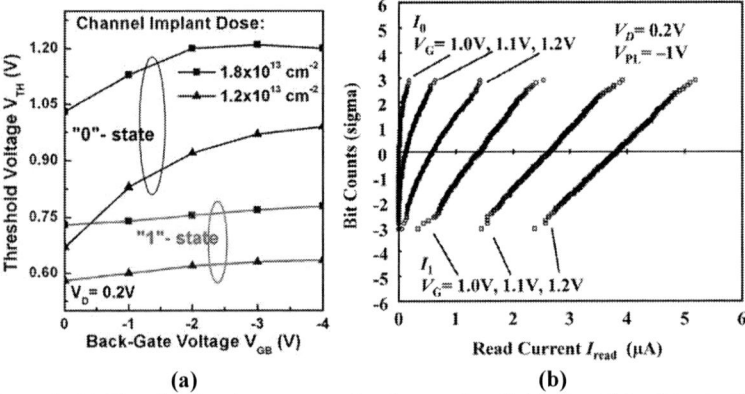

(a)   (b)

Figure 5. (a) Threshold voltage versus back-gate (i.e. field plate) bias V_{GB} in "0" and "1" states of a fully-depleted 1T-DRAM for two different channel dopings (adapted from reference [4]). (b) Statistical distribution of the memory cell 0-state I_0 and 1-state I_1 read currents of 1000 transistors in a 96kb memory array plotted for several front-gate voltages V_{GF} (from [4]).

The RDF effect and the junction leakage current are attenuated if the channel doping decreases. This is beneficial for the memory reliability and stability as well as for retention time improvement. However, when the doping is lowered, ΔV_{TH} and therefore the current sense margin decrease. It is not clear if by optimizing other technological parameters one can simultaneously take advantage of a low doping and keep a decent current sense margin.

In PD SOI MOSFETs, the threshold variation ΔV_{TH} can be approximated by ΔV_B x (C_D/C_{OX}) [4], where the body potential variation ΔV_B is proportional to the change in hole density. This model remains valid for FD SOI MOSFET if the back channel is in accumulation mode. To increase the depletion layer capacitance C_D and hence ΔV_{TH}, the depletion layer thickness T_D has to become thinner (Fig.4.a). As explained previously, the channel doping in PD MOSFETs can be increased. In FD devices, this can be achieved by decreasing the silicon film thickness T_{SI} ($\approx T_D$) while the back interface remains accumulated by applying a negative voltage V_{GB} on the back gate (Fig.4.b). It can be observed that, as V_{GB} becomes more negative, the 1-state threshold shift ΔV_{TH1} is nearly constant whereas the 0-state ΔV_{TH0} increases and finally saturates when the back interface enters in strong accumulation regime ($V_{GB} < -2V$, Fig. 5.a). Remark that if the back interface is depleted, then C_D is connected in series with the buried oxide capacitance, hence the resulting C_D is much lower and ΔV_{TH} is drastically reduced.

In FD devices (Fig.4.b), the channel doping can be decreased to the 'intrinsic' value while keeping an excellent current sense margin as long as a sufficiently negative V_{GB} is applied. Since I_{LEAK} and the RDF are reduced, the retention time is enhanced. As the silicon film thickness decreases, I_{LEAK} reduction also originates from the smaller effective junction area ($I_{LEAK} \alpha T_{Si}$). When a P^+ doped back gate or a P^+ field plate is used with FD devices (Fig. 4.b and Fig.6.a [7,9]), the maximum ΔV_{TH} (i.e. the maximum signal margin) can be further improved by increasing the plate doping [4]. It is also obvious that the optimal $|V_{GB}|$ can be lowered to a less negative bias if a thinner BOX is used. Finally, the amplification factor (C_D/C_{OX}) of ΔV_{TH} can be enlarged with thicker front-gate oxide T_{OX}.

These considerations naturally lead to adopt FD SOI as the solution to overcome the short-channel effect and improve the retention time and the reliability of the 1T-DRAM by reducing the silicon film thickness T_{SI} and the channel doping. However, in fully depleted SOI, the use of a common back-gate voltage V_{GB} different from 0V induces compatibility problems with the peripheral circuits and requires special implantations under the back gate (Fig.6.a). On the other hand, a common V_{GB} limits the possible combinations of the different 0 and 1-state programming methods. To the best of our knowledge, only the second generation BJT method (Fig.1.c) allows keeping a common grounded V_{GB} for all the memory cells embedded in the same array while enabling the capacitive coupling method which is the most efficient 0-state programming technique without disturb issues. This is made possible because the front inversion channel is not required during the reading and programming periods (Fig.2.a).

In FD SOI, as the silicon film thickness T_{SI} decreases, the electric field at the gate-to-drain overlap region is enhanced, deteriorating the retention time of 0-state by enabling parasitic B2B tunneling generation due to the negative V_{GB}. However, it is demonstrated [9] that the FD SOI 1T-DRAM can be scaled down while keeping the electric field strength and the electrical characteristics constant. In this case, the operation bias and the thicknesses of the front oxide, back gate oxide and silicon film should be reduced. In order to keep the electric field constant, special technological "improvers" can also be used, for example lightly doped drain LDD [7] or gate non-overlap regions [11]. These techniques tend to reduce the parasitic B2B tunneling generation and consequently extend the retention time.

All previous considerations highlight that the ultimate technological step to be considered for the 1T-DRAMs is their implementation within double-gate (DG) devices

[3,6,10,13,15]. Indeed, with FD devices, in order to improve memory cell performances and achieve low bias operation, it is necessary to use thinner silicon film and BOX. In addition, as a negative V_{GB} is required to secure the hole storage, separated wells have to be implanted below the back gates in order to ensure that the other surrounding circuits work properly [7]. Finally, in double-gate devices, the back gate of each transistor can be controlled individually, which enables to get rid of additional technological steps and take full advantage of the different programming methods [6,10,15].

Best Candidates to Scaling

We have evoked the advantages of the different programming techniques and the scaling effects on the SOI 1T-DRAM performance. We now review the best combinations of the programming methods and technological process which highlight the most promising candidates to scaling. In the following, examples of 1T-DRAMs using FD, PD and double-gate devices are briefly described. The performance in terms of programming bias, retention time, and current sense margin are provided quantitatively, for various programming methods.

Toshiba approach

In the first example (Fig.6.a, [7,9]), the feasibility of a FD SOI 1T-DRAM cell was experimentally demonstrated for 90 nm technology node in a 128 Mb memory array and predicted by numerical simulations for 32 nm technology. The impact ionization and the forward bias techniques are used respectively for the 1 and 0-state programming. To improve the retention time and the current signal margin, the CMOS process includes LDD, a relatively high channel doping ($3\times10^{17}cm^{-3}$) and P-doped field plate. In addition, Co salicide and Cu wirings are used to reduce the parasitic source and drain resistances which usually degrade the current signal. The gate length, the front and buried gate oxide and the silicon film thicknesses are 145, 6, 25 and 43 nm, respectively. The 1 (0)-state programming voltages are +1.5 V (-2.3 V) for the front gate and +2.2 V (-1.5 V) for the drain. The plate bias V_{GB} is fixed at -2.5 V. A ΔV_{TH} of 420 mV and a retention time of 70 ms were achieved for a temperature of 85°C.

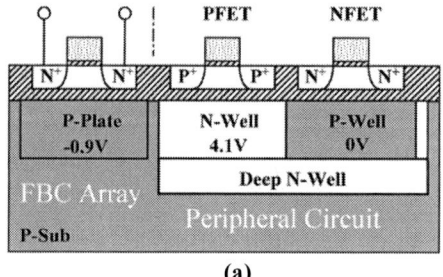 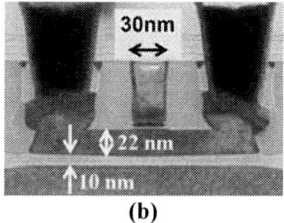

(a)   (b)

Figure 6. Cross-section of (a) well structure implanted for compatibility purpose between the memory array and the peripheral circuits (from [7]) and (b) very thin silicon film 1T-DRAM using a 45 nm logic technology (TEM picture from [8]).

Intel approach

A stand-alone FD SOI 1T-DRAM memory cell (Fig.6.b, [8]) was fabricated with a 45 nm logic technology. Simulations were used to demonstrate the viability of the memory for 16 nm technological node. No details were provided on the programming methods and bias levels. Nevertheless, since (i) the worst-case disturb condition occurs at high drain voltage and lowers the retention time due to residual B2B tunneling generation and (ii) the 0-state current level has the same order of magnitude as the 1-state, it is reasonable to assume that the impact ionization and the forward bias programming methods were applied. To improve the retention time, low doped SD implants were used. A high-k dielectric was implemented. The gate length, width, and the thicknesses of the BOX and Si film were 55, 65, 10 and 22 nm, respectively.

Since the silicon film is very thin, the channel can be undoped. This allows suppressing the RDF effects and lowering significantly the junction leakage current. A ΔV_{TH} of 400mV can be achieved with a V_{GB} of -2V and the maximum retention time at the worst disturb condition is 25ms for a temperature of 85°C. Notice that, if the disturb condition is not taken into account, the hold retention time could reach about 100s at 85°C. It is due to the fact that the main physics phenomenon responsible for the retention time degradation is the SRH recombination/generation.

ZRAM

PD SOI 1T-DRAM were fabricated with a 100 nm design rules and tested in the memory cell array condition [10-12]. The second generation BJT effect and the capacitive coupling techniques were respectively used to program the 1 and 0-states (Fig.1.c and 3.b). To improve the retention time, a non-overlap structure was implemented. It allows decreasing the maximum electric field and hence reducing the leakage current during the holding and reading operations. Indeed, the maximum B2B tunneling generation peak usually occurs at the gate-to-drain and -source overlap regions. Moreover, the lower electric field drastically decreases the SRH generation rate by 4 orders of magnitude compared with conventional devices. Notice that a high channel doping (10^{18}cm^{-3}) is needed in order to increase the impact ionization generation during the 1-state programming. This allows reducing the programming time and drain voltage.

The gate length, front gate oxide and Si film thicknesses are 55, 5 and 80 nm, respectively. The gate voltage (-1 V) and drain voltage (3 V) are the same for 1 and 0-state programming, whereas a source voltage of 0.5V is used only for the 0-state programming. The back gate is kept grounded. For this particular programming technique, a ΔV_{TH} larger than 500mV and an excellent I_1/I_0 current ratio (10^4-10^6) are achieved. Thanks to the underlap regions, a retention time of 70ms can be obtained at 85°C.

These programming modes, previously proposed by Okhonin et al [10], can be applied to advanced technologies like MuGFET devices (Fig.2.a). For a fin width of 11nm and a gate length of 50nm, a retention time of 1ms at 125°C was reported [12].

MSDRAM

The Meta-Stable DRAM (MSDRAM) memory cell physics principles are based on the MSD hysteresis effect (Fig. 7.a) [20]. Note the very wide memory window and the current ratio I_1/I_0 which exceeds 6 orders of magnitude. The MDRAM [15] is the first 1T-DRAM taking full advantage of double-gate operation which allows combining the low consumption and superior reliability (thanks to the B2B tunneling for 1-state programming) with the efficiency of the 0-state programming by dynamic gate coupling. This preliminary demonstration used large area FD MOSFETs fabricated with unoptimized technology, which explains the disproportionate bias levels. It is clear that specific source and drain architectures [16] allow enhancing the retention time while reducing the programming voltage and/or time.

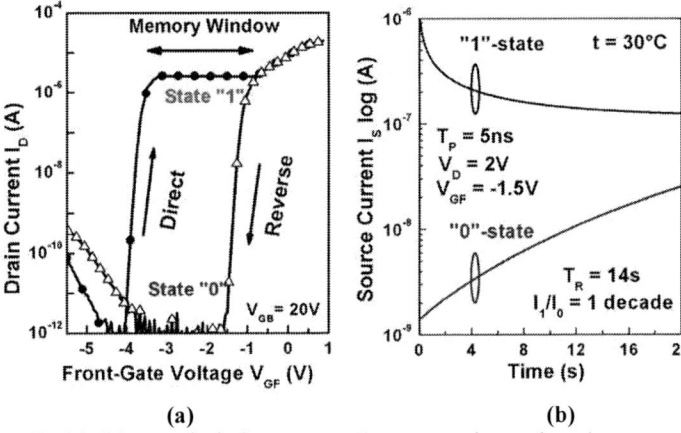

(a)   (b)

Figure 7. (a) Measured drain current I_D versus decreasing (reverse scan) and increasing (direct scan) front-gate bias V_{GF}. MSD effect is observed for direct scan where I_D saturates. The applied back-gate V_{GB}, source V_S, and drain V_D voltages are 30V, 0V and 0.1V, respectively. This n-MOSFET had 400nm thick BOX, 6nm thick gate oxide, 80nm thick Si film and 5×10^{16} cm^{-3} doping. The gate length and width are respectively 1.5μm and 20μm. (b) Simulated short-channel DG-MSDRAM: drain current I_D versus time during the "0"-state and "1"-state reading cycle. V_{GB} and V_D are 0.25V and 0.1V and the silicon film thickness and the channel length are 40nm and 50nm respectively. The front and back-gate oxide thicknesses are 3nm and 6nm and the body doping is 10^{16} cm^{-3}. A retention time T_R of 14s with $I_1/I_0=10$ is achieved at 30°C. During the 1-state writing by B2B tunnelling, V_D and V_{GF} are equal to 2V and -1.5V with a programming time T_P equal to 5ns.

The MSDRAM scalability was demonstrated by 2D simulations of a 50 nm long channel memory cell with DG configuration and ultra thin BOX (Fig.7.b). The gate and drain voltages used for programming, reading, and holding have been restricted to less than 2 V, without affecting the MSD effect. The performance is remarkable: $I_1/I_0=10^3$ at the reading onset, 14 s retention time for $I_1/I_0=10$ and 5 ns programming time.

Recent measurements confirm that the MSD memory effect is maintained in small MOSFETs with gate area of 0.1 μm^2 (0.35 x 0.35 μm) [21].

Conclusion

The floating-body SOI memory is not an intellectual exercise, it can compete in the DRAM market. There are several options, the principles of which have been discussed and compared. Among the basic mechanisms, we prioritize B2B tunneling for 1-state programming and capacitive coupling for 0-state. In terms of CMOS scaling, double-gate transistors are best suited devices. This is why we believe that the MSDRAM has remarkable assets to succeed in the 1T-DRAM competition.

Acknowledgements

Part of this work has been supported by the European Union programs EUROSOI+ and NANOSIL.

References

1. S. Okhonin, M. Nagoga, J.M. Sallese and P. Fazan, IEEE Int. SOI Conf., pp. 153-154 (2001).
2. S. Okhonin, M. Nagoga, J.M. Sallese and P. Fazan, EDL, Vol.23, No.2, pp. 85-87 (2002).
3. C. Kuo, T.-J. King and C. Hu, TED, Vol.50, No.12, pp. 2408-2416 (2003).
4. T. Shino, T. Ohsawa, T. Higashi, K. Fujita, N. Kusunoki, Y. Minami, M. Morikado, H. Nakajima, K. Inoh, T. Hamamoto and A. Nitayama, TED, Vol.52, No.10, pp. 2220-2226 (2005).
5. S. Okhonin, M. Nagoga, P. Fazan, L. Mathew, B. –Y. Nguyen, H., -H. Chen and T. Stephens, ICMTD, pp. 63-65 (2005)
6. I. Ban, U. E. Avci, U. Shah, C. E. Barns, D. L. Kencke and P. Chang, IEDM Technical Digest. pp.1-4 (2006).
7. T. Hamamoto, Y. Minami, T. Shino, N. Kusunoki, H. Nakajima, M. Morikado, T. Yamada, K. Inoh, A. Sakamoto, T. Higashi, K. Fujita, K. Hatsuda, T. Ohsawa and A. Nitatama, TED, Vol.54, No.3, pp. 563-571 (2007).
8. U.E. Avci, I. Ban, D.L. Kencke and P.L/D. Chang, IEEE Int. SOI Conf., pp. 29-30 (2008).
9. T. Hamamoto and T. Ohsawa, ESSDERC, pp. 25-29 (2008).
10. S. Okhonin, M. Nagoga, E. Carman, R. Beffa and E. Faraoni, IEDM Technical Digest. pp.925-928 (2007).
11. K.-W. Song, H. Jeong, J.-W. Lee, S.I. Hong, N.-K. Tak, Y.-T. Kim, Y.L. Choi, H.S. Joo, S.H. Kim, H.J. Song, Y.C. Oh, W.-S Kim, Y.-T. Lee, K. Oh and C. Kim, IEDM Technical Digest. pp.797-800 (2008).
12. S. Okhonin, M. Nagoga, C.-W. Lee, J.-P. Colinge, A. Afzalian, R. Yan, N. D. Akhavan, W. Xiong, V. Sverdlov, S. Selberherr and C. Mazure, IEEE Int. SOI Conf., pp. 157-158 (2008).
13. T. Tanaka, E. Yoshida and T. Miyashita, IEDM Technical Digest. pp.919-922, (2004).
14. E. Yoshida and T. Tanaka, TED, Vol.53, No.4, pp. 692-697 (2006).
15. M. Bawedin, S. Cristoloveanu and D. Flandre, EDL, Vol.29, No.7, pp. 795-798 (2008).
16. M. Bawedin, S. Cristoloveanu, D. Flandre, C. Renaux and C. Crahay, Patent No. EP2009/050031

17. H. Furuhashi, T. Shino, T. Ohsawa, F. Matsuoka, T. Higashi, Y. Minami, H. Nakajima, K. Fujita, R. Fukuda, T. Hamamoto and A. Nitayama, IEEE Int. SOI Conf., pp. 33-34 (2008)
18. M. Blagojevic, M. Kayal, M. Pastre, L. Harik, M.J. Declercq, S. Okhonin and P. Fazan, J. Solid-State Circuits, Vol. 41, No. 6, pp. 1463-1470 (2006).
19. T. Ohsawa, K. Fujita, K. Hatsuda, T. Higashi, T. Shino, Y. Minami, H. Nakajima, M. Morikado, K. Inoh, T. Hamamoto, S. Watanabe, S. Fujii and T. Furuyama., J. Solid-State Circuits, Vol. 41, No. 1, pp. 135-145 (2006).
20. M. Bawedin, S. Cristoloveanu, J.G. Yun and D. Flandre, SSE, Vol.49, No.9, pp. 1547-1555 (2005).
21. A. Hubert, S. Cristoloveanu, M. Bawedin and T. Ernst, Accepted in IEEE ULIS Conference (2009).

ECS Transactions, 19 (4) 257-264 (2009)
10.1149/1.3117416 ©The Electrochemical Society

65nm Low Power (LP) SOI Technology on High Resistivity (HR) Substrate for WLAN and Mmwave SOCs

C. Raynaud[1,2], S. Haendler[1], G. Guegan[2], F. Gianesello[1],
B. Martineau[1], P. Touret[1,2], N.Planes[1]

[1] STMicroelectronics, 850 rue Jean Monnet, 38926 Crolles Cedex, France,
[2] CEA-LETI Minatec, 17 rue des Martyrs, 38054 Grenoble Cedex 9, France

> **Abstract.** We present a 65nm RF SOI CMOS technology, targeted as Low Power (LP) to serve mobile applications. The integration has been made on High Resistive (HR) back substrate 300mm SOI wafers from SOITEC to improve performances in high frequency range, compared to bulk [1, 2]. For the first time, low leakage SRAM (Isb< 10pA at 0.9V, 25°C, for 0.62µm^2 and 0.52µm^2 cells) are integrated on these HR wafers, and the paper reports a 30% power reduction in operation for a given maximum speed, compared to similar SRAM design on bulk. Furthermore, we have demonstrated a 21% measured power-delay product reduction compared to bulk, at 125°C, on loaded ring oscillators.

Introduction

There have been many publications on 65nm LP bulk technologies for mobile applications [3] but very few on 65nm LP SOI [4], as CMOS technologies with a thin SOI film (< 0.2µm) have been commercialized up to now with Partially Depleted devices for high performance applications, with a higher cost and complex process, and with a quite high leakage in standby mode, not well suited for mobile applications. Fully Depleted SOI is a good candidate for low leakage but it requires a high k/metal gate stack with a too complex and high cost process for Low Power products in 65nm or 45nm nodes.

Technology Description

We present a 65nm RF SOI CMOS technology, targeted as Low Power (LP) to serve mobile applications. The integration has been made on 300mm SOI wafers with a High Resistive (HR) back substrate (> 1kOhm-cm) to improve RF performances [1,2]. The initial SOI film thickness is 70nm on top of a 145nm thick oxide (see cross section in Figure 1). The technology is a double gate oxide process: 50A oxide for 2.5V IO and 18.5A oxide for 1.2V triple Vt (HVt, SVt, LVt) Core devices, with a 0.025A/cm^2 max gate current density. High Vt (HVt) floating body device characteristics are presented in Figure 2. The Source-Drain to Body junctions have been optimized to reduce the leakage (Figure 3) w/o the need of an asymmetric implant. The back-end process is similar to bulk one: 6 Copper metal levels and low-K dielectrics.

257

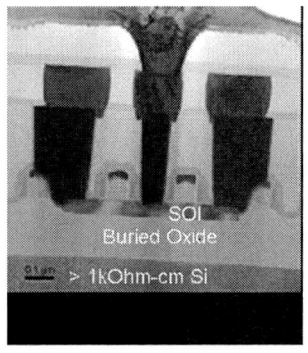

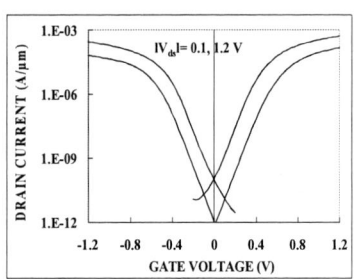

Figure 2. log Id(Vg) @ Vds = 0.1V, 1.2V, L = 60nm, W = 1μm, floating body HVt

Figure 1. TEM SRAM cross-section

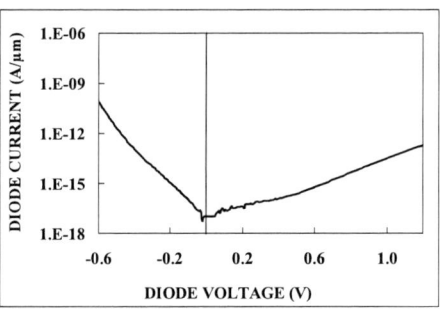

Figure 3. Source-drain to body diode I-V in SOI NFETs

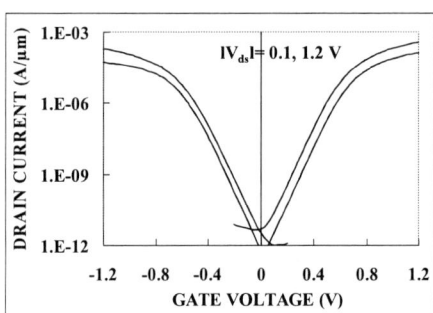

Figure 4. log Id(Vg) @ Vds = 0.1V, 1.2V, L = 60nm, W = 1μm, body contacted HVt

Figure 4 shows that the leakage can be further reduced with a Body Contact (BC) to get low leakage HVt devices. The body contact is made by a specific layout, but w/o any added mask (Figure 5).

Figure 5. Body Contacted NMOS top layout and cross-section

Figures 6 and 7 show Ioff and Ion at 1.2V for NMOS and PMOS respectively. Body Contacted (BC) devices can be used to switch power off in Core Logic. By stacking body contacted (BC) and floating body (FB) devices in Core Logic, designers can take advantage of FB devices for speed and cut leakage with these BC devices, w/o any process extra cost.

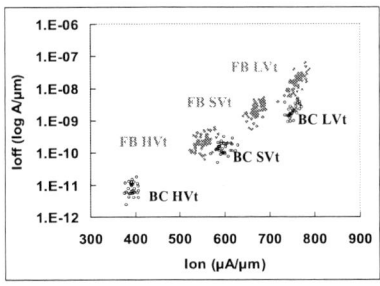
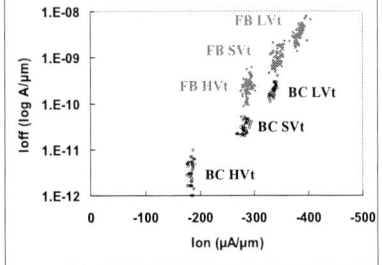

Figure 6. Ioff-Ion @ Vds = 1.2V, NMOS, L = 60nm

Figure 7. Ioff-Ion @ Vds = -1.2V, PMOS, L = 60nm

Power Delay Product Measurements

Figures 8 and 9 show a 19% and 21% measured power-delay product reduction compared to bulk, at 25°C and 125°C respectively, for loaded (FO3) ring oscillators. The improvement is due to a gate to body capacitive coupling effect, which increases the performance by a dynamic forward body bias, as the body is floating and follows the gate voltage.

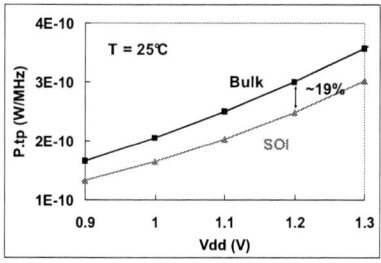 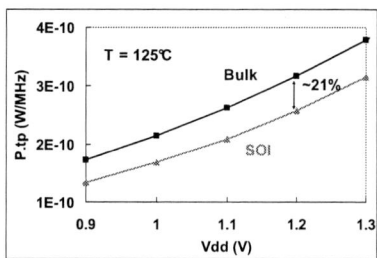

Figure 8. Power-delay product for loaded ring oscillators (FO3) @ 25°C

Figure 9. Power-delay product for loaded ring oscillators (FO3) @ 125°C

Ultra Low Power SRAM

The measured standby leakage (Isb) of the $0.62\mu m^2$ and $0.52\mu m^2$ SRAM cells, at 25°C and 125°C, is reported in Figures 10, 11 and Figures 12, 13 respectively.

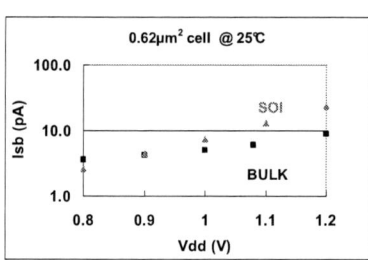 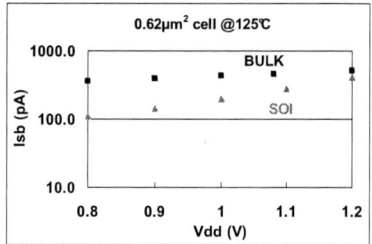

Figure 10. Isb for the $0.62\mu m^2$ bitcell @ 25°C

Figure 11. Isb for the $0.62\mu m^2$ bitcell @ 125°C

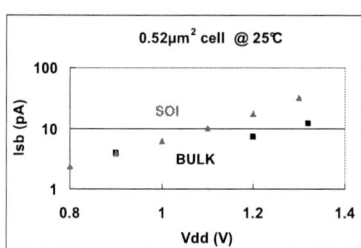 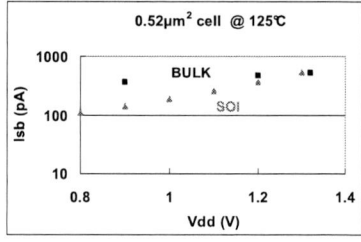

Figure 12. Isb for the $0.52\mu m^2$ bitcell @ 25°C

Figure 13. Isb for the $0.52\mu m^2$ bitcell @ 125°C

Due to the floating body effect, the leakage dependence on the supply voltage is different in case of SOI: it is slightly higher than bulk at 1.2V, 25°C, but similar to bulk and lower than 10pA/μm at 0.9V, 25°C. At 125°C, the leakage is lower in case of SOI, whatever the supply voltage is (Figures 11, 13). This is due to the reduced impact ionization at high temperature and also to the reduced junction leakage in case of SOI (no vertical SD/ bulk junction).

The SRAM Static Noise Margin is similar to bulk at low supply voltage: ~150mV at 0.9V, 25°C (Figure 14).

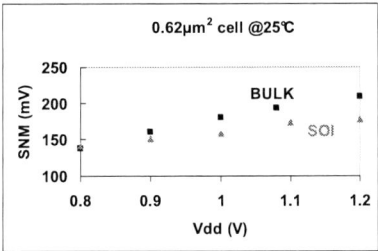

Figure 14. Static Noise Margin for the $0.62\mu m^2$ SRAM bitcell @ 25°C

The dynamic current has been measured on a 128K SRAM cut as a function of frequency (I = CV*f) and CV (A/MHz) is compared (Figure 15) to similar SRAM on bulk wafers. It is 14% lower in case of SOI due to a reduced equivalent capacitance by the dynamic forward floating body bias, as for the ring oscillators. The maximum access frequency f_{max} of SRAM is 40% higher in case of SOI at 1.2V (Figure 16). As the same f_{max} than bulk is reached at a lower supply voltage (~1.04V instead of 1.2V), ~30% power consumption (CV*V in W/MHz) is saved with SOI, for a given targeted speed. The capability to reach same yield as an industrial 65nm LP bulk technology has also been demonstrated on the same test vehicle, which gives clear indications of the easy manufacturability of this process.

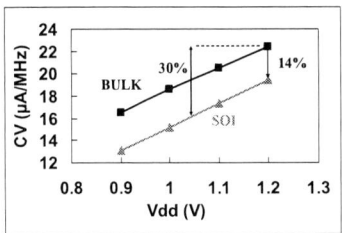

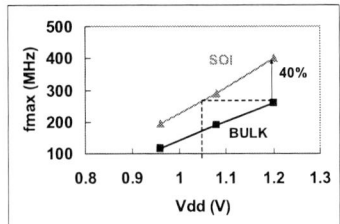

Figure 15. CV measured on HVt 128K SRAM @ 25°C (with 0.62μm2 cells)

Figure 16. Maximum access frequency measured on HVt 128K SRAM @ 25°C (with $0.62\mu m^2$ cells)

High f_T, f_{max} and Low NF_{min} floating body devices

For the LNA, high f_T, f_{max} and low minimum Noise Figure devices are required. For same gate length and same trans-conductance than bulk, SOI and bulk active devices have same f_T, f_{max} and fit the ITRS roadmap (see Figures 17, 18). Measured f_T and f_{max} (Figure 19) on a 60nm gate length FB LVt NMOS are 160GHz and 200GHz respectively. Figure 20 shows the measured NF_{min} vs. frequency. As only 0.2dB is measured @ 6GHz, these devices are well suited for LNA circuits [2].

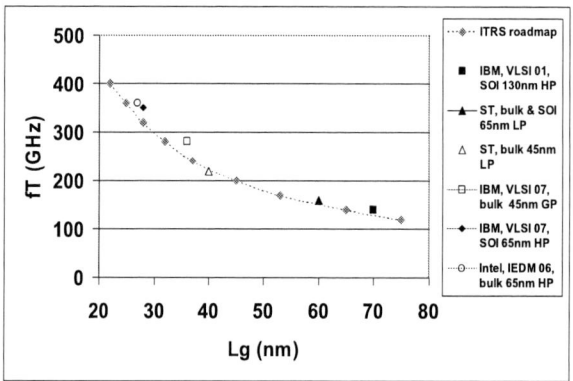

Figure 17. f_T as a function of gate length

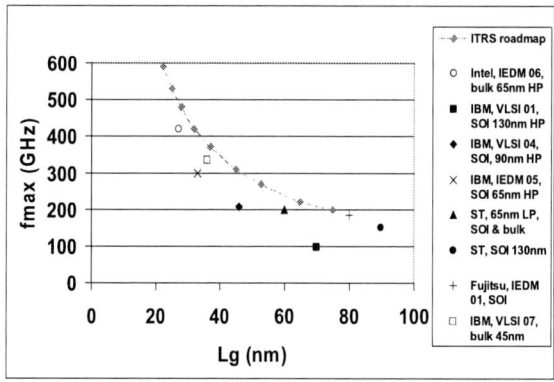

Figure 18. f_{max} as a function of gate length

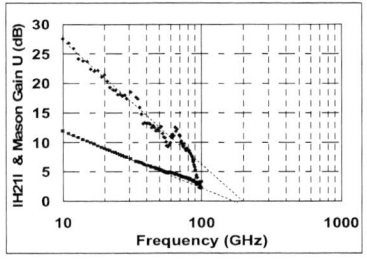
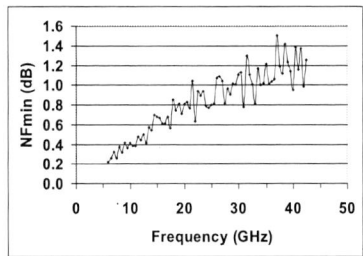

Figure 19. H21 and Mason Gain vs frequency L = 60nm, W = 64μm, floating body LVt NMOS

Figure 20. NFmin @ Vds =1.2V, L = 60nm, W = 100μm, Ids =123μA/μm, floating body LVt NMOS

Body Contacted devices can be used for VCO circuits thanks to a lower frequency noise (Figure 21).

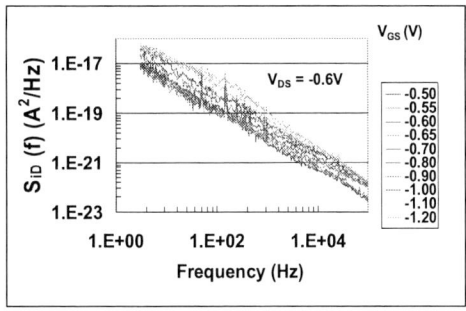

Figure 21. Low Frequency Noise vs frequency for LVt Body Contacted PMOS, L= 0.5μm, W=5μm

High Q inductors

Figure 22 shows a measured 30% higher Q factor for a 0.82nH inductor, with larger bandwidth, compared to bulk, thanks to the high resistive SOI wafers. This is obtained with a standard back-end process, but with a specific layout. Such high Q factor is not feasible in a 65nm LP bulk technology without a thick Cu special back-end process, which adds cost and complexity.

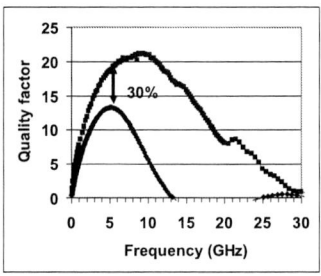

Figure 22. Quality factor of a 0.82nH inductor on HR 300mm SOI wafers, compared to bulk; high resistivity can't be used in bulk, due to the latch-up issue.

Conclusion

For the first time, low leakage and ultra low power SRAM (Isb < 10pA/bitcell at 0.9V and 30% power saving) have been measured on 300mm thin SOI wafers with a High Resistive back substrate. This SOI material is similar to a III-V substrate in high frequency range (reduced losses and improved isolation) but is better suited for VLSI integration with a lower cost. Therefore, this 65nm LP SOI process is a good candidate for a co-integration of low power analog-RF [1, 2, 5] and digital circuits.

Acknowledgments

The authors would like to thank M. Guillermet, JP. Oddou, F. Di Zanni, F. Leverd, R. Beneyton, L. Pinzelli, C. Laviron, A. Torres, JD Chapon, C. Perrot, D. Neira, M. Fournier, O. Gonnard for their help in the 65nm SOI process development, Y. Espinoux, T. Valentin, for their help in the SRAM characterization and SRAM yield analysis, A. Lachater for LFN characterization, J. Pretet and Patrick Scheer for SOI devices modeling and optimization.

References

[1] F. Gianesello, S. Montusclat, B. Martineau, D. Gloria, C. Raynaud, S. Boret, G. Dambrine, S. Lepilliet, R. Pilard , " 65 nm HR SOI CMOS Technology : emergency of Millimeter-Wave SoC ", pages 555 – 558, *IEEE RFIC* (June 2007)

[2] B. Martineau, A. Cathelin, F. Danneville, A. Kaiser, G. Dambrine, S. Lepilliet, F. Gianesello, D. Belot, "80 GHz Low Noise Amplifiers in 65 nm CMOS SOI", pages 348-351, *ESSCIRC* (2007)

[3] Y. Wang, H. Ahn, U. Bhattacharya, T. Coan, F. Hamzaglu, W. Hafez, C.H. Jan, P. Kolar, S. Kulkarni, J. Lin, Y. Ng, I. Post, L. Wei, Y. Zhang, K. Zhang, M. Bohr, "A 1.1GHz 12μA/Mb-Leakage SRAM Design in 65nm Ultra-Low-Power CMOS with Integrated Leakage Reduction for Mobile Applications", pages 324-325, *ISSCC* (2007)

[4] J.Cai, A. Majumdar, D. Dobuzinsky, T H. Ning, S. J. Koester, and W. E Haensch, "Ultra-Low Leakage Silicon-on-Insulator Technology for 65nm Node and Beyond", pages 908-910, *IEDM* (2007)

[5] O.Bon, O. Gonnard, L. Boissonnet, F. Dieudonné, S. Haendler, C. Raynaud and F. Morancho, "RF Power NLDMOS Technology Transfer Strategy from the 130nm to the 65nm node on thin SOI", pages 61-62, *IEEE International SOI Conference* (October 2007)

ECS Transactions, 19 (4) 265-270 (2009)
10.1149/1.3117417 ©The Electrochemical Society

Performance of Common-Source, Cascode and Wilson Current Mirrors Implemented with Graded-Channel SOI nMOSFETs in a Wide Temperature Range

M. de Souza[a*], D. Flandre[b] and M. A. Pavanello[a,c]

[a] Department of Electrical Engineering, Centro Universitário da FEI,
São Bernardo do Campo, Brazil
*e-mail: michelly@fei.edu.br
[b] Microelectronics Laboratory, Université catholique de Louvain, Louvain-la-Neuve, Belgium
[c] LSI/PSI/USP, University of São Paulo, São Paulo, Brazil

This work presents an experimental comparative analysis of the behavior of current mirrors implemented with standard uniformly doped and Graded-Channel (GC) SOI nMOSFETs as a function of the temperature. Three different current mirror architectures were used, Common-source, Wilson and Cascode. The experimental results show that the use of Graded-Channel transistors promotes not only the increase of the output swing, but also the increase of the output resistance in all evaluated architectures, in comparison to the standard uniformly doped counterpart. Despite some degradation observed with the temperature reduction, current mirrors with GC transistors still present better performance than those implemented with standard SOI transistors.

Introduction

Current mirrors (CMs) are basic transistor stages of great importance for analog systems (1) and are typically used to actively bias or load analog circuit branches. In this building block, the input current (I_{IN}) is mirrored to the output (I_{OUT}) for any value of voltage applied to the drain of the output transistor ($V_{D,OUT}$). Ideally, current mirrors present mirroring precision independent of the temperature (T) and the output voltage ($V_{D,OUT}$), high output resistance (R_{OUT}), and low saturation voltage (V_{SAT}) (1). The simplest implementation of a current mirror is the Common-source, represented in Figure 1(A). Despite its simplicity, Common-source current mirrors present low output resistance. This parameter can be increased by using different circuit architectures, such as the Wilson and Cascode ones, which include more transistors, as shown in Figure 1(B) and (C), respectively.

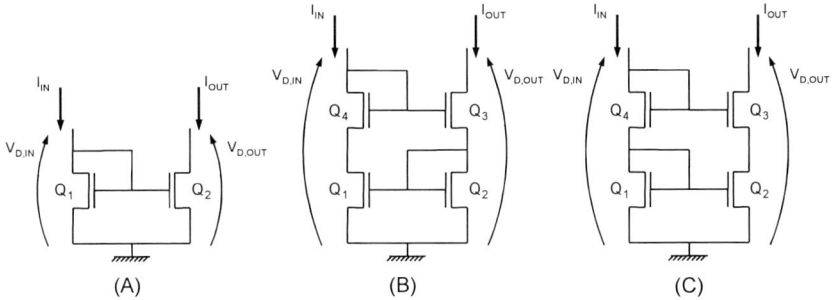

Figure 1 – Common-source (A), Wilson (B) and Cascode (C) current mirrors architectures.

265

Additionally to the use of different architectures, previous works reported that the use of Graded-Channel (GC) SOI nMOSFET (2) can further improve the characteristics of CMs operating at room temperature (3, 4). The Graded-Channel (GC) SOI nMOSFET is an asymmetric channel device that efficiently improves the SOI analog characteristics (2). In a GC transistor, through a simple mask arrangement in the standard SOI processing, a region with length L_{LD} near the drain is kept with the natural wafer doping concentration during the ion implantation performed to adjust the threshold voltage, V_T. This lightly doped region acts as an extension of the drain below the gate, reducing the effective channel length ($L_{eff} \cong L - L_{LD}$, L being the mask gate length), promoting the increase of the transconductance and drain current (2, 3). Furthermore, the presence of the undoped region near the drain substantially diminishes the electric field in this region, reducing the impact ionization, causing the breakdown voltage increase (2, 3).

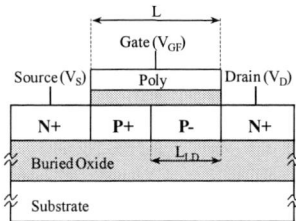

Figure 2 – Cross-section of a Graded-Channel SOI nMOSFET.

The low temperature operation of MOS transistors constitutes an efficient way to improve devices characteristics without scaling down the devices dimensions. The combination of increased carrier mobility, improved saturation velocity and reduced subthreshold slope and leakage current, provided by the temperature roll-off, results in better device performance in comparison to room temperature operation (5). It can be particularly interesting for analog circuits, in which, in some cases, devices cannot be scaled down to the minimum dimensions, to avoid undesirable effects, such as worsening of matching properties (6) and noise (7).

This work presents, for the first time, the evaluation of different current mirror architectures implemented with GC SOI nMOSFETs in the temperature range from 300K down to 150 K and compare it to the same circuits implemented with standard uniformly doped SOI transistors. The mirroring precision, output resistance, saturation voltage and breakdown voltage are addressed.

Results and Discussion

Current mirrors implemented with uniformly doped and Graded-Channel SOI nMOSFETs were fabricated starting from a SOI wafer with doping concentration of 10^{15} cm^{-3} and buried oxide thickness of 390 nm, with a 31 nm-thick gate oxide in a silicon layer with final thickness of 80 nm (3). The p-type threshold voltage ion implantation increases the body concentration level to about 6×10^{16} cm^{-3}. CM transistors have L= 2μm, channel width (W) of 20 μm for the Cascode and Wilson CMs and 18 μm for the Common-source ones. In the case of GC CMs, different L_{LD}/L ratios were fabricated and measured. The effective L_{LD}/L ratio has been experimentally derived from drain current

(I_{DS}) versus drain voltage (V_{DS}) curves with $V_{GT}=V_{GF}-V_T=200mV$ (V_{GF} being the applied gate voltage) (8) of standard and GC SOI transistors present in the same chips than the current mirrors.

Current mirrors were measured at fixed output voltage ($V_{D,OUT}=1.5V$) and sweeping the input voltage, $V_{D,IN}$. From the obtained curves, the mirroring precision (I_{OUT}/I_{IN}) was extracted as a function of the normalized input drain current ($I_{IN}/(W/L_{eff})$) for CMs with different transistors and architectures, and is presented in Figure 3, at T=150 K and 300 K. As observed in these curves, the mirroring precision is better (closer to unity) in CMs implemented with GC transistors for all architectures and in all operational regimes, due to the larger value of the Early voltage in GC SOI MOSFET devices (4). Also it is weakly temperature-dependent in strong inversion. As devices move to weak inversion, CMs mirroring precision departs from the unity due to the threshold voltage mismatching between devices, which is dominant in this inversion regime (1).

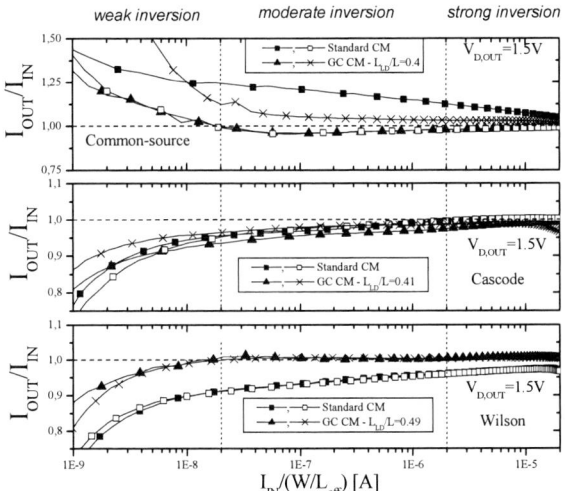

Figure 3 – Mirroring precision extracted at $V_{D,OUT}=1.5V$, at 150K (closed symbols) and 300K (open symbols).

The studied current mirrors were also measured at constant input current and sweeping the output voltage. Figure 4 presents the output current as a function of the output voltage for the studied current mirrors with a fixed bias current of 1µA, at 150 K and 300 K. From these curves one can see that, for a given temperature, the use of GC SOI promotes the reduction of the saturation voltage (V_{SAT}) and increase of the breakdown voltage (BV_{CM}). On the other hand, the temperature reduction causes the opposite effect, i.e. the increase of the saturation voltage and reduction of the breakdown voltage, both for current mirrors implemented with standard uniformly doped and graded-channel transistors, independently of the architecture. This behavior can be clearly seen in Figure 5(A) that presents V_{SAT} and BV_{CM} values. These values were extracted by using the methodology presented in (9). This method uses $\left[\dfrac{d(1/g_D)}{dV_{D,OUT}} \times g_D \right]$ curves as a function of $V_{D,OUT}$, measured at constant

I_{IN} (in this case $I_{IN}=1\mu A$), with g_D being the output conductance. This curve presents a positive peak, which marks the onset of the saturation region and a negative one that defines the breakdown voltage. The distance between these two peaks gives the output swing ($V_{OS}=BV_{CM}-V_{SAT}$).

The presented results show the expected increase of V_{SAT} and BV_{CM} promoted by the Cascode and Wilson structures in comparison to Common-source (1), resulting in the enhancement of the output swing, as shown in Figure 5(B).

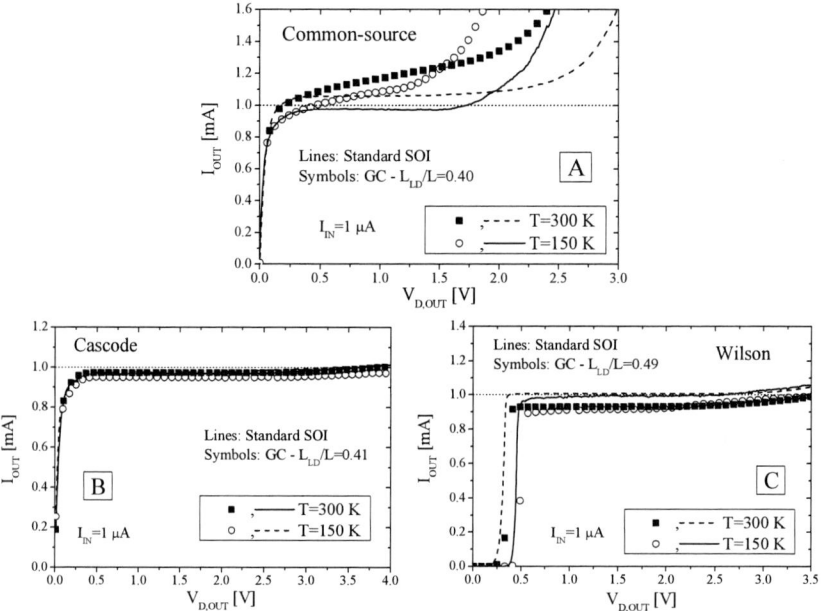

Figure 4 – Measured $I_{OUT} \times V_{D,OUT}$ curves, biased at $I_{IN}=1\mu A$ for Common-source (A), Cascode (B) and Wilson (C) current mirrors implemented with standard and GC SOI.

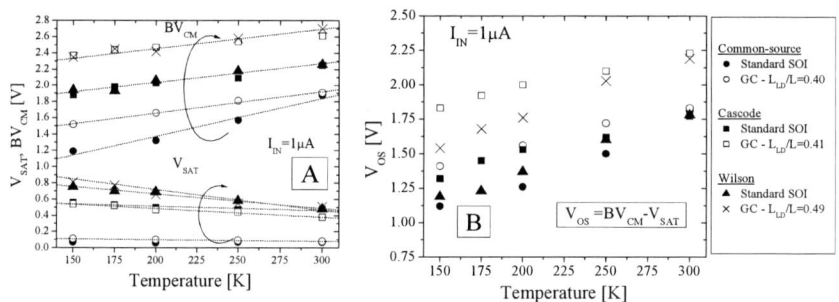

Figure 5 – Saturation voltage and breakdown voltage as a function of the temperature for different CMs with $I_{IN}=1\mu A$ (A) and output swing as a function of the temperature (B).

The results presented in Figure 5(B) allow for noting that, for a given CM architecture, the use of GC transistors promotes the increase of V_{OS}, due to the larger values of BV_{CM} and smaller V_{SAT} shown in Figure 5(A). This increase is larger in Cascode and Wilson CMs, and reaches up to 510 mV. As the temperature is reduced V_{SAT} increases and BV_{CM} reduces, resulting in V_{OS} reduction in all architectures, independent of the channel engineering. However, for all configurations, GC CMs present improved V_{OS} at all temperatures.

From the slope of the saturation region of I_{OUT} *vs.* $V_{D,OUT}$ curves measured at $I_{IN}=1\mu A$, the values of R_{OUT} were extracted and are presented in Figure 6, for the three CM architectures, implemented with standard and GC nMOSFETs, as a function of the temperature.

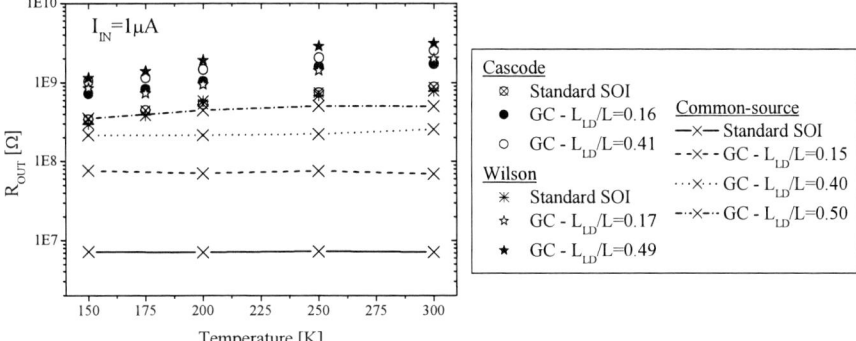

Figure 6 – Output resistance as a function of the temperature for different CMs with $I_{IN}=1\mu A$.

As presented in this figure, the output resistance is increased by the use of Wilson and Cascode architectures both for standard and GC transistors. While the output resistance of Common-source current mirrors is determined by the output resistance of the output transistor only (Q_2, in Figure 1), for the Cascode and Wilson architectures R_{OUT} is given by eqns. [1] and [2], respectively.

$$R_{out} = \left(g_{m3}r_{d2}\right)r_{d3} \qquad [1]$$

$$R_{out} = \left(g_{m1}r_{d1}\right)\left(\frac{g_{m3}}{g_{m2}}\right)\cdot r_{d3} \qquad [2]$$

where g_{m1}, g_{m2} and g_{m3} are the transconductance of transistors Q1, Q2 and Q3, respectively, r_{d1}, r_{d2} and r_{d3} are the output resistances of transistors Q1, Q2 and Q3, respectively (see Figure 1). From these equations one can note the increase of R_{OUT} promoted by Cascode and Wilson CMs with respect to Common-source ones.

As the output conductance g_D of GC transistors reduces with L_{LD}/L increase, R_{OUT} increases in Common-source CMs with L_{LD}/L, and this increase can be as large as 70 times in comparison to the standard one ($L_{LD}/L=0.50$ @ 300K). In this particular case, the output resistance can become as large as the R_{OUT} of Cascode and Wilson CMs with standard nMOSFETs, by simply changing the channel architecture. In the case of

Cascode and Wilson CMs, the maximum obtained increase of R_{OUT} provided by GC use is in the order of 3 and 4 times, respectively.

The temperature reduction promotes the output conductance degradation of the CM transistors (10), worsening the output resistance in all architectures. However, despite the decrease that may be seen, the R_{OUT} increase promoted by the use of GC transistors remains virtually constant with temperature in all architectures.

Conclusions

In this work it has been shown that the use of GC transistors allows for obtaining increased output swing of current mirrors in three different architectures while simultaneously increasing R_{OUT} in comparison to the standard counterpart. In Common-source configuration the improvement can reach 70 times while it is of 3 and 4 times for Cascode and Wilson configurations, respectively. Despite the temperature reduction tends to worsen these characteristics, GC CMs still present improved performance in comparison to the same architecture implemented with standard SOI transistors.

Acknowledgments

M. de Souza and M. A. Pavanello would like to acknowledge the Brazilian research-funding agencies FAPESP, CAPES and CNPq for the financial support for this work.

References

1. K. R. Laker and W. M. C. Sansen, *Design of Analog Integrated Circuits and Systems*, Mcgraw-Hill, Inc. (1994).
2. M. A. Pavanello, J. A. Martino, V. Dessard, D. Flandre, *Solid-State Electronics*, **44**, 1219 (2000).
3. M. A. Pavanello, J. A. Martino, D. Flandre, Solid-State Electronics, 46, 1215 (2002).
4. A. A. Santos, D. Flandre and M. A. Pavanello, in *22nd Symposium on Microelectronics Technology and Devices - SBMicro2007*, J. A. Martino, M. A. Pavanello, and C. Claeys, Editors, PV 9-1, p. 441, The Electrochemical Society Proceeding Series, Pennington, NJ (2007).
5. E. A. Gutierrez, M. J. Deen and C. L. Claeys, *Low Temperature Electronics: Physics, Devices, Circuits and Applications*, Academic Press, San Diego (2001).
6. P. R. Kinget, *IEEE Solid-State Circuits*, **40**, 1212 (2005).
7. C. Hu, G. P. Li, in: *53rd Annual Device Research Conference*, p. 16 (1995).
8. M. A. Pavanello, J. A. Martino, V. Dessard, D. Flandre, in: *Silicon-On-Insulator Technology and Devices 1999*, Pennington (EUA): the Electrochemical Society, p.293 (1999).
9. R. S. Ferreira, M. A. Pavanello, in *19th Symposium on Microelectronics Technology and Devices - SBMicro2004*, E. J. P. Santos, R. P. Ribas, J. Swart, Editors, PV 2004-3, p. 45, The Electrochemical Society, Inc., NJ (2004).
10. M. de Souza, M. A. Pavanello and D. Flandre, in *Proc. of the 7th WOLTE*, p. 57 (2006).

CHAPTER 12

CIRCUIT TECHNOLOGY II

ECS Transactions, 19 (4) 273-282 (2009)
10.1149/1.3117418 ©The Electrochemical Society

Highly Reliable SRAM Circuit Technology Using FinFETs

Shin-ichi O'uchi, Kazuhiko Endo, Takashi Matsukawa, Yongxun Liu, Yuki Ishikawa,
Junichi Tsukada, Hiromi Yamauchi, Kunihiro Sakamoto, and Meishoku Masahara

Nanoelectronics Research Institute, National Institute of AIST,
1-1-1 Umezono, Tsukuba, Ibaraki 305-8568, Japan

This paper introduces a flex-pass-gate SRAM (Flex-PG SRAM),
which is a FinFET-based SRAM to enhance both the read and
write margins independently. The flip-flop in the Flex-PG SRAM
consists of usual FinFETs, while the pass gates consist of double-
"independent"-gate FinFETs, i.e., "four-terminal"- (4T-) FinFETs.
By optimizing the current drivability in the 4T-FinFET pass gates
according to operational conditions of read and write, both the read
and write margins are enhanced. TCAD simulations revealed that
the Flex-PG SRAM increases the read margin by 71 mV without
the cell size penalty and decrease in the write margin, even when
its 6σ tolerance is ensured. Also, a fabricated device proved its
feasibility.

Introduction

A static random access memory (SRAM) is an important element as a working
memory device integrated together with logic circuits in system-on-a-chip's (SoCs). The
device scaling has increased the memory capacity of SRAM in the limited chip die size.
However, the device scaling will also cause a serious problem of variability in the device
performance and thus a severe degradation in the SRAM cell stability which affects the
yield of LSIs. Extending the operational stability of the SRAM is an effective means to
improve their chip yield. So far, the read and write margins (RM and WM) have been
optimized by arranging the gate width of MOSFETs. In the scaled CMOS technology,
this method will not be sufficient anymore, and some novel circuit approaches to enhance
both the RM and WM is helpful. Plural circuit schemes using dynamic voltage control
were proposed so far [1-5]. In addition, 7- and 8-transistor SRAM cells were proposed as
an alternative of a 6-transistor cell [6, 7].
Also, an approach based on advanced device technology is expected [8-10]. For
example, fin-type-FET (FinFET) technology [9, 10] is a promising candidate for the
scaled SRAM. A FinFET uses an intrinsic body. It greatly suppresses the device-
performance variability caused by the fluctuation in the number of dopant ions, while a
planar-bulk MOSFET requires a heavily doped channel, which causes serious process
variability.
In this paper, we introduce a FinFET-based 6-transistor SRAM which enhances both
the RM and WM by using variable-threshold-voltage pass-gates[11, 12], and its
performance with respect to the enhancement of operational stability is demonstrated
through TCAD simulation [13]. Further, the device fabrication results are presented to
validate its feasibility.

273

Flex-PG SRAM Cell and Array

Figure 1 shows the circuit topology of the proposed SRAM cell and schematics of devices comprising the cell. In this SRAM cell, the flip-flop is composed of 3-terminal fin-type field-effect transistors (3T-FinFETs) [9] while the pass gates (PGs) are composed of 4-terminal FinFETs (4T-FinFETs) [14, 15]. The 4T-FinFET has four terminals of the first gate, second gate, source, and drain. This gives us a function of variable threshold voltage controlled by the second gate voltage, V_{G2}. Using this flexibility in the pass gates, the RM and WM are optimized in the proposing SRAM cell. During the read operation, in general, a large current drivability ratio of the driver to the pass gate (MN1/MN3 and MN2/MN4) is preferred to reduce read disturbance and enhance the RM. This is realized by raising the V_{th} of the pass gates. On the other hand, during the write operation, a strong pass gates are desired to easily flip the potential at the memory node and to enhance the WM. This requirement is also attained by lowering the V_{th} of the pass gates. These requirements for larger RM and WM contradict each other in the fixed-V_{th} technology. The V_{th} flexibility in the pass gate of the 4T-FinFET frees the SRAM cell from this contradiction. We named this SRAM a flex-pass-gate (Flex-PG) SRAM. Figure 2(a) and (b) show an example of simulated butterfly curves for read and write operations, and the dependency of the RM and WM on V_{G2} is summarized in Fig. 2(c). Here, RM is equivalent to the static noise margin (SNM) [16] in the read operation, and WM is valued by the counterpart of the read SNM in this figure. The effect of the Flex-PG appears as follows. By lowering V_{G2}, V_{th} of the pass gates is raised and the RM is enhanced as shown in Fig. 2(a). On the other hand, by raising V_{G2}, V_{th} is lowered and the WM is enhanced as shown in Fig. 2(b). It is clear that the trade-off relationship between the RM and WM is controlled by V_{G2}, as shown in Fig. 2(c).

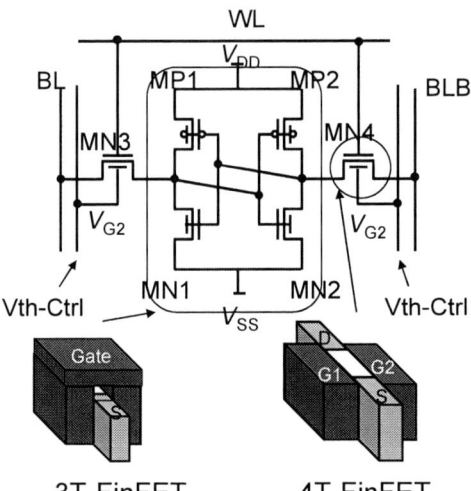

Figure 1. Circuit topology of the Flex-PG SRAM. The flip-flop consists of 3T-FinFETs and the pass gates consist of 4T-FinFETs.

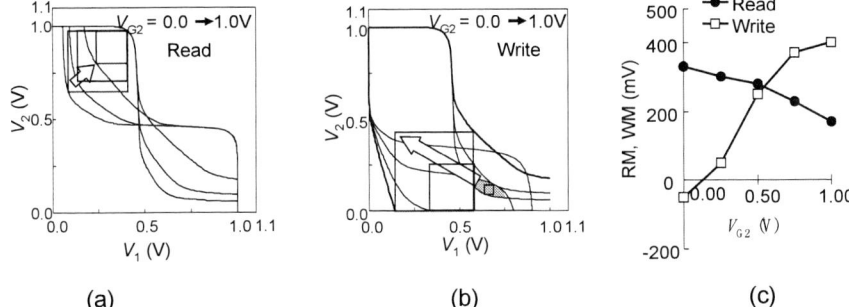

(a)　　　　　　　　　　(b)　　　　　　　　　　(c)

Figure 2. (a) Simulated butterfly curves for the read operation and (b) those for the write operation, where V_{G2} of pass gates is varied from 0.0 to 1.0 V. The RM and WM are summarized in the graph (c) as a function of V_{G2}. Simulation condition: gate length, $L_G = 20.0$ nm; fin thickness, $t_{si} = 9.0$ nm; first and second oxide thickness, $t_{ox1} = t_{ox2} = 1.2$ nm; gate work function, $\Phi_m = 4.71$ eV for both the p- and n-channel devices; power supply voltage, $V_{DD} = 1.0$ V.

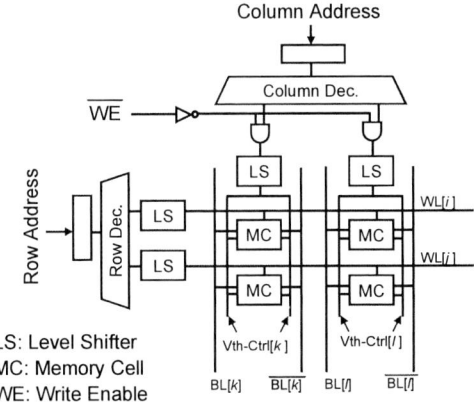

Figure 3. Block diagram of the Flex-PG SRAM array. The V_{th}-control lines, which are configured in parallel with the bit lines, provide the second gate voltage for the pass gates.

A topology to provide V_{G2}'s to all the pass gates in an array should be designed carefully so as to maintain the cell stability in both the accessed and unaccessed cells. In our scheme, the V_{th}-control lines to provide V_{G2} are configured in parallel with the bit lines (BLs and BLBs) to adopt column-by-column V_{th} control, as shown in Fig. 3. The level shifter (LS) in each column enables to provide $V_{G2,R}$ and $V_{G2,W}$, according to the write enable (WE) and the column address.

Here, in the unaccessed cells, which are located in the selected column for the write operation, the increase in the leakage current through the pass gates by the lowered V_{th}

should be managed for low-power operation. There is a contradiction between extending the WM and reducing the leakage currents in the BL sharing cells. The higher $V_{G2,W}$ for increasing the WM leads to an unacceptable leakage current in the datum-holding pass gate. There are two solutions to this problem. One is an introduction of the negative voltage to the low level of the signal on WLs, and the other is the optimization of the second gate oxide thickness. Applying a negative voltage to the low level of WL signal, $V_{WL,L}$, of the unselected row is effective for suppressing the pass-gate leakage current. In order to apply the negative $V_{WL,L}$, an LS is needed in each row, as shown in Fig. 3. When the negative gate voltage is applied to the fist gates of pass gates, the gate induced drain leakage (GIDL) may be increased. To compensate the GIDL, the gate-source/drain underlap structure is effective [17, 18]. On the other hand, a second gate oxide that is thicker than the first gate one helps to suppress the leakage current because the thicker second gate oxide improves the subthreshold slope (S-slope) and enhances on/off-current ratio of the 4T-FinFET. We have experimentally demonstrated the effect of the thicker second gate oxide in a previous study [19]. It will be possible to form the thicker second gate oxide only in the pass-gate devices by using this method together with a masking process.

The cell size of the Flex-PG SRAM for the half-pitch- (hp-) 32-nm technology is estimated to be $120F^2$, which is comparable to that of the planar-bulk SRAM cell having a β-ratio of the driver to pass gate of 2, for the same technology node. Here, F denotes half of the first-layer metal wiring pitch. Figure 4 compares the layouts of the Flex-PG and planar-bulk SRAM cells. In this estimation, three rules for the hp-32-nm technology were assumed based on the ITRS [20]: (i) the first- and intermediate-metal-wiring pitches are $2F$, (ii) contacts and vias are arranged in stagger configurations along with the $2F$-wiring pitch, and (iii) side-by-side contacts or vias are arranged in the pitch of $3F$. In addition, (iv) the length of the metal reservoir, which is the wire extension prolonged beyond a contact or a via to prevent the electromigration breakdown, is assumed to be $F/2$. According to this estimation, the Flex-PG SRAM improves both the RM and the WM independently without increasing the cell area size.

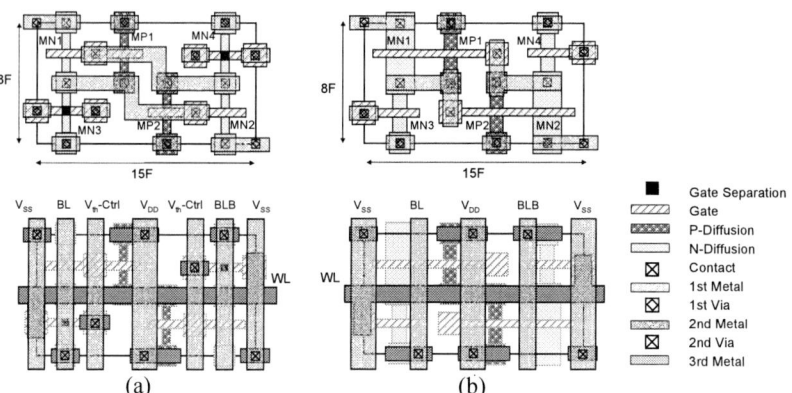

Figure 4. Layouts of (a) Flex-PG SRAM cell and (b) planar bulk SRAM cell having β-ratio of 2.

Variability Tolerance

We demonstrated the SNM enhancement of the Flex-PG SRAM for the hp-32-nm-technology low-standby-power (LSTP) applications by employing TCAD simulation. The FinFET model used in this study is summarized in Table I together with the corresponding planar-bulk MOSFET model. The FinFET model adopted the gate-source/drain underlap structure used to suppress short channel effects (SCEs) [17]. Together with the gate-source/drain underlap length of 10 nm, the source/drain-doping decay profile was assumed as a Gaussian profile with a standard deviation of $\sigma = 3.0$ nm. The model satisfies the on- and off-current requirements for the 32-nm LSTP. The planr-bulk MOSFET model was decided to realize the same off current as the FinFET model.

We performed a TCAD-based Monte Carlo simulation taking into account the process-induced random variation: (i) the gate length, L_G, (ii) the fin thickness, t_{si}, and (iii) the number of dopant ions in the channel, N. The variabilities in L_G and t_{si} are assumed to have identical Gaussian distributions of $3\sigma = 0.12$ $L_G = 2.4$ nm [20], and the variation in N was simulated by a Poisson distribution. These random variations cause the distributions of V_{th} in the 3T-FinFET and planar-bulk MOSFET, as shown in Fig. 5. It is found from Fig. 5 that the 3T-FinFET has narrower V_{th} distributions than that of the planar-bulk MOSFET. This is a contribution of the intrinsic bodies of FinFETs.

TABLE I. List of Device Model Parameters

Parameters	FinFET	Planar-Bulk MOSFET
Physical Gate Length, L_G (nm)	20	
Effective Oxide Thickness, t_{ox} (nm)	1.2	
Fin Thickness, t_{Si} (nm)	9.0	---
Gate Work Function for PMOS / NMOS, Φ_m (eV)	4.71 / 4.71	4.94 / 4.53
Channel Doping Conc., N_{chan} (cm^{-3})	1.0x10^{16}	3.0x10^{18}
S/D Doping Conc., N_{SD} (cm^{-3})	1.0x10^{20}	
Power Supply Voltage, V_{DD} (V)	1.0	

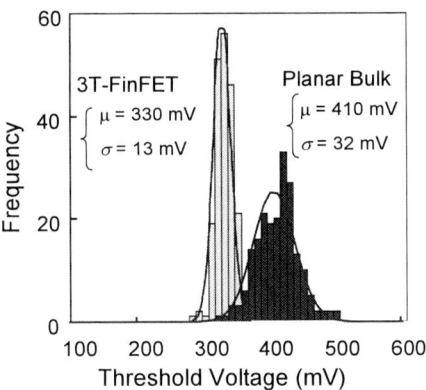

Figure 5. Comparison of V_{th} fluctuation between 3T-FinFET and planar-bulk MOSFETs. The number of samples is 200.

The 100-sample Monte Carlo simulation gave histograms of the RM and WM of the Flex-PG, 3T-FinFET and planar-bulk SRAMs, as shown in Fig. 6. Here, V_{DD} = 1.0 V. Both two references were calculated based on the assumption that they have the β-ratio of 2. As for the Flex-PG SRAM, $V_{G2,W}$ = 1.0 V and $V_{G2,R}$ = −1.0 V. The pass gates in the Flex-PG SRAM were optimized by choosing the second gate oxide thickness of them, t_{ox2} = $4t_{ox1}$, according to the procedure presented in [11]. It was found from Fig. 6 that the Flex-PG SRAM showed the enhancement of both the RM and WM, although these distributions are broader than that of the 3T-FinFET SRAM. On the quantitative comparison of the SNMs, an excess margin ensuring the tolerance for 6σ variation, $\mu -$ 6σ, gives a good criterion. Here, μ and σ are the average and standard deviation respectively. Based on this measure, the RM of the Flex-PG SRAM is larger than that of the 3T-FinFET SRAM and the WM of the Flex-PG SRAM is still comparable to that of the 3T-FinFET SRAM. The increase in the RM by the Flex-PG is estimated to 92 mV on average, and 71 mV taking account of the 6σ tolerance. Consequently, the Flex-PG SRAM enhances the RM without reducing the WM, even when the process-induced random variation is taken into account. Note that the planar-bulk SRAM has a broadest distribution in both the RM and the WM. In particular, there is no WM for ensuring 6σ tolerance. This disadvantage is brought about by the variability in the number of heavily doped ions necessary for the planar-bulk MOSFET (3.0×10^{18}/cm^3 in this simulation) to suppress the SCEs.

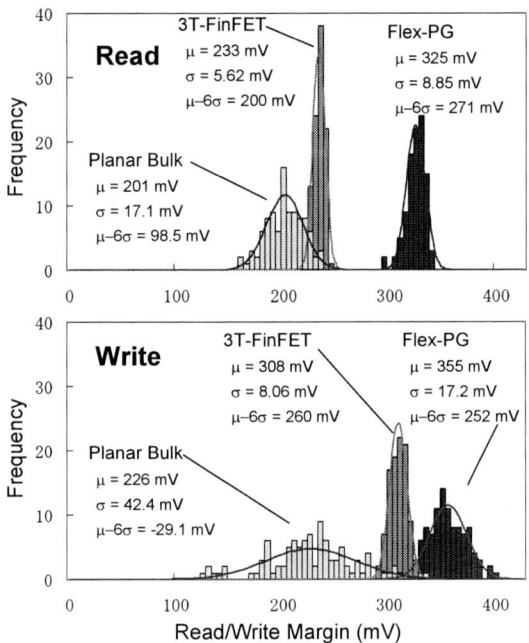

Figure 6. Histogram of the read and write margins given by a 100-sample TCAD-based Monte Carlo simulation.

Device Fabrication

We have also demonstrated the feasibility of the Flex-PG SRAM by experiment [21, 22]. We fabricated a Flex-PG SRAM cell to verify the controllability of the RM and WM. The fin structure was formed on a (100) oriented SOI wafer by a conventional reactive ion etching (RIE). Its channel surface has the (110) orientation. After the Si-fin etching, a 2.5-nm-thick gate oxide was formed at 850°C and then covered with TiN metal gate and n^+ poly-Si. The gate electrode was then patterned by the electron beam (EB) resist and the SiO_2 hard mask. To form the source/drain, a shallow ion-implantation (As, BF_2) was performed. Finally, the gate electrodes of the 4T-FinFET were separated by selective etching back process. The second gate oxide thickness was set to the same as that of the first gate.

The cross-sectional STEM images of the 3T- and 4T-FinFETs are shown in Fig. 7. The gate-separation of the 4T-FinFET was clearly confirmed and rectangular shaped 3T- and 4T-FinFETs with TiN metal gates were successfully co-integrated. As a result, the Flex-PG SRAM cell was successfully integrated as shown in Fig. 8.

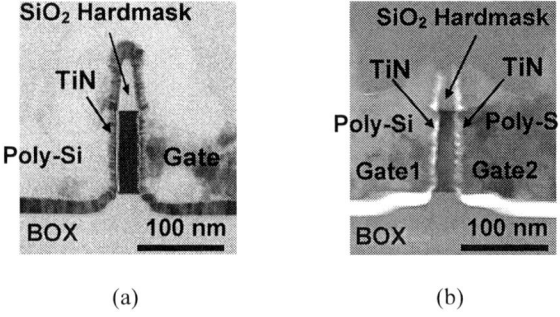

(a)   (b)

Figure 7. Cross-sectional STEM images of the (a) 3T- and (b) 4T-FinFETs.

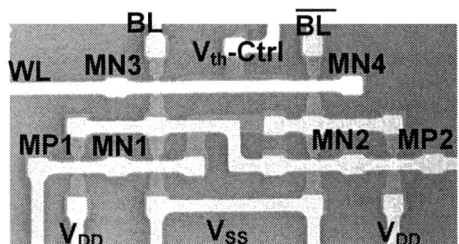

Figure 8. Top view of the fabricated Flex-PG SRAM cell.

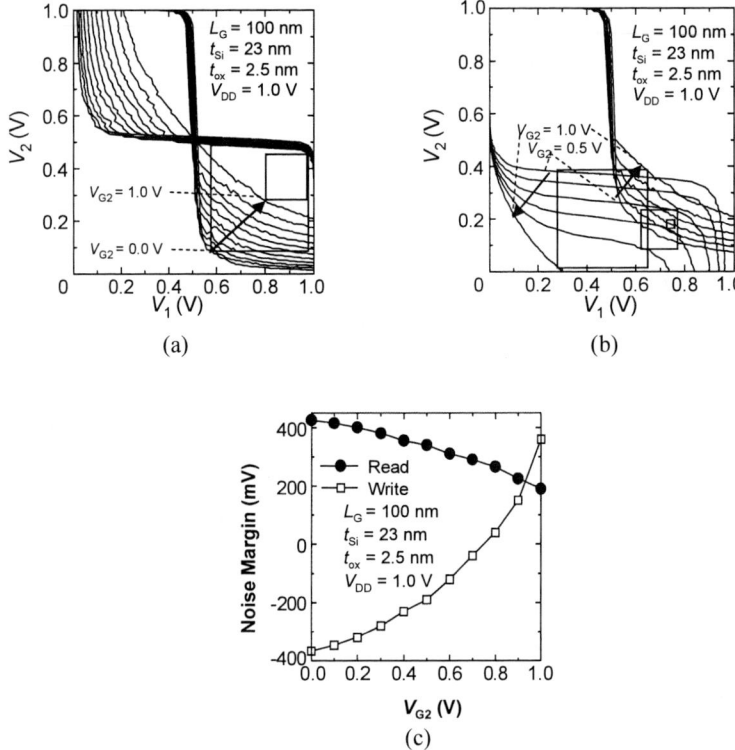

Figure 9. Experimental butterfly curves of the Flex-PG SRAM for (a) the RM evaluation and (b) the WM evaluation with various V_{G2} of the pass gates. The RM and WM are summarized in the graph (c) as a function of V_{G2}.

We measured the butterfly curves of the fabricated Flex-PG SRAM cell for RM and WM evaluation. These results are summarized in Fig. 9. As for the RM shown in Fig. 9(a), the RM decreased when V_{G2} was raised. Simultaneously, the WM was increased. This is summarized in Fig. 9(c) as a function of V_{G2}. It is clear that the RM and WM dependency on V_{G2} well coincides with the simulated results, although the device dimensions are different from the hp-32nm condition. Thus, it is apparent that the feasibility of the Flex-PG SRAM was successfully verified by this experiment.

Conclusion

In this paper, we introduced the Flex-PG SRAM. The Flex-PG SRAM enhances both the RM and WM using only 6 transistors in the comparable cell area with the conventional planar-bulk SRAM. The operational-stability enhancement is due to the variable-V_{th} pass gates composed of the 4T-FinFET, with the column-by-column V_{th} control. TCAD-based Monte Carlo simulation revealed that the Flex-PG SRAM designed for the hp-32-nm-technology LSTP, achieves a 92-mV increase in the RM over the usual 3T-FinFET SRAM, on average. Even when the 6σ tolerance for the process-induced random variation is ensured, a 71-mV increase in the RM is realized without decreasing the WM. We suggest that the Flex-PG SRAM is an attractive means to improve the SRAM yield by co-design of the device and circuit.

Acknowledgments

This work was partly supported by METI's Innovation Research Project on Nanoelectronics Materials and Structures.

References

1. A.J. Bhavnagarwala, S.V. Kosonocky, S.P. Kowalczyk, R.V. Joshi, Y.H. Chen, U. Srinivasan, and J.K. Wadhwa., in *Digest of Technical Papers of 2004 Symposium on VLSI Circuits*, p. 292 (2004).
2. M. Yamaoka N. Maeda, Y. Shinozaki, Y. Shimazaki, K. Nii, S. Shimada, K. Yanagisawa, and T. Kawahara, in *Digest of Technical Papers of 2005 IEEE International Symposium on Solid-State Circuits Conference (2005 ISSCC)*, p.480 (2005).
3. K. Zhang U. Bhattacharya, Z. Chen, F. Hamzaoglu, D. Murray, N. Vallepalli, Y. Wang, B. Zheng, and M. Bohr, in *Digest of Technical Papers of 2005 IEEE International Symposium on Solid-State Circuits Conference (2005 ISSCC)*, p. 474 (2005).
4. Y. Morita H. Fujiwara, H. Noguchi, K. Kawakami, J. Miyakoshi, S Mikami, K. Nii, H. Kawaguchi, and M. Yoshimoto, in *Digest of Technical Papers of 2006 Symposium on VLSI Circuits*, p. 13 (2006).
5. S. Ohbayashi M. Yabuuchi, K. Nii, Y. Tsukamoto, S. Imaoka, Y. Oda, M. Igarashi, M. Takeuchi, H. Kawashima, H. Makino, Y. Yamaguchi, K. Tsukamoto, M. Inuichi, K. Ishibashi, and H. Shinohara, in *Digest of Technical Papers of 2006 Symposium on VLSI Circuits*, p. 17 (2006).
6. L. Chang, D.M. Fried, J. Hergenrother, J.W. Sleight, R.H. Dennard, R.K. Montoye, L. Sekaric, S.J. McNab, A.W. Topol, C.D. Adams, K.W. Guarini, and W. Haensch, in *Digest of Technical Papers of 2005 Symposium on VLSI Technology*, p. 128 (2005).
7. K. Takeda, Y. Hagihara, Y. Aimoto, M. Nomura, Y. Nakazawa, T. Ishii, and H. Kobatake, IEEE J. Solid-State Circuits, **41**, 113 (2006).
8. M. Yamaoka, R. Tsuchiya, and T. Kawahara, in *Proceedings of 2005 Asian Solid-State Circuits Conference*, p. 109 (2005).
9. D. Hisamoto, W.C. Lee, J. Kedzierski, H. Takeuchi, K. Asano, C. Kuo, E. Anderson, T.J. King, J. Bokor, and C. Hu, IEEE Trans. Electron Devices, **47**, 2320 (2000).
10. H. Kawasaki, K. Okano, A. Kaneko, A. Yagishita, T. Izumida, T. Kanemura, K. Kasai, T. Ishida, T. Sasaki, Y. Takeyama, N. Aoki, N. Ohtsuka, K. Suguro, K.

Eguchi, Y. Tsunashima, S. Inaba, K. Ishimaru, and H. Ishiuchi, in *Digest of Technical Papers of 2006 Symposium on VLSI Technology*, p. 86 (2006).

11. S. O'uchi, M. Masahara, K. Sakamoto, K. Endo, Y.X. Liu, T. Matsukawa, T. Sekigawa, H. Koike, and E. Suzuki, in *Proceedings of IEEE 2007 Custom Integrated Circuits Conference (CICC 2007)*, p. 33 (2007).

12. S. O'uchi, T. Matsukawa, T. Nakagawa, K. Endo, Y.X. Liu, T. Sekigawa, J. Tsukada, Y. Ishikawa, H. Yamauchi, K. Ishii, E. Suzuki, H. Koike, K. Sakamoto, and M. Masahara, in *Technical Digest of 2008 International Electron Devices Meeting (IEDM 2008)*, p.709, (2008).

13. *ATLAS User's Manual*, Silvaco International.

14. Y. X. Liu, M. Masahara, K. Ishii, T. Tsutsumi, T. Sekigawa, H. Takashima, H. Yamauchi, and E. Suzuki, in *Technical Digest of 2003 IEEE International Electron Devices Meeting (IEDM 2003)*, p. 986 (2003).

15. L. Mathew, Y. Du, A. V.-Y. Thean, M. Sadd, A. Vandooren, C. Parker, T. Stephens, R. Mora, R. Rai, M. Zavala, D. Sing, S. Kalpat, J. Hughes, R. Shimer, S. Jallepalli, G. Workman, W. Zhang, J.G. Fossum, B.E. White, B.-Y. Nguyen, and J. Mogab, in *Proceedings of 2004 IEEE International SOI Conference*, p. 187 (2004).

16. E. Seevinck, F. J. List, and J. Lohstroh, IEEE J. Solid-State Circuits, **SC-22**, 748 (1987).

17. V.P. Trivedi, J.G. Fossum, and M.M. Chowdhury, IEEE Trans. Electron Devices, **52**, 56 (2005).

18. K. Tanaka, K. Takeuchi, and M. Hane, in *Technical Digest of 2005 IEEE International Electron Devices Meeting (IEDM 2005)*, p. 1001 (2005).

19. M. Masahara, R. Surdeanu, L Witters, G. Doornbos, V.H. Nguyen, G. Van den Bosch, C. Vrancken, K. Devriendt, F. Neuilly, E. Kunnen, M. Jurczak, and S. Biesemans, IEEE Electron Device Lett., **28**, 217 (2007).

20. International Technology Roadmap for Semiconductors 2005 Edition (ITRS 2005), http://public.itrs.net.

21. K. Endo, S. O'uchi, Y. Ishikawa, Y.X. Liu, T. Matsukawa, M. Masahara, K. Sakamoto, J. Tsukada, K. Ishii, H. Yamauchi, and E. Suzuki, in *Proceedings of 38th European Solid-State Device Research Conference*, p. 146 (2008).

22. K. Endo, S. O'uchi, Y. Ishikawa, Y.X. Liu, T. Matsukawa, K. Sakamoto, J. Tsukada, K. Ishii, H. Yamauchi, E. Suzuki, and M. Masahara, in *Technical Digest of 2008 International Electron Devices Meeting (IEDM2008)*, p. 857 (2008).

ECS Transactions, 19 (4) 283-288 (2009)
10.1149/1.3117419 ©The Electrochemical Society

Optimizing Spacer–to–Straggle Ratio in Underlap Double Gate MOSFETs for Low Voltage Analog and Digital Circuits

A. Kranti, Rashmi, and G.A. Armstrong

Semiconductors and Nanotechnology Group,
School of Electronics, Electrical Engineering and Computer Science,
Queen's University Belfast, Belfast BT9 5AH, Northern Ireland, UK

In this work, we report on the significance of non–overlap channel architecture in nanoscale double gate (DG) FETs to improve performance of low–voltage analog and digital circuits. It is shown that the low–voltage operation of a cascoded Operational Transconductance Amplifier (OTA) and 6-T SRAM cell can be considerably enhanced using underlap gate–source/drain design. The present work provides new opportunities for realizing future low–voltage analog and digital circuits with underlap DG FETs.

Introduction

Silicon–on–Insulator (SOI) based low–voltage design has been a topic of considerable interest because of its suitability for highly integrated circuits for logic and analog/rf applications. Double Gate (DG) SOI MOSFETs (fig. 1(a)) are well suited for low–voltage operation due to the inherent suppression of short channel effects (SCEs), reduced drain–induced barrier lowering, excellent scalability and 'volume–inversion' (VI) phenomena [1-2]. For low–voltage analog operation, weak inversion (W.I.) region offers relatively high open loop voltage gain along with lowest power dissipation and lower harmonic distortion [3]. The drawbacks of operating in W.I. – SCEs and speed degradation, can be effectively controlled by using DG FETs and underlap channel design (fig. 1(b)) [4-5]. Undoped DG FETs, free from random dopant fluctuations, and independent gate control [6], are considered very promising in minimizing variability and leakage in SRAM cells. The leakage current in DG FETs can be further reduced by underlap channel design [7-8]. Variability studies on DG based SRAMs have shown much lower statistical variation than bulk cells [9]. However, the improvement of SRAM base performance at low voltages is still a crucial issue.

In this work, we report 0.6 V operation of low–voltage analog/digital circuits – Operational Transconductance Amplifier (OTA) (fig. 2(a)) and 6T SRAM cell (fig. 2(b)) based on underlap DG SOI MOSFETs. For analog applications, underlap design [5] has been applied to evaluate open loop intrinsic voltage gain and unity gain frequency of the OTA whereas underlap parameters have been optimized for enhanced read stability, write-ability and low leakage performance of SRAM cell.

Simulations

OTA [10] and SRAM [6] circuits based on undoped DG FETs were simulated using 2D Mixed–mode module in ATLAS with Lombardi mobility model [11]. Mixed–mode simulation avoids the requirement for detailed SPICE models and is a very useful predictive tool for nanoscale technologies. Device parameters used for the study are given in Table 1. As analog/RF technology has trailed high performance logic CMOS by two generations, a longer gate length was used for OTA design. Gaussian source/drain (S/D) doping profile was modeled [4] as $N_{SD}(x) = (N_{SD})_{peak} \exp(-x^2/\sigma^2)$, where $(N_{SD})_{peak}$ is the peak S/D doping, σ is the lateral straggle that defines S/D doping roll-off [5] in the channel direction as $\sigma = \sqrt{2sd/\ln(10)}$, s is the spacer width and d is the S/D doping gradient. SRAM metrics evaluated are Retention Noise Margin (*RNM*), Static Noise Margin (*SNM*), Write–ability current (I_{wr}) and Leakage current (I_{leak}). As S/D doping gradient should be scaled with gate length [1], a nominal target value of $d = 3$ nm/dec is selected for SRAM design whereas a slightly gradual profile with $d = 5$ nm/dec is used for the OTA. The results are analyzed in terms of spacer-to-straggle ratio [5], as it is the most significant parameter for the design of underlap devices [5]. In this work, s/σ values have been generated by increasing the spacer widths at a constant d. While higher (lower) s/σ values imply longer (shorter) effective channel length (L_{eff}, in weak/moderate inversion) [5] and low (high) leakage, they also signify wider spacer regions which result in series resistances for above threshold operation and reduces the current [5]. Underlap S/D profile with $s/\sigma < 2$ results in a gate–overlap design whereas $s/\sigma \sim 2$ result in L_{eff} equal to L_g. Underlap design with $s/\sigma \geq 2.1$ yields $L_{eff} > L_g$.

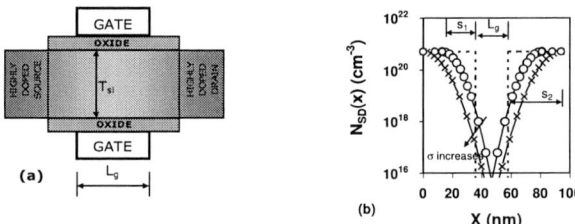

(a)

(b)

Figure 1. (a) Schematic of underlap DG SOI MOSFET. (b) Variation of source/drain (S/D) doping profile along the channel. Notations: o—o: $\sigma = 10$ nm ($s = s_1 \sim 23$ nm), ×—×: $\sigma = 12.5$ nm ($s = s_2 \sim 36$ nm) ——: Abrupt S/D regions.

Results and Discussion

Operational Transconductance Amplifier (OTA) Design

Fig. 3(a) shows the gain–bandwidth behavior of OTA circuit [10] designed with underlap and non–underlap (abrupt S/D junction) DG FETs. High values of open loop intrinsic voltage gain (A_{VO_OTA}) and unity gain frequency (f_{T_OTA}) of ~ 41 dB and 13 GHz respectively, can be achieved with non–underlap DG devices due to enhanced gate controllability and suppressed SCEs. These values can be further improved by using gate–underlap design in DG devices. As illustrated in fig. 3(b), an increase of 15 dB is obtained for A_{VO_OTA} whereas f_{T_OTA} improves by a factor of 3 in OTA designed with $s/\sigma = 3.5$. In a conventional non–underlap S/D design, it is not possible to attain a simultaneous improvement in A_{VO_OTA} and f_{T_OTA}, as an increase in A_{VO_OTA} requires a longer L_g (assuming a linear dependence of Early voltage on L_g), which would compromise f_{T_OTA} because of a reduction in g_m (g_m is approximately inversely proportional to L_g). OTA designed with a larger s/σ value (> 3.5) shows degraded performance due to the additional series resistance associated with very wide spacers.

The optimal spacer–to–straggle value $((s/\sigma)_{\text{optimal}})$ for achieving significant improvement in both $A_{\text{VO\_OTA}}$ and $f_{\text{T\_OTA}}$ is ~ 3.3.

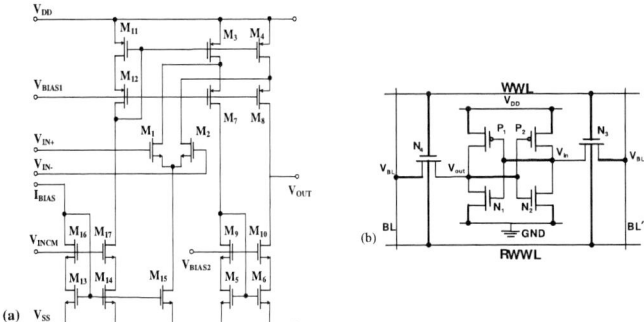

Figure 2. (a) Schematic diagram of cascoded OTA [10]. Parameters: $V_{\text{DD}} = 0.6V$, $V_{\text{SS}} = -0.6V$, M_1-M_2: $W/L_g = 176$, M_3-M_4: $W/L_g = 60$, M_5-M_6: $W/L_g = 4.5$, M_7-M_8: $W/L_g = 120$, M_9-M_{10}: $W/L_g = 27$, M_{11}: $M_3/3$, M_{13}-M_{14}: $W/L_g = M_{15}/4$, M_{16}-M_{17}: $M_1/4$, M_{15}: $W/L_g = 35$. (b) DG FET based Two WordLine (2WL) 6–T SRAM cell [6]. WWL = V_{DD} in write mode and 0.3V in read mode, RWWL= V_{DD} in read/write modes. All n and p–type transistors were simulated with identical structural parameters.

Parameter		OTA	SRAM
Gate length (L_g)	(nm)	60	22
Silicon film thickness (T_{si})	(nm)	30	13
Gate oxide/high–κ thickness (T_{ox})	(nm)	1.3	3.5
Gate oxide/high–κ permittivity	(F/m)	$3.9\varepsilon_o$	$15\varepsilon_o$
Doping gradient (d)	(nm/dec)	5	3
Gate workfunction (Φ_m)	(eV)	4.72	
Spacer width (s)	(nm)	$13 - 52$	$8 - 25$
Lateral straggle (σ)	(nm)	$7.5 - 15$	$4.5 - 8$
Spacer–to–straggle ratio (s/σ)		$2 - 3.5$	$1.7 - 3$
Source/Drain resistance (R_{SD})	($\Omega\mu$m)	120	

Table 1: Values of device parameters analyzed in the work.

The gain of the OTA can be expressed [10] as

$$A_{\text{VO\_OTA}} \cong G_M R_{\text{OUT}} \qquad [1]$$
$$G_M \cong g_{m1} \left(g_{m7}/(g_{m7} + g_{ds1} + g_{ds3}) \right) \qquad [2]$$
$$R_{\text{OUT}} \cong (r_{ds8} (1 + g_{m8}/(g_{ds2} + g_{ds4}))) (r_{ds10} (1 + g_{m10}/g_{ds6})) \qquad [3]$$

where r_{ds} is the output resistance ($1/g_{ds}$), g_{ds} is the output conductance and g_m is the transconductance of individual devices. The improvement in $A_{\text{VO\_OTA}}$ and $f_{\text{T\_OTA}}$ can be related to the enhancement of important analog parameters (g_m and g_{ds}) of individual devices. Therefore, for both underlap and non–underlap devices, we evaluate analog metrics such as Early voltage (V_{EA}), transconductance (g_m) and gate capacitance (C_{gg}).

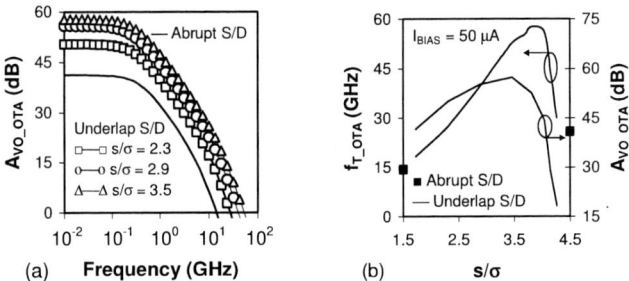

(a) Frequency (GHz) (b) s/σ

Figure 3. (a) Gain–bandwidth behavior of OTA at $I_{BIAS} = 50$ μA. (b) Variation of A_{VO_OTA} and f_{T_OTA} of OTA. Filled symbols (■) on the y–axis represent OTA metrics for DG devices designed with abrupt S/D junctions.

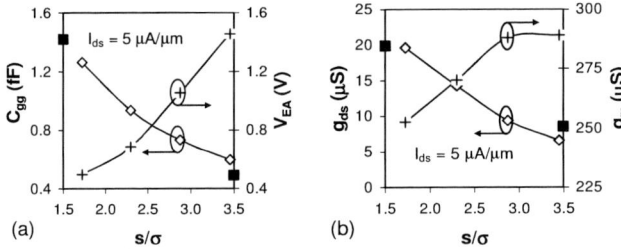

(a) s/σ (b) s/σ

Figure 4. (a) Dependence of total gate capacitance (C_{gg}) and Early voltage (V_{EA}) and (b) output conductance (g_{ds}) and transconductance (g_m) on s/σ ratio for underlap and non–underlap devices. Filled symbols (■) on the y–axis represent the values of DG devices designed with abrupt S/D junctions.

As underlap devices are designed to operate in weak/moderate inversion, a reduction in internal fringing capacitance results in substantial reduction (fig. 4(a)) by a factor of 2 in gate capacitance (C_{gg}). An increase in s/σ within the range 2.3–3.5 improves the gate control and increases g_m (for a fixed I_{ds}) (fig. 4(b)). Although g_m does not increase beyond $s/\sigma = 2.8$, its value is still higher than that achieved by abrupt S/D devices. The most significant aspect of increasing s/σ is the 3 times reduction of g_{ds} (fig. 4(b)) leading to the substantial improvement in V_{EA} (fig. 4(a)) at $s/\sigma = 3.5$ due to the reduction of peak electric field at the gate edge. This improvement in V_{EA} and g_m along with a reduction in C_{gg} leads to improved A_{VO_OTA} and f_{T_OTA} values as shown in fig. 3.

6T – SRAM Cell Design

As shown in fig. 5(a), underlap architecture is beneficial for low–voltage SRAM design as SNM and RNM increase with s/σ values. A design with $s/\sigma < 2$ ($L_{eff} < L_g$) results in SNM and RNM values lower than those of cells based on abrupt S/D junctions (shown by filled symbols on y-axes) due to a gate–overlap design instead of the desirable underlap architecture. A two fold reduction in I_{wr} with s/σ increasing from $2 - 3$, alongside two decade reduction in I_{leak} can be seen in fig. 5(b). Any increase in s/σ beyond 2.8 will result in severe write–ability degradation as I_{wr} reduces significantly. Therefore, s/σ between 2.0 and 2.8 is optimal for balancing the read/write requirements as it results in a maximum SNM and RNM of 145 mV and 245 mV respectively, and maximum I_{leak} of 1 nA along with minimum I_{wr} of 25 µA at $V_{DD} = 0.6$ V

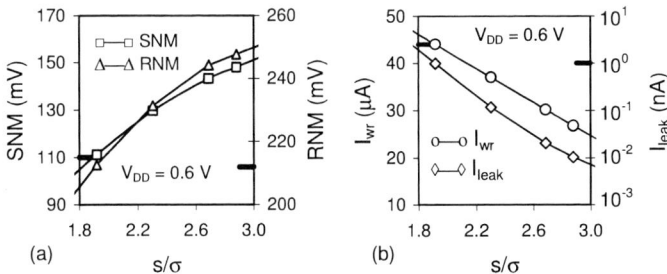

(a) s/σ   (b) s/σ

Figure 5. Dependence of (a) SNM and RNM (b) I_{wr} and I_{leak} on s/σ of 6T–SRAM cell. Lines (——) on the y–axis represent SRAM metrics for DG FETs with abrupt S/D regions.

The proposed SNM value ~ 130 mV at 0.6 V is comparable to those published in the literature (122 – 172 mV) for DG/Fin FETs [12-13] for a wide range of gate lengths (20 – 40 nm). SNM being the most cited SRAM performance metric, should always be compared along with I_{leak}, RNM and I_{wr} to evaluate the optimal design and minimize the read/write trade–offs. Unfortunately, data corresponding to I_{leak}, RNM and I_{wr} for DG/Fin FETs is not available in the literature for comparison. Nevertheless, a 6T SRAM design with SNM ~ $0.22V_{DD}$ along with ultra low leakage (I_{leak} ~ 0.12 nA) and write–ability current of 37 µA clearly illustrates the significance of underlap channel architecture for low–voltage SRAM design.

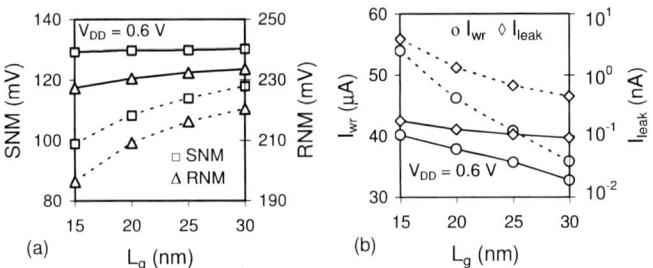

(a) L_g (nm)   (b) L_g (nm)

Figure 6. Variation of (a) SNM and RNM (b) I_{wr} and I_{leak} with L_g for SRAM cell based on abrupt and underlap S/D profiles at $s/\sigma = 2.3$. Notations: ——— Underlap S/D design – – – Abrupt S/D junctions.

As shown in figs. 6(a)-(b), values of SNM, RNM, I_{wr} and I_{leak} can be maintained relatively independent of L_g in range of 15–30 nm for a SRAM cell designed with $s/\sigma = 2.3$. Gate-underlap design offers increasing relative improvement in SNM, RNM and I_{leak} with L_g reduction, as compared to an abrupt S/D design, in which any advantage associated with ~ 50% increase in I_{wr} for two fold L_g reduction is nullified by a corresponding order of magnitude rise in I_{leak}, along with a strong degradation of *SNM* and *RNM*. Analysis suggests the validity of a well defined scaling relationship ($s/\sigma \sim 2.3$) as a simple yet effective parameter for the design of low–voltage SRAM cells.

Conclusions

The enormous potential of underlap architecture in nanoscale DG FETs for enhancing low–voltage analog and digital metrics has been presented in a circuit environment. The ability to control S/D extension region parameters provides an additional degree of freedom, alongside device and circuit parameters to improve OTA and SRAM metrics. Underlap DG FETs operated at low currents are particularly useful for analog circuits as 30% improvement in gain (in dB) and 3 times higher unity gain frequency can be achieved. At a spacer–to–straggle ratio of ~ 2.3, *SNM* of 130 mV and *RNM* of 230 mV can be achieved without compromising write–ability current ($\sim 30~\mu A$) or leakage current (~ 0.1 nA). For analog applications, a larger spacer–to–straggle (s/σ) ratio of 3.3 is appropriate for achieving enhanced performance metrics whereas for SRAM design, s/σ of 2.3 is more useful. In both cases, the significance of optimal device design for enhancing performance metrics in a circuit environment has been demonstrated.

Acknowledgments

This work was funded by Engineering and Physical Sciences Research Council, UK.

References

1. International Technology Roadmap for Semiconductors (www.itrs.net).
2. F. Balestra, S. Cristoloveanu, M. Benachir, J. Birni and T. Elewa, *IEEE Elect. Dev. Lett.*, **8** 410 (1987).
3. T. M. Hollis, D. J. Comer and D. T. Comer, *IEEE Trans. Circuits Syst. II, Exp. Briefs*, **52**, 545 (2005).
4. A. Kranti and G.A. Armstrong, *IEEE Elect. Dev. Lett.*, **28**, 139 (2007).
5. A. Kranti and G.A. Armstrong, *IEEE Trans. Elect. Dev.*, **54**, 3308 (2007).
6. O. Thomas, M. Reyboz and M. Belleville, *IEEE Int. Symp. Circuits and Systems*, 2778 (2007).
7. V.P. Trivedi and J.G. Fossum, *IEEE SOI Conference*, 192 (2004).
8. A. Kranti and G.A. Armstrong, *Semi. Sci. and Tech.*, **21**, 409 (2006).
9. A. Yagishita, *IEEE Int. Conf. on Integrated Circuit Design and Technology*, 1 (2007).
10. D.M. Binkley, *Int. Conf. Mixed Design of Integrated Circuits and Systems*, 47 (2007).
11. ATLAS Users Manual, SILVACO, 2006.
12. A. Nackaerts *et al.*, *IEDM Tech. Dig.*, 269 (2004).
13. H. Kawasaki et al., *VLSI Tech. Dig.*, 70 (2006).

ECS Transactions, 19 (4) 289-294 (2009)
10.1149/1.3117420 ©The Electrochemical Society

Harmonic Distortion Analysis of SOI Triple Gate FinFETs Applied to 2-MOS Balanced Structures

R. T. Doria[a], J. A. Martino[a], A. Cerdeira[b], M. A. Pavanello[a,c]

[a] LSI/PSI/USP, University of Sao Paulo, Sao Paulo, Brazil
e-mail: rdoria@lsi.usp.br

[b] Seccion de Eletrónica del Estado Sólido (SEES), CINVESTAV, Mexico

[c] Department of Electrical Engineering, Centro Universitário da FEI, São Bernardo do Campo, Brazil

> This work presents an evaluation of the non-linearities exhibited in 2-MOS resistive structures composed by triple gate FinFETs with several fin widths down to 30 nm. The harmonic distortion has been analysed in terms of its third order component (HD3) as a function of the gate voltage, the input amplitude voltage and the fin width. The linearity has also been analysed with respect to the on-resistance, which constitutes a key parameter in such circuits. Along the harmonic distortion evaluation, the non-linearity causes are pointed out. At lower gate voltages, wider devices present smaller HD3 with respect to the narrower ones, while the contrary occurs at higher gate voltages.

Circuit and Devices Characteristics

Several analog circuits such as continuous time filters require the use of resistors, which are responsible for a large die area and do not allow for on-chip tuning to compensate for process and temperature variations. As an alternative, long-channel transistors operating in the linear region can be used (1), since on-resistances (R_{ON}) of a few hundred kΩ are usually required. However, the extremely non-linear behavior exhibited in MOSFET characteristics hinder these devices to be widely applied. Thus, resistive balanced structures composed by at least two transistors (2-MOS) can be used aiming the non-linearity (or harmonic distortion) reduction (2). This work presents a study of the harmonic distortion of FinFETs applied to 2-MOS structures, varying the fin widths.

The 2-MOS structures are responsible for the suppression of the odd order harmonics exhibited in the output of single transistor circuits. As a result, the second order harmonic distortion (HD2), usually the preponderant non-linearity component, becomes less influent and the third harmonic distortion (HD3) represents the major non-linearity source. Figure1(A) shows the 2-MOS structure under analysis where Vo is the DC bias (set to 0 V), V_G is the gate bias (common to both transistors) and Va is the amplitude of a sinusoidal signal applied symmetrically to the drain of the devices (The input signal has been considered as DC voltage bias (V_o) associated to a sinusoidal-like signal of amplitude Va.sin(x) with $-\pi/2 < x < \pi/2$). Along this work Va has been varied up to 0.5 V, which implies in a peak-to-peak input amplitude of 1 V.

The triple gate SOI nMOS FinFETs analyzed in this work were fabricated on an active silicon layer with p-type doping concentration of 10^{15} cm^{-3} in which the threshold voltage is controlled by the gate material workfunction, as described in ref. (3). The measured

289

devices present effective gate oxide thickness equal to 2 nm (1 nm of SiO_2 and 2 nm of HfO_2), channel length (L) of 10 μm, fin height (H_{fin}) of 60 nm and several fin widths (W_{fin}) from 120 nm down to 30 nm (3). The channel width (W) is estimated as $W = 2H_{fin} + W_{fin}$ due to the presence of three gates. A schematic view of the studied devices is presented in Figure 1(B).

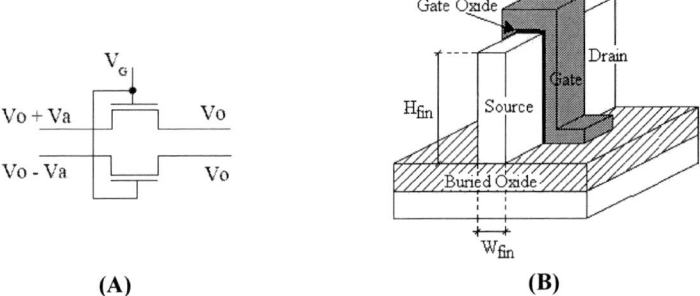

(A)   (B)

Figure 1. (A) 2-MOS balanced structure scheme and (B) schematic view of a triple gate FinFET device.

HD3 has been determined through the Integral Function Method (IFM) (4) since this technique allows the non-linearity extraction directly from the DC characteristics of the devices, without the need of an AC characterization as in Fourier based methods.

Harmonic Distortion Evaluation

Initially, the distortion was extracted as a function of the gate overdrive voltage (V_{GT}) as presented in Figure 2 with two different amplitudes Va = 0.1 V and 0.25 V, respectively. According to the curves, an improvement on HD3 is obtained as V_{GT} is increased for both analysed input amplitudes, moving the saturation voltage to higher values, except for the peaks region where a better linearity is attained. Similar behaviour with respect to V_{GT} is exhibited for all analysed amplitudes. However, for the smaller V_{GT} a lower harmonic distortion is attained in the right side of the linearity peaks.

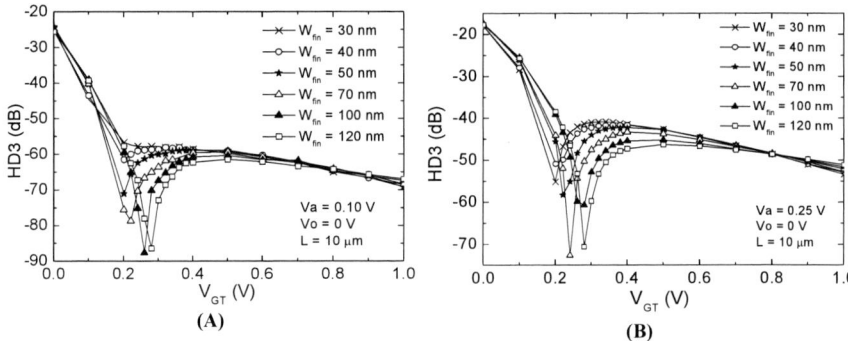

(A)   (B)

Figure 2. Extracted HD3 vs. V_{GT} curves for 2-MOS balanced structures with 10 μm-long FinFETs at (A) Va = 0.10 V and (B) Va = 0.25 V.

According to references (5,6), the non-linear behaviour presented by 2-MOS structures is resultant from two different parameters: the body effect and the mobility degradation. In the left side of the linearity peaks (lower V_{GT}), HD3 is influenced by the body effect (7), while the mobility degradation governs the distortion in the right side of the peaks. In the peaks region, a compensation of the non-linearities generated by each of these effects takes place.

Although a higher linearity is attained, biasing the devices at the peaks region is unpractical since process and temperature variations can change the HD3 peaks position along V_{GT}. Besides that, in this region the distortion is dominated by the fifth order harmonic and the peaks are attenuated if total harmonic distortion is observed (4). For lower V_{GT} the devices start to operate in saturation and a poor linearity is obtained. Thus, from the linearity perspective, the optimal bias stands for V_{GT} larger than 0.4 V. For this region of the curves in Figure 2, one can observe a slight influence of W_{fin} in HD3. For $V_{GT} = 0.4$ V, the wider device presents the lower distortion, while for $V_{GT} = 1.0$ V the better linearity is obtained for the narrower one. Figure 3 illustrates this changing on the tendency of HD3 as V_{GT} is varied for Va = 0.25 V.

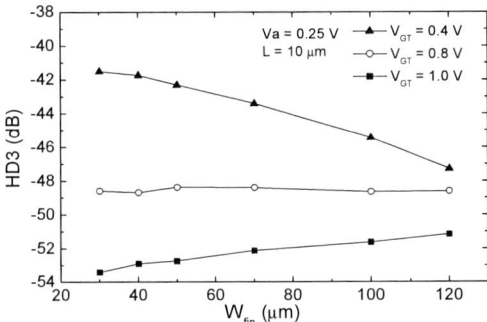

Figure 3. Extracted HD3 vs. W_{fin} curves for 2-MOS balanced structures with 10 μm-long FinFETs at Va = 0.25 V.

As one can observe, all the measured devices present similar linearity at $V_{GT} = 0.8$ V. At $V_{GT} = 0.4$ V the wider FinFET exhibits 6 dB better HD3 than the device of $W_{fin} = 30$ nm, while at $V_{GT} = 1.0$ V an advantage of 2 dB is obtained in favour of the narrower transistor. The variation of HD3 as a function of W_{fin} for different V_{GT} can be explained through the mobility degradation. As described in (8,9), wider devices present stronger electron mobility degradation in relation to the narrower ones just after attaining the mobility peak, which is translated as a higher linearity. For higher V_{GT}, the mobility degradation of wider devices reduces with respect to the narrower ones, resulting in a lower linearity. According to reference (5), for low amplitudes the HD3 behaviour can be essentially described by the devices transconductance (g_m) and its second order derivative as expressed in equation [1].

$$HD3 = \frac{1}{4}.Va^2.\frac{\frac{\partial^2 g_m}{\partial V_{GT}^2}}{6g_m} \qquad [1]$$

The transconductance is strongly dependent on the devices mobility (μ_n) and varies proportionally to μ_n. Admitting a quadratic model for the mobility degradation as in equation [2] where θ_1 and θ_2 are the mobility degradation coefficients and μ_0 is the low field mobility, HD3 equation can be rewritten as a function of the mobility degradation coefficients.

$$\mu_n = \frac{\mu_0}{1 + \theta_1 . V_{GT} + \theta_2 . V_{GT}^{~2}} \tag{2}$$

Considering $g_m \propto \mu_n$, substituting [2] in [1] and simplifying the expression, HD3 can be described as a function of θ_2 as exhibited in equation [3].

$$HD3 \approx \frac{1}{4} . Va^2 . \frac{3.\theta_2^{~2} . V_{GT}^{~2} - \theta_2}{3\left(\theta_2 . V_{GT}^{~2} + 1\right)^2} \tag{3}$$

Aiming to confirm the dependence between HD3 and the mobility degradation, the parameters μ_0 and θ_2 were extracted for the measured devices and the curves of μ_n/W have been plotted as a function of V_{GT} as shown in Figure 4. According to this figure, at lower V_{GT} narrower devices present a higher normalized mobility. However, these devices exhibit a larger mobility degradation as V_{GT} is increased and an inversion on the curves of the normalized mobility can be observed around $V_{GT} = 0.6$ V as observed in Figure 2 for HD3 curves.

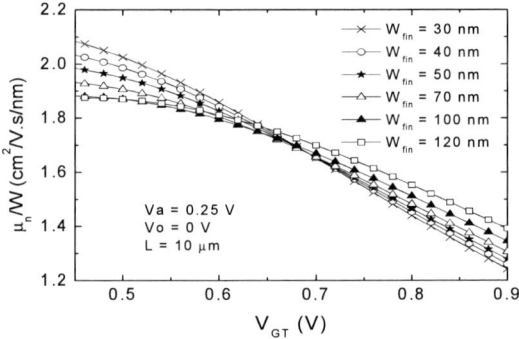

Figure 4. Extracted μ_n/W vs. V_{GT} curves for 2-MOS balanced structures with 10 μm-long FinFETs at Va = 0.25 V.

In Figure 5 the third order distortion is shown as a function of the input signal amplitude for devices of several W_{fin} at $V_{GT} = 1.0$ V. According to the figure, the smaller distortion exhibited by the FinFET of $W_{fin} = 30$ nm extends through the whole input signal range. Furthermore, when reducing the input signal amplitude a higher linearity is reached. For an input signal peak-to-peak amplitude of 0.1 V HD3 is in the order of −70 dB while a − 40 dB harmonic distortion is obtained for a peak-to-peak signal of 1 V.

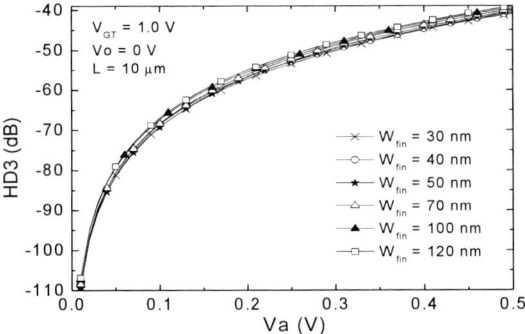

Figure 5. Extracted HD3 vs. Va curves for 2-MOS balanced structures with 10 μm-long FinFETs at $V_{GT} = 1$ V.

Usually, in balanced structures the device to be used is chosen based on the targeted on-resistance. Thus, Figure 6 presents HD3 as a function of R_{ON} multiplied by the total device width at the input amplitudes Va of (A) 0.10 V and (B) 0.25 V. When Va = 0.1 V is applied, a 15 dB HD3 improvement is obtained for the whole $R_{ON}.W$ range with respect to the distortion observed for the higher input bias. At 30 kΩ.μm the 2-MOS FinFET biased at Va = 0.25 V with W_{fin} = 120 nm exhibits 5 dB better HD3 with respect to the 30 nm-W_{fin} FinFET while 3 dB better HD3 is obtained at Va = 0.1 V. At R_{ON} = 35 kΩ.μm, the better HD3 shown by the wider device is related to the linearity peaks region (V_{GT} = 0.3) and, therefore, is unpractical to be reached. Nevertheless, the lower distortion presented by the narrower transistor at $R_{ON}.W$ = 20 kΩ.μm corresponds to the HD3 obtained at $V_{GT} \cong 0.95$ V in Figure 2. For $R_{ON}.W$ between 25 kΩ.μm and 30 kΩ.μm, all the devices are biased for V_{GT} between 0.4 V and 0.6 V and, as exhibited in Figure 2, the wider device present the lower distortion.

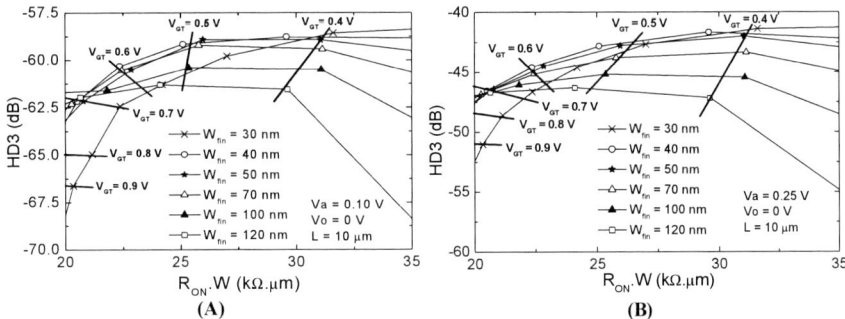

Figure 6. Extracted HD3 vs. $R_{ON}.W$ curves for 2-MOS balanced structures with 10 μm-long FinFETs at (A) Va = 0.10 V and (B) Va = 0.25 V.

An important feature of FinFETs consists in the higher value of R_{ON} obtained without degrading significantly the distortion when compared to planar devices but reducing the area. When multiplying $R_{ON}.W$, the analyzed devices have presented on-resistance between 100 kΩ and 250 kΩ, which are values interesting for 2-MOS structures applications.

Conclusions

This work evaluated the non-linearities exhibited in 2-MOS structures composed of triple gate FinFETs in terms of gate and input amplitude bias, fin width and on-resistance. According to the analysis carried out, the optimal structures V_{GT} from the distortion perspective takes place for voltages higher than 0.4 V. Although linearity peaks are observed for lower voltages, biasing the devices at these peaks is unpractical since process and temperature variation can influence on their position. For V_{GT} higher than 0.4 V, a linear improvement on HD3 can be observed for devices of several W_{fin}. For V_{GT} between 0.5 V and 0.8 V, HD3 is lower for wider fins, while for V_{GT} larger than 0.8 V the distortion increases with W_{fin}. This behavior is related to the mobility degradation observed in devices of different fin widths. The evaluation of HD3 as a function of the input amplitude demonstrated that for lower amplitudes a huge improvement on HD3 is obtained for all the devices analyzed and at $V_{GT} = 1$ V a 2 dB advantage in favor of the 30 nm-W_{fin} is obtained for the whole input range. Finally, when HD3 is analyzed as a function of R_{ON}, the wider device presents a lower HD3 for almost the whole studied range and propitiates the reduction of HD3 up to 5 dB in relation to the narrower device at $R_{ON}.W = 30$ k$\Omega.\mu$m.

Acknowledgments

The authors thank Prof. Cor Claeys from IMEC for supplying the devices for characterization and aknowlegde the Brazilian research-funding agencies FAPESP and CNPq for the financial support.

References

1. Cerdeira, A.; Alemán, M. A.; Pavanello, M. A.; Martino, J. A.; Vancaillie, L.; Flandre, D.; *IEEE Transactions on Electron Devices*, **52**, 967(2005).
2. Banu. .M.; Tsividis, Y.; *J. Solid-State Circuits*, **SC-18**, 644(1983).
3. Collaert. N.; Demand, M.; Ferain, I.; Lisoni, J.; Singanamalla, R.; Zimmerman, P.; *et al. In: Symposium on VLSI technology digest of technical papers*, 108(2005).
4. Cerdeira, A.; Alemán, M. A., Estrada, M.; Flandre, D.; *Solid-State Electronics*, **48**, 2225(2004).
5. Groenewold, G.; Lubbers, W. J.; *IEEE Transactions on Circuits and Systems II. Analog Digit. Signal Process.*, **41**, 569(1994).
6. Vancaillie, L.; Kilchytska, V.; Alvarado, J.; Cerdeira, A.; Flandre, D.; *IEEE Transactions on Electron Devices*, **53**, 263(2006).
7. Akarvardar, K.; Mercha, A.; Cristoloveanu, S.; Gentil, P.; Simoen, E.; Subramanian, V.; Claeys, C.; *IEEE Transactions on Electron Devices*, **54**, 767(2007).
8. Rudenko, T.; Collaert, N.; De Gendt, S.; Kilchytska, V.; Jurczak, M.; Flandre, D.; *Microelectronic Engineering*, **80**, 386(2005).
9. Chowdhury, M. M.; Fossum, J. G.; *IEEE Transactions on Electron Devices*, **53**, 263(2006).

CHAPTER 13

DEVICE TECHNOLOGY III

ECS Transactions, 19 (4) 297-302 (2009)
10.1149/1.3117421 ©The Electrochemical Society

Comparison of Short-Channel Effects in SOI and ssOI Triple-Gate MOSFETs

K-I Na [a, b], S. Cristoloveanu [b], Y-H Bae [c], W. Xiong [d], C. R. Cleavelin [d],
P. Patruno [e], and J-H Lee [a]

[a] Electrical Eng. and Computer Sci., Kyungpook National University,
Daegu, South Korea.
[b] IMEP, Grenoble Polytechnic Institute, Minatec, Grenoble, France
[c] Division of Electronics Engineering, Uiduk University at Gyeongju, South Korea
[d] Texas Instruments, USA
[e] SOITEC USA.

We have compared the short-channel effects (SCE) in triple-gate
MOSFETs fabricated on SOI and strained SOI (ssOI) substrates.
The triple-gate ssOI MOSFETs exhibit high gain in mobility but
show earlier threshold voltage roll-off than unstrained devices.
This indicates that the low effective mass and the narrow band-gap,
due to tensile stress in ssOI substrate, increase the junction and
gate tunneling leakage current and hence enhance the SCE. The
off-current of ssOI device at $V_g = 0$ is one order of magnitude
higher than that of SOI device, decreasing the on/off current ratio.
The subthreshold slope of ssOI device is also affected. The
stronger short-channel effect in ssOI device is correlated with a
more intense channel-to-substrate coupling.

INTRODUCTION

Multiple-gate transistors such as double-, triple-, and quadruple-gate MOSFETs
fabricated on silicon on insulator (SOI) substrate may become the key devices for
advanced CMOS technology because of their high performance and tolerance of short-
channel effects (SCE) (1-4). Furthermore, strained-SOI (ssOI) techniques, which benefit
from the mobility improvement, are very attractive for increasing the current drive (5, 6).

In this paper, we present an experimental analysis of triple-gate MOSFETs by
comparing strained and unstrained devices in terms of carrier mobility and short-channel
effects which affect the threshold voltage, ON/OFF properties, and subthreshold slope.
We also investigate the relation between short-channel effects and 3-dimensional
coupling.

DEVICE FABRICATION AND MEASUREMENT

Conventional Unibond SOI and strained-SOI (ssOI) substrates fabricated by Smart-
Cut technology were used as starting materials. The ssOI strained silicon layer,
transferred from a Si/SiGe stack, had an initial biaxial strain of 1.5 GPa. The thicknesses
of the buried oxide (BOX) and silicon film were 145 and 60 nm, respectively. The top
silicon layer defined the fin height. The wafers were patterned to make fins and
subsequently annealed in a hydrogen atmosphere to smoothen the sidewalls of the fin for

reducing the corner effect. The fins were undoped ($\sim 2\times10^{16}$ cm^{-3}) to preserve high carrier mobility. The gate stack was composed of a thin oxide layer (t_{ox} = 1.8 nm) formed by *in-situ* wet oxidation and mid-gap gate material (TiSiN), grown by LPCVD. To investigate the dimensional effects, triple-gate MOSFETs with variable fin widths were prepared. We will refer to the physical fin width which is typically 35-45 nm smaller than the designed fin width. A resist trim etch was used to shrink the fin width by 30-40 nm, and H$_2$ anneals for an additional 5 nm reduction. Figs. 1(a) and (b) show the schematic and the TEM image of a final triple-gate MOSFET with 12 nm fin width.

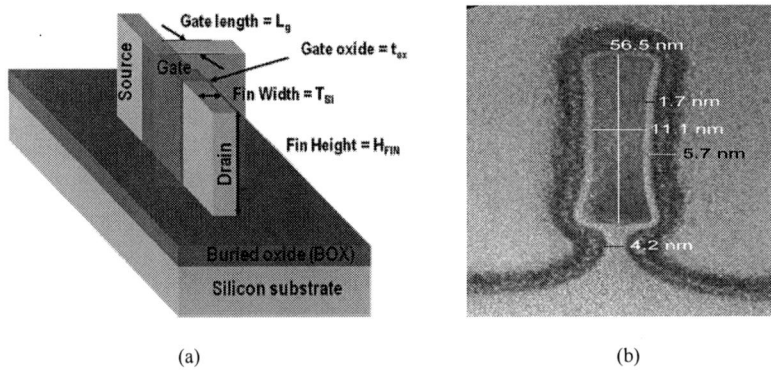

(a)   (b)

Figure 1. (a) Schematics of a triple-gate MOSFET on SOI substrate with 1.8 nm gate oxide (t_{ox}). The gate length (L_g) ranges from 85 to 10 μm. (b). Cross-sectional TEM of a triple-gate MOSFET with 12 nm fin width after trimming and H$_2$ annealing.

The drain current versus gate voltage $I_D(V_g)$ characteristics were measured, at 50 mV drain voltage, with Agilent 4155C semiconductor parameter analyzer. The electrical parameters such as threshold voltage and mobility were extracted using the Y-function, $Y(V_g) = I_D/g_m^{1/2}$ (7).

SHORT-CHANNEL EFFECTS

Fig. 2(a) shows the comparison of electron mobility between SOI and sSOI triple-gate MOSFETs as a function of gate length. The strain results in a remarkable gain in mobility. In long transistors, the electron mobility is high (500 cm^2/Vs) even in unstrained MOSFETs. It reaches an excellent value (700 cm^2/Vs) for sSOI devices. In shorter devices the mobility degrades but still exhibits reasonable values. The key point is that the benefit of strain is maintained in 85-nm-long sSOI FinFETs where μ = 300 cm^2/Vs.

Fig. 2(b) shows the linear transconductance $g_m(V_g)$ curves of n- and p-channel triple-gate MOSFETs with sSOI and SOI substrate. These devices feature 1 μm gate length, 20 fins in parallel, and {110}/<110> fin surface/direction. The g_m peak of n-channel MOSFET increases from 70 to 110 μS, however, that of p-channel MOSFET decreases slightly from 40 to 35 μS. This is so because the tensile stress improves the electron mobility and decreases the hole mobility via the modulation of the effective mass and surface scattering rate.

Fig. 2(b) also shows that the threshold voltage for long n-channel triple-gate MOSFETs with sSOI substrate is 50 ~ 100 mV lower than for the control (unstrained) devices. This is explained by the conduction-band shift induced by strain, which enables more electrons in the inversion layer for same gate bias (8, 9). The material properties of strained-silicon on SiGe layer can be summarized as follows:

$$\chi_{SS} = 4.17 + 0.67x \qquad [1]$$

$$E_{g,SS} = 1.08 - 0.4x \qquad [2]$$

$$\varepsilon_{r,SS} = 11.8 \qquad [3]$$

where χ is the affinity, E_g is the bandgap energy, and x is the germanium content in the initial $Si_{1-x}Ge_x$ template used to grow the sSi layer (which was subsequently transferred on BOX). These equations indicate that the bandgap energy of the strained Si film is lower than in unstrained Si.

For the p-channel triple-gate MOSFETs, less than 25 mV reduction in V_T is observed in sSOI (vs. SOI) because the strain-induced valence-band shift is smaller than the conduction-band shift (8, 9).

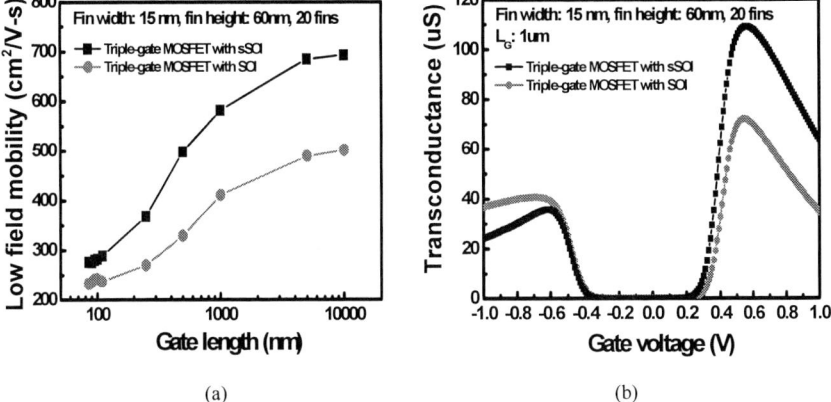

(a)   (b)

Figure 2. (a) Low-field electron mobility as a function of gate length and (b) transconductance versus gate bias in n- and p-channel triple-gate MOSFETs fabricated on sSOI and SOI substrates. These devices feature a fin width of 15 nm and a fin height of 60 nm. $V_D = 50$ mV.

Fig. 3 shows the variation of threshold voltage as a function of gate length in n-channel triple-gate MOSFETs fabricated on SOI and sSOI substrate. The triple-gate sSOI MOSFETs exhibit earlier threshold voltage roll-off than triple-gate SOI MOSFETs. The primary effect of strain is to enhance the transport properties. In principle, the transistor electrostatics is not significantly affected by strain. However, our results indicate that the SCE are amplified in sSOI, presumably in relation with the increase in leakage current. On one hand, the lower effective mass and the narrower band-gap (Eq.(2)), due to tensile stress, increase the junction and gate tunneling leakage currents. On the other hand, increased leakage means less gate control, hence more severe SCE.

Fig. 4 shows the off-current leakage current (measured at $V_g = 0$) that is almost 10 times higher in sSOI than in SOI. High off-current also degrades the on/off current ratio (Fig. 3), although the on-current was improved via the mobility gain. These results suggest that triple-gate MOSFETs fabricated on sSOI substrate may be less attractive for low standby power applications than for high performance circuits, where a superior on-current is the primary request.

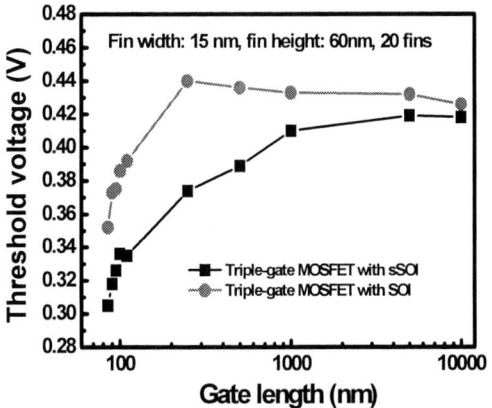

Figure 3. Threshold voltage as a function of gate length for triple-gate MOSFETs with SOI and sSOI substrates. Fin width, fin height, and the number of fins are 15, 60, and 20 nm, respectively.

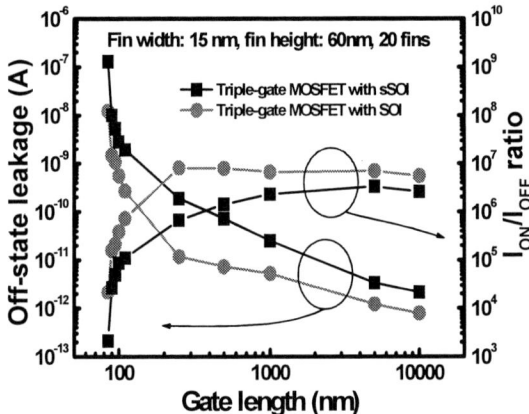

Figure 4. Off-state leakage ($V_g = 0$) and I_{ON} ($V_g = 1$)/I_{OFF} ratio versus gate length for triple-gate MOSFETs with SOI and sSOI substrates. Fin width, fin height, and the number of fins are 15, 60, and 20 nm, respectively.

The subthreshold slope is equally excellent (60 mV/decade) in long-channel SOI and sSOI devices Fig. 5(a). However, as a consequence of SCE below 100 nm length, the swing increases more rapidly in sSOI than in SOI FinFETs.

COUPLING EFFECTS

Figure 5(b) illustrates the channel-to-substrate coupling effect. The threshold voltage was measured as a function of substrate (back-gate) bias and the coupling coefficient was defined as $C = \Delta V_T / \Delta V_{BACK}$ (10). In narrow fins, the lateral gates control the potential at the fin-BOX interface (lateral coupling) and inhibit the back-gate effect (vertical coupling) (11). This is why the coupling coefficient is very small in long and narrow FinFETs.

In shorter devices, the fringing fields from source and drain tend to bias the body (longitudinal coupling via the BOX), lowering the threshold voltage (12). This short-channel effect (DIVSB) deteriorates the gate control over the channel, letting the substrate action increase. Since the SCE effect is stronger in sSOI, the vertical coupling coefficient increases more rapidly in short sSOI devices than in SOI (Fig. 5(b)). This is an example of 3D coupling effects, typical to FinFETs.

Accordingly, the relevance of the back-gate on both the channel-to-substrate coupling and the charge sharing is accentuated as the channel length reduces.

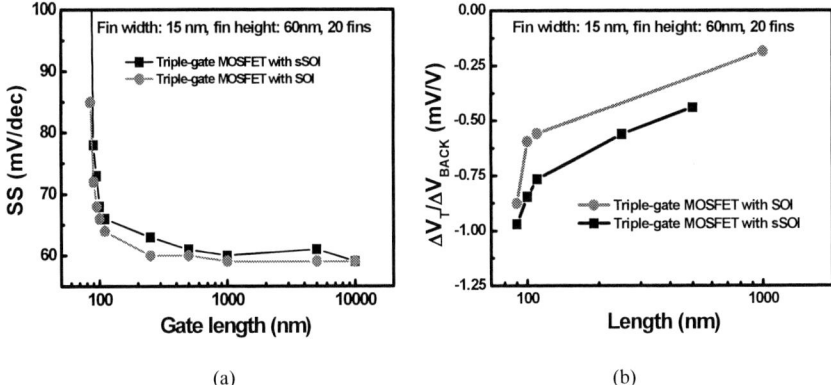

(a)   (b)

Figure 5. (a) Subthreshold slope and (b) vertical coupling coefficient ($\Delta V_T / \Delta V_{BACK}$) versus gate length in triple-gate MOSFETs with SOI and sSOI substrates. Parameters are as in Fig. 4.

CONCLUSION

The effects of strain on the performance of SOI triple-gate MOSFET with {110}/<110> surface-orientation/current-direction have been investigated. Substrate-induced strain was found to remarkably improve {110} electron mobility, although {110} hole mobility was slightly degraded. However, the short-channel effects (threshold voltage roll-off, off-current leakage, subthreshold slope, etc) of strained triple-gate MOSFETs are more severe compared to unstrained devices. An unexpected consequence is that the 3D coupling effects are stronger in sSOI FinFETs.

Acknowledgments

This research was supported by Brain Korea 21 (BK21) of the Ministry of Education, the Korea Science and Engineering Foundation (KOSEF) through the National Research Lab. Program (No. M10600000273-06J0000-27310), the World Class University (WCU) Program funded by the Ministry of Science and Technology, and the European Union programs EUROSOI+ and Nanosil.

References

1. E. Bernard, T. Ernst, B. Guillaumot, N. Vulliet, T. C. Lim, O. Rozeau, F. Danneville, P. Coronel, T. Skotnicki, S. Deleonibus, and O. Faynot, *IEEE Electron Device Letters*, vol. 30, no. 2, pp. 148-151, 2009.
2. J. Cai, A. Majumdar, D. Dobuzinsky, T.H. Ning, S.J. Koester, and W.E. Haensch, *IEEE International Electron Devices Meeting*, pp. 907-910, 2007.
3. J.P. Colinge, *Microelectronics Engineering*, vol. 84, pp. 2071-2076, 2007.
4. S. Cristoloveanu, R. Ritzenthaler, A. Ohata, and O. Faynot, *International Journal of High Speed Electronics and System*, vol. 16, no. 1, pp. 217-230 (2006).
5. T. Numata, T. Irisawa, T. Tezuka, J. Koga, N. Hirashita, K. Usuda, E. Toyoda, Y. Miyamura, A. Tanabe, N. Sugiyama, S. Takagi, *IEEE Transactions on Electron Devices,* vol. 53, pp. 1030-1038, 2006.
6. K.J. Chui, K.W. Ang, H.C. Chin, C. Shen, L.Y. Wong, C.H. Tung, N. Balasubramanian N., M.F. Li, G.S. Samudra, Y.C. Yeo, *IEEE Electron Device Letters,* vol.27, pp. 778-780, 2006.
7. S. Cristoloveanu, S.S. Li, *Electrical characterization of silicon-on-insulator materials and devices*, Kluwer Academic Publishers, 1995.
8. J.S. Goo, Q. Xiang, Y. Takamura, F. Arasnia, E.N. Paton, P. Besser, J. Pan, and M.R. Lin, *IEEE Electron Device Letters*, vol. 24, pp. 568-570, 2003.
9. S. Thompson, G. Sun, K. Wu, J. Lim, and T. Nishida, *IEEE International Electron Devices Meeting*, pp. 221-224, 2004.
10. K.I. Na, J.H. Lee, S. Cristoloveanu, Y.H. Bae, P. Patruno, W. Xiong, *International Journal of High Speed Electronics and Systems,* vol. 18. No. 4, pp. 773-782, 2008.
11. R. Ritzenthaler, M. Gaillardin, K. Akarvardar, O. Faynot, C. Jahan, S. Cristoloveanu, ECS Transactions, vol. 6, no. 4, pp. 89-94, 2007.
12. T. Ernst, R, Ritzenthaler, O. Faynot, S. Cristoloveanu, IEEE Trans. Electron Devices, vol. 54, no. 6, pp. 1366-1375, 2007.

CHAPTER 14

POSTER SESSION

ECS Transactions, 19 (4) 305-310 (2009)
10.1149/1.3117422 ©The Electrochemical Society

Thermal Resistance Model for Multi-finger Trench-Isolated Bipolar Transistors on SOI Substrate

Rashmi[a], A. Kranti[a], G.A. Armstrong[a], S. Nigrin[b]

[a] School of Electronics, Electrical Engineering and Computer Science, Queen's University Belfast, BT9 5AH, N. Ireland.
[b] MHS-Electronics UK Ltd., Cheney Manor, Swindon SN2 2QW, UK.

A simple predictive model to estimate the thermal resistance of a trench isolated multi-finger bipolar device on SOI substrate is proposed. The model shows very good agreement with 3-D electro-thermal simulations over a wide range of device parameters. The simulations have been verified with measurements. The model is virtually independent of critical device parameters such as trench depth, oxide layer thickness and emitter dimensions and serves as a useful tool to predict the thermal resistance of a multi-emitter device reasonably accurately, from the known thermal resistance of the corresponding single-emitter device.

Introduction

Complementary bipolar technologies play a very important role in high performance analogue circuits such as low noise RF amplifiers, high slew rate wide band operational amplifiers and highly linear power amplifiers. Operation at large power densities usually results in considerable self-heating and can degrade the DC and AC performance [1] of bipolar transistors. The limit of electro-thermal stability of a bipolar transistor is defined as the snapback point where the collector current reaches a critical value, which is a strong function of the thermal resistance (R_{th}) of the device [2,3]. A lattice temperature rise of typically ~17 K above ambient temperature will be sufficient to make the device electro-thermally unstable [4]. Technological advancements such as shrinking photolithography capabilities, self aligned double polysilicon architectures and introduction of deep trench isolation (DTI) schemes [5] have improved bipolar electrical performance considerably. A combination of DTI method with silicon-on-insulator (SOI) substrates has emerged as a desirable technique as it results in reduced collector-substrate capacitances, latch-up immunity, radiation hardness, substrate noise immunity and transistor size reduction [6]. However, due to the very low thermal conductivity of thermal oxide inside the trenches and buried oxide of the SOI substrates, being nearly two orders of magnitude lower than thermal conductivity of Silicon (Si), DTI+SOI based bipolar devices show much higher thermal resistance. Furthermore, use of multiple emitter fingers in high power bipolar transistors to avoid current inhomogeneities and hot spots [7] results in mutual thermal coupling between emitters, leading to a temperature rise of all fingers to well above their individual temperature values and thus a higher overall device thermal resistance compared with the equivalent single emitter device [8] of same current density. It is therefore extremely important to optimize the thermal resistance of multiple emitter finger trench-isolated bipolar transistors on SOI substrate.

While the thermal resistance of a single emitter bipolar transistor can be determined using analytical models [9-10] or device simulation, the thermal coupling

305

associated with multi-finger structures is not amenable to analytical models and requires extensive 3-D electro-thermal simulation. To the best of our knowledge, no published models are available to easily estimate R_{th} of a multi-emitter device. In this work, we propose a simple empirical expression for R_{th}, based on power law dependence on number, N_e and length, L_e of emitter fingers and provide comparisons with fully coupled 3D electro-thermal simulations [11] of multiple emitter finger bipolar devices. The simulations have been verified with experimental measurements.

Device Structure and Simulation

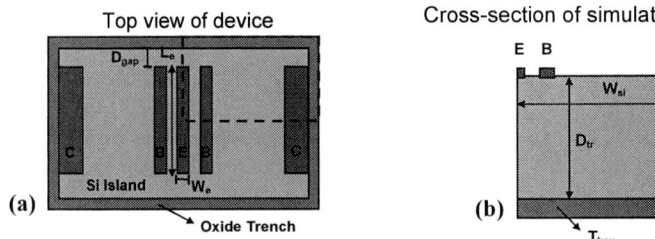

Figure 1. (a) Schematic diagram of the top-view of simulated symmetric single-emitter device with two collector and two base contacts. (b) Vertical cross-section of the simulated quarter of the device, as shown in (a) by dotted lines.

Fig. 1(a) shows the schematic diagram of a single-emitter device with trench oxide of thickness T_{trox} surrounding the Si island on all four sides and a buried oxide layer of thickness T_{box} at the bottom. Due to the inherent symmetry of the layout, we simulate one-quarter of the structure, as shown in fig. 1(b), to reduce computation time. The emitter width, W_e is 0.4 μm with its length, L_e varying from 4 – 20 μm. The emitter finger is separated from the trench oxide by a distance D_{gap} of 1.6 μm. The trench is therefore, (L_e+2D_{gap}) long, $2W_{si}$ wide and D_{tr} deep. The thermal resistance, R_{th} was evaluated as the ratio of maximum temperature rise to applied power [12] as

$$R_{th} = \frac{(T_{MAX} - T)}{V_{CE}I_C} \tag{1}$$

where V_{CE} is the collector-emitter voltage, I_C is the collector current, T_{MAX} is the maximum lattice temperature and T is the ambient substrate temperature, applied below the buried oxide and beyond the trench as a fixed isothermal boundary condition.

TABLE I. Comparison with measured R_{th} values ($W_e = 0.4$ μm, $D_{tr} = 10.5$ μm, $T_{trox} = 0.5$ μm, $T_{box} = 1.0$ μm, $S_e = 4.6$ μm, $D_{gap} = 1.6$ μm)

L_e (μm)	Measured R_{th} (K/W)	Simulated R_{th} (K/W)	Modelled R_{th} (K/W)($\alpha = 0.3$, $\beta = -0.42$)
\multicolumn{4}{c|}{$N_e = 4$: Symmetric Design}			
4	600	557	631
48	227	185	222
\multicolumn{4}{c|}{$N_e = 1$: Asymmetric Design}			
4	1360	1378	1371

3-D electro-thermal simulations were performed for bipolar devices with up to five emitter fingers and L_e in range 4 – 20 μm with $T_{box} = 1$ μm, $T_{trox} = 0.5$ μm and trench depths, D_{tr} of 5.5 and 10.5 μm. The spacing between emitter fingers, S_e was 4.6 μm and the Si island width, W_{si} of simulated quarter for single-finger device was 13.4 μm, based on design rules and mask layout. Simulations were verified with measured thermal resistance values for four emitter finger devices with emitter lengths of 4 and 48 μm. An asymmetric single-finger design with one emitter, one collector and two base contacts was also measured and simulated. In order to confirm that measurement technique of thermal effects is accurately reproduced in simulation, the variation of base-emitter voltage, V_{BE} with ambient temperature was simulated for a fixed emitter current, I_E of 0.1 mA, resulting in a simulated dV_{BE}/dT value of −1.56 mV/K, which is in close agreement with the measured value of −1.52 mV/K. Simulated thermal resistance values also agree well with measured R_{th} values, as shown in Table 1, the difference may be attributed to the use of simplified boundary conditions which ignore the thermal resistance of surrounding substrate.

Analytical Model for R_{th} of Multiple-Emitter Bipolar Transistor

On the basis of a known thermal resistance value for a single emitter device, the thermal resistance of a multiple emitter finger device can be evaluated by the expression

$$R_{th} = R_{th0} \frac{1}{N_e^{\alpha}} \left(\frac{L_e}{L_{e0}} \right)^{\beta} \qquad [2]$$

where N_e is the number of emitter fingers, L_e is emitter length, L_{e0} is a reference length, R_{th0} is thermal resistance of the single-finger reference device with $L_e = L_{e0}$ and α, β are the model parameters defined below. L_{e0} has been chosen arbitrarily as 12 μm for the present work, compatible with typical aspect ratio (L_e/W_e) of ~ 20–50 associated with traditional bipolar technology. β is a non-linearity factor associated with emitter length dependence of R_{th} while α is a measure of the mutual thermal coupling between emitter fingers in a multi-finger design [12], evaluated as

$$\alpha = \frac{\ln(R_{th0} / R_{th})}{\ln(N_e)} \bigg|_{L_e = L_{e0}} \qquad [3]$$

$$\beta = \frac{\ln(R_{th} / R_{th0})}{\ln(L_e / L_{e0})} \bigg|_{N_e = 1} \qquad [4]$$

The lower the value of α, the stronger is the mutual coupling between the multiple emitter fingers.

Results and Discussion

Fig. 2(a)-(b) compare the simulated and modelled dependence of R_{th} on length and number of emitter fingers for trench depths of 5.5 μm and 10.5 μm respectively. The thermal resistance decreases by ~ 45% with increase in L_e from 4 μm to 20 μm, due to changes in the relative proportions of heat dissipating orthogonal to the finger and increased vertical and horizontal heat dissipation due to larger Si volume available. Similarly, increasing the number of emitter fingers from 1 to 5 results in nearly 40% reduction in R_{th}. The model shows good agreement with simulations over the range of L_e and N_e, achieving the best overall fit with $\alpha = 0.3$ and $\beta = -0.42$ for $L_{e0} = 12$ μm at $S_e = 4.6$ μm, independent of the choice of D_{tr}. The model is also valid for the asymmetric single-finger device, the results for which have not been shown here.

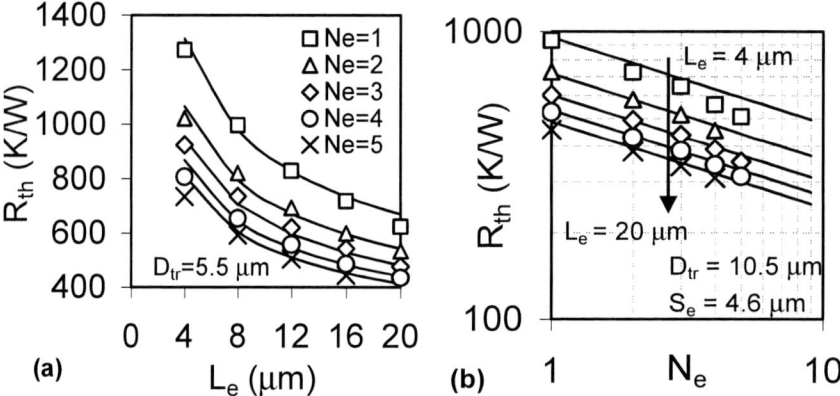

Figure 2. Variation of thermal resistance with L_e and N_e for (a) $D_{tr} = 5.5$ μm, (b) $D_{tr} = 10.5$ μm. Lines: Model, Symbols: 3-D Simulation. All notations are same in (a) and (b).

It should be noted that the present simulation results were based on symmetric devices where W_{si} changes with N_e, from 13.4 μm for single-finger device to 23.4 μm for $N_e = 5$ for S_e of 4.6 μm. Thus, the observed decrease in thermal resistance of a multi-emitter device is in fact a combined effect of (i) increasing the number of emitter fingers and (ii) the associated increase in Si island volume. The mutual thermal coupling, α can therefore be expressed as:

$$\alpha = \alpha_{Coupling} + \alpha_0 \qquad [5]$$

where $\alpha_{Coupling}$ is the dominant term and represents the effect of N_e on R_{th} for a fixed W_{si} and α_0 is the effect of consequent increase in W_{si} on R_{th0}. Simulating one and five finger devices, both with same W_{si} of 23.4 μm models the effect of multiple fingers alone and gives an approximately 25% lower value of α (= $\alpha_{Coupling}$), indicating stronger thermal coupling between the emitter fingers. Thus, analysing α instead of $\alpha_{Coupling}$ underestimates the thermal coupling between emitter fingers in a multi-emitter device.

As the degree of coupling between emitter fingers would depend on the distance between the fingers, it is important to see the effect of S_e on $\alpha_{Coupling}$. Simulation of a four-finger device with a fixed W_{si} of 23 μm suggests that $\alpha_{Coupling}$ increases from 0.22 to 0.3 and 0.37 respectively as S_e increases from 4.6 μm to 6.9 μm and 9.2 μm. Thus, increase in emitter finger spacing reduces the mutual thermal coupling between the fingers.

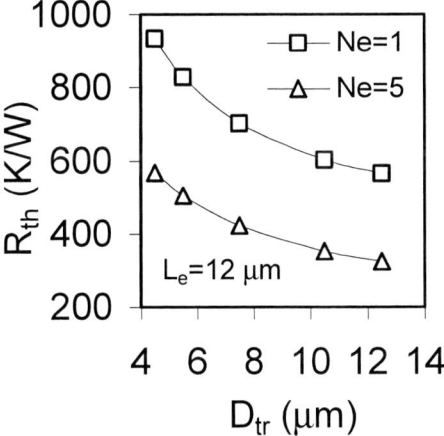

Figure 3. Simulated dependence of thermal resistance on trench depth for $N_e = 1$ and 5 for $L_e = 12$ μm, $S_e = 4.6$ μm.

Fig. 3 shows the effect of trench depth, D_{tr} on thermal resistance for $N_e = 1$ and 5. Increase in D_{tr} from 5.5 μm to 10.5 μm significantly reduces R_{th} by nearly 32% due to increased Si island volume. However D_{tr} does not affect the model coefficients α, β significantly as any change in trench depth affects both R_{th} and R_{th0} similarly and R_{th0}/R_{th} remains virtually unchanged. Similarly, any change in oxide thicknesses or emitter dimensions does not affect the model parameters significantly.

Conclusion

The proposed model serves as a useful tool to predict the thermal resistance of a multiple emitter finger trench-isolated bipolar device reasonably accurately, if measured value of thermal resistance for a single-finger test device is known. The model is relatively independent of the depth of trench and the thickness of trench oxide and buried oxide layers and is valid for both, symmetric and asymmetric designs over a wide range of device parameters.

Acknowledgments

This work was co-funded by UK Technology Strategy Board under Project TP/8/ADM/6/I/Q2102J.

References

1. R. M. Fox, S. G. Lee and D. T. Zweidinger, *IEEE J. Solid State Circuits*, **SC28**, 678 (1993).
2. N. Rinaldi, V. d'Alessandro and F. M. de Paola, *In Proc. BCTM*, 2.3 (2006).
3. T. Vanhoucke and G. A. M. Hurkx, *In Proc. BCTM*, 37 (2005).
4. N. Nenadovic, V. d'Alessandro, L. K. Nanver, N. Rinaldi, H. Schellevis and J. W. Slotboom, *In Proc. BCTM*, 24 (2002).

5. M. C. Wilson, P. H. Osborne, S. Nigrin, S. B. Goody, J. Green, S. J. Harrington, T. Cook, S. Thomas, A. J. Manson and A. Madni, *In Proc. BCTM*, 164 (1998).
6. M. Mastrapasqua, P. Palestri, A. Pacelli, G. K. Celler, M. R. Frei, P. R. Smith, R. W. Johnson, L. Bizzarro, W. Lin, T. G. Ivanov, M. S. Carroll, I. C. Kizilyalli and C. A. King, *IEEE Elec. Device Lett.*, **23**, 145 (2002).
7. M. Pfost, P. Brenner and R. Lachner, *In Proc. BCTM*, 100 (2004).
8. D. J. Walkey, D. Celo, T. J. Smy and R. K. Surridge, *IEEE Trans. on Elec. Devices*, 49, 1375 (2002).
9. I. Marano, V. d'Alessandro and N. Rinaldi, *Solid-State Electronics*, **52**, 730 (2008).
10. T. Vanhoucke and G. A. M. Hurkx, *IEEE Trans. on Elec. Devices*, **53**, 1379 (2006).
11. ATLAS Users Manual, Silvaco, 2008.
12. S. Nigrin, G. A. Armstrong and A. Kranti, *Solid-State Electronics*, **51**, 1221 (2007).

Fabrication of Thick-film Silicon-on-Insulator by Separation by Implanted Oxygen Layer Transfer

Xing Wei, Aimin Wu, Bo Zhang, Miao Zhang, Xi Wang, and Chenglu Lin

State Key Laboratory of Functional Material for Informatics, Shanghai Institute of Microsystem and Information Technology, Chinese Academy of Sciences, Shanghai, 200050, China

In this paper, two approaches combined the Separation by IMplanted OXygen (SIMOX) layer transfer (SLT) process with Si epitaxy are proposed to fabricate thick-film Silicon-on-Insulator (SOI). Spectroscopic ellipsometry indicates thick-film SOI wafers with the top Si layers of 1430.86 ± 19.8 nm and 1476.44 ± 18.5 nm are achieved. Atomic-scale sharp interfaces between the top Si layer and buried oxide layers are observed by high-resolution transmission electron microscopy. Using atomic force microscopy, surface morphology of the SOI wafer is also investigated.

INTRODUCTION

Thick-film silicon-on-insulator (SOI) is mainly applied in MEMS [1], high voltage power devices [2], automobile electronics, and optical waveguides [3]. In most of the applications, bulk silicon quality of the top Si and thermal oxide properties of the buried oxide (BOX) are required. Currently, SOI wafers are normally produced with bonding [4] and SIMOX [5] techniques. Bonding offers SOI wafers with high quality and wide thickness flexibility of both top Si and BOX layers. However, the drawback of bonding SOI is the inability of producing top Si layer with good thickness uniformity, which restricts the applications of bonding SOI wafers. In contrast, SIMOX offers a top Si layer with good thickness uniformity, but it suffers from limited thickness of the BOX layer. Furthermore, the BOX quality of SIMOX is not as good as thermally oxidized SiO_2. It is reported that Si islands [6] and pinholes [7] in the BOX layer led to degradation of breakdown voltage and leakage current, respectively.

Recently, there is a newly developed approach for fabricating SOI wafers, referred as SIMOX layer transfer (SLT) [8] [9], which have the advantages of both the bonding and SIMOX techniques. The SIMOX donor wafer used in the SLT process provides thickness uniformity of the top Si. The thermal growth of oxide and wafer bonding provide high quality and uniformity of the BOX. Thus, the SLT process opens up an innovative way of synthesizing SOI wafers for the electronics industry.

In this paper, two approaches involving different SIMOX layer transfer/epitaxy sequences are proposed to fabricate thick-film SOI wafers: the 1st approach, denoted as process A, is formation of thick-film SIMOX donor wafer by Si epitaxy before SLT process; the 2nd approach, denoted as process B, is formation of thin-film seed wafer by SLT process before Si epitaxy. In addition, the authors investigate the cross-sectional structure and surface morphology of the thick-film SOI wafers by transmission electron microscopy (TEM) and atomic force microscopy (AFM), respectively.

EXPERIMENTAL

Process A: a 5 inch SIMOX donor wafer with 190 nm top Si was used as the seed wafer. Subsequently, a 1.5 μm-thick epitaxial Si layer was deposited onto the seed wafer by chemical vapor deposition (CVD). The thick-film SIMOX donor wafer was then oxidized in a wet oxygen atmosphere at 1100 °C. Both the as-oxidized donor wafers and the handle wafers were cleaned in modified solutions of SC-1 (NH_4OH: H_2O_2: H_2O = 1: 6: 30) and SC-2 (HCl: H_2O_2: H_2O = 1: 1: 30), followed by rinsing in deionized water and spin drying. Afterwards, the SIMOX donor wafer was bonded upside down to the Si handle wafer at room temperature. The bonded wafer pair was annealed at 1150 °C in wet oxygen for 3 h. The annealed wafer pair was ground from the backside of the donor wafer until a 3-μm Si layer was left. A 5% tetramethyl-ammonium hydroxide (TMAH) solution was used to remove the remaining Si layer of the donor wafer at 55 °C, and then the etch-stop layer was removed with 5% hydrofluoric (HF) acid solution. Finally, a thick-film SOI wafer with 1430.86 ± 19.8 nm top Si layer on top of a 3.5 μm BOX layer was obtained.

Process B: a 5-inch thin-film SOI wafer was obtained by transferring the top Si of the SIMOX donor wafer upside down onto thermally oxidized Si handle wafer, and more details of the process can be found in reference 8. After cleaning and spin drying, the top Si of the thin-film SOI wafer was thickened with the similar epitaxy process used in Process A for fabricating the thick-film SIMOX donor wafer. Finally, thick-film SOI wafer with 1476.44 ± 18.5 nm top Si and 1 μm BOX layer was fabricated.

RESULTS AND DISSCUSSION

The thick-film SOI wafers were measured by spectroscopic ellipsometry (SE) at 49 locations over the entire 5-inch wafer. The thickness distribution of the top Si is shown in Figure 1. The mean top Si thickness of the SOI fabricated by process A is 1430.86 nm, and the thickness variation is ± 19.8 nm. Correspondingly, the top Si thickness and its variation of the SOI fabricated by process B are 1476.44 nm and ± 18.5 nm, respectively. Compared to the original donor wafers used in both process A and B, the thinner top Si layers after the layer transfer are attributed to partially remove the transition region near the top Si/BOX interface [8]. It is clear that the top Si with excellent thickness uniformity can be achieved by both process A and process B.

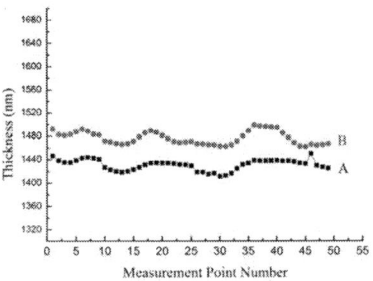

Figure 1. SE measurement results of the top Si thickness: Curve A is the top Si thickness distribution of the SOI fabricated by process A; Curve B is the top Si thickness distribution of the SOI fabricated by process B.

Cross-sectional transmission electron microscopy (XTEM) was employed to investigate the microstructure of the thick-film SOI wafers. As shown in Figure 2, sandwich-like structure including the top Si, the BOX and Si substrate can be easily distinguished, and the sharp interfaces can be observed. The results reveal the perfect structure of the thick-film SOI.

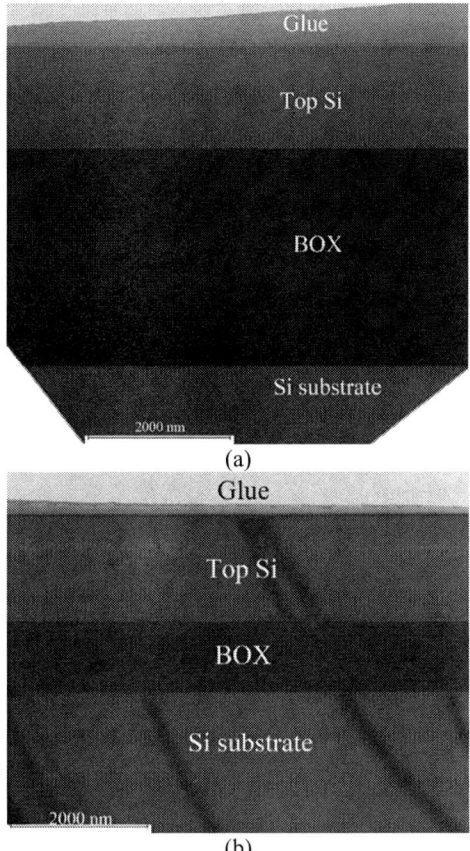

Figure 2. XTEM image of the thick-film SOI structure: (a) sample fabricated by process A; (b) sample fabricated by process B.

High-resolution TEM images of the interfacial region between the top Si and the BOX are shown in Figure 3. Atomic-scale sharp interfaces without undulation are clearly demonstrated. No discernable defects can be observed, indicating high crystal quality of the top Si.

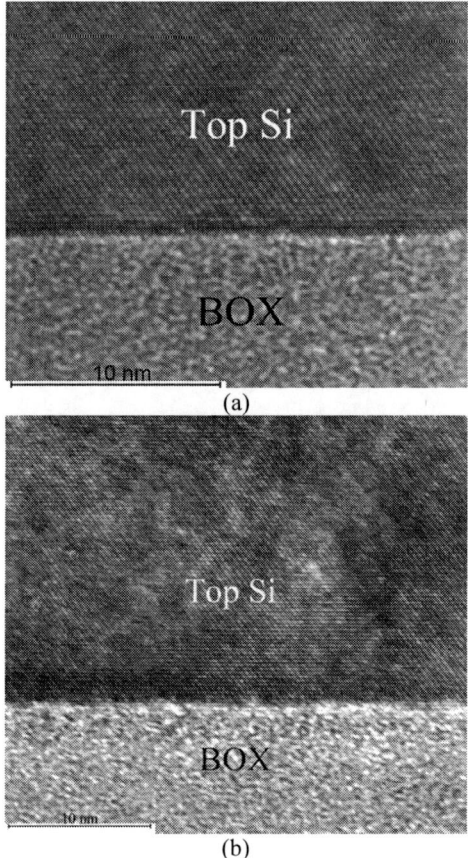

(a)

(b)

Figure 3. High-resolution TEM image of the interfacial region between the top Si layer and the BOX layer: (a) sample fabricated by process A; (b) sample fabricated by process B.

Surface morphology of the thick-film SOI wafer fabricated by process A is shown in Figure 4. Clearly, as was seen on the surface of the thin-film seed wafer [9], square pit morphology is also observed. Most of the pits have square surface and depth of less than 1×1 μm^2 and 10 nm, respectively. The edges of the square pits, which reflects the crystal orientation of the top Si, are along the [110] crystal orientation. The average root-mean-square (rms) roughness (calculated from a 5×5 μm^2 scan area) is 2.383 nm. The increased surface roughness of SOI wafers leads to a higher thermal oxide defect density which further causes a lower breakdown voltage of MOSFET's gate oxide [10]. As a result, chemical mechanical polishing (CMP) is needed to improve surface roughness of the top Si. It is worth noting that the material removed by CMP is originated from the top Si/BOX interfacial region, which contains numerous embedded defects [11]. Therefore, crystal quality of the device layer can be further improved.

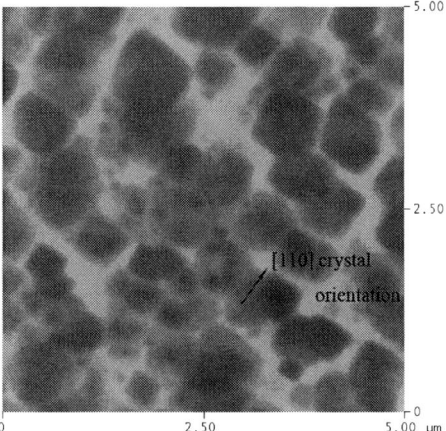

Figure 4. AFM image of the surface morphology of the SLT SOI obtained by process A.

Compared to the thick-film SOI wafer fabricated by process A, a polycrystalline silicon layer is observed at the edge exclusion area of the SOI fabricated by process B, which is possibly due to the Si epitaxy. This is because a 4 mm exclusion area was formed at the edge of the thin-film seed wafer after the layer transfer, i.e. the BOX layer was exposed at the site. During the Si epitaxy process, Si layer was deposited onto the BOX layer. Therefore, the polycrystalline silicon layer was formed at the exclusion area. Since the potential influence of the polycrystalline silicon layer on the CMOS process can not be conclusively ruled out, process A is more frequently used in industrial production.

CONCLUSIONS

In summary, top Si layers of thick-film SOI wafers with excellent thickness uniformity have been achieved by the SLT process. Structure characterization shows sharp interfaces and defect-free top Si layer, revealing the perfect structure of the thick film SOI wafers. Therefore, the SLT process for fabricating high-quality thick-film SOI was demonstrated and it will be a competitive candidate for mass production.

ACKNOWLEGEMENTS

This work was financially supported by the National Science Foundation of China (Grant No. 60721004) and the Shanghai Institute of Microsystem and Information Technology Fund for Young Scholars.

REFERENCES

1. A. M. Wu, J. Chen, and X. Wang, *Proceedings of the 2007 IEEE International SOI Conference*, Indian Wells, CA, 1-4 Oct. 2007, p.151-152.
2. R. P. Zingg, *Microelectro. Eng.*, **59**, 461 (2001).
3. Y. J. Wang, Z. L. Lin, X. L. Cheng, C. S. Zhang, F. Gao, and F. Zhang, *Appl. Phy. Lett.*, **85**, 3995 (2004).

4. W. P. Maszara, *J. Appl. Phys.,* **64**, 4943, (1988).

5. S. Nakashima and K. Izumi, *J. Mater. Res.,* **8**, 523, (1993).

6. K. Kawamura, T. Yano, I. Hamaguchi, S. Takayama, Y. Nagatake, and A. Matsumura, *Jpn. J. Appl. Phys.,* **38**, 2477 (1999).

7. A. G. Revesz, G. A. Brown, and H. L. Hughes, *J. Electrochem. Soc.,* **140**, 3222 (1993).

8. X. Wei, A. M. Wu, M. Chen, J. Chen, M. Zhang, X. Wang, C. L. Lin, *J. Vac. Sci. Technol.,* **26**, L45 (2008).

9. X. Wei, B. Zhang, M. Chen, M. Zhang, X. Wang, C. L. Lin, *Proceedings of the 2008 IEEE International SOI Conference*, New Paltz, NY, 6-9 Oct. 2008, p. 81-82.

10. W. M. Huang, *IEDM Technical Digest*, p.735, (1993).

11. A. Ogura, T. Tatsumi, T. Hamajima, and H. Kikuchi, *Appl. Phy. Lett.,* **69**, 1367 (1996).

The "U" Shape Behavior of GIFBE in Function of Back Gate Bias in FinFETs

Paula Ghedini Der Agopian [1*], João Antonio Martino[1], Eddy Simoen[2] and Cor Claeys[2, 3]

[1]LSI/ PSI/USP - University of Sao Paulo, Brazil
Av. Prof. Luciano Gualberto, trav. 3 n° 158, 05508-900 – SP – Brazil.
*biba@lsi.usp.br
[2]IMEC, Kapeldreef 75, B-3001 Leuven, Belgium
[3] KU Leuven, Electr.Eng. Dept., B-3001 Leuven, Belgium

In this work the gate induced floating body effect (GIFBE) was analysed in FinFET devices in function of the substrate bias. This analysis was performed in strained and unstrained devices that were fabricated with and without selective epitaxial grow (SEG). For all evaluated devices, this analysis results in a "U" shape behavior of GIFBE onset. Strained devices present earlier GIFBE due to the band gap variation as well as the presence of SEG structure.

Introduction

As devices are scaled down, the short-channel control becomes more complicated for planar devices. In order to avoid the threshold reduction, the high channel doping concentration is necessary but it is responsible for the carrier mobility degradation. FinFETs have been considered as promising structures to replace the planar devices due to the high drive capability and short channel immunity [1].

Besides the short-channel effects, another consequence of the device reduction is the gate tunneling current. The dominant tunneling mechanism for the gate oxide smaller than 3nm is the direct one [2]. This tunneling is composed for three components: ECB (electron conduction band), EVB (electron valence band) and HVB (hole valence band). Although the gate tunneling current is composed of several components, the two smaller components called electron valence band (EVB) and hole valence band (HVB), which flows between the gate and the body, are responsible for the appearance of a floating body effect in partially depleted (PD) SOI MOSFETs and in fully depleted (FD) SOI when the back interface is accumulated [3].

The FinFETs has been extensive studied due to the strong coupling among the three gates of the transistor that results in a better control of short channel effect (SCE). These structures, usually fully depleted (FD), may also suffer from the Gate Induced Floating Body Effect (GIFBE) when the back interface is accumulated [4]. In this work the GIFBE behavior is analyzed as a function of the substrate bias for devices with and without strain (uniaxial and biaxial).

Devices Characteristics

The measured FinFETs devices were fabricated at Interuniversity Microelectronics Centre (IMEC) with and without Selective Epitaxial Growth (SEG). Besides the presence or absence of SEG, three types of structures were evaluated. The structure used as reference was the conventional FinFET and the others were FinFETs submitted to either an uniaxial or a biaxial stress.

The studied FinFETs characteristics are: effective gate oxide (EOT) and buried oxide (t_{oxb}) thicknesses of 1.5nm and 145nm, the silicon finger height (H_{Fin}) of 60nm the channel doping concentration is $1x10^{15}$ cm^{-3}. All measurements were performed at room temperature.

Analysis and Discussions

Figure 1A, the transconductance (gm) is evaluated for small negative back gate bias. The results show a lateral shift in the transconductance (gm) curves when the threshold voltage (Vth) increases due to a negative back gate bias. When the focus is on the gm second peak, it is possible to note that, as the substrate bias become more negative, the gm second peak amplitude increases and the onset of GIFBE occurs for lower gate voltages. GIFBE occurs when the back interface reaches accumulation, whereby the gate tunneling current is able to modulate the body potential.

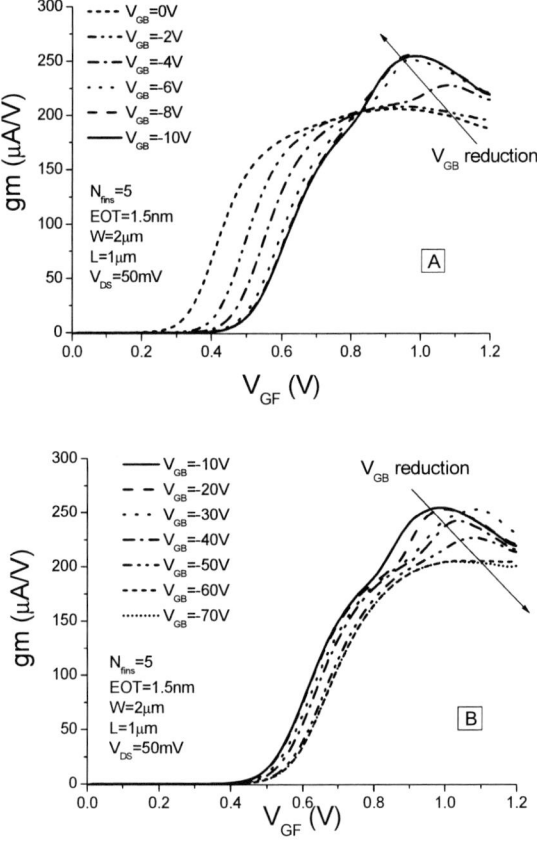

Figure 1: Experimental transconductance behavior as a function of gate voltage for a back gate bias from 0V to –10V (A) and from –10V to –70V (B)

In triple gate FinFET devices, the gate coupling is stronger and when a high negative bias is applied at the back gate (figure 1B), the accumulated region expansion causes a higher recombination at the back interface. It is also known that the recombination process increase results in a smaller body potential variation that, in turn, is not only responsible for the second peak amplitude reduction, but also for the shift of the gm second peak towards a higher gate voltage [3].

Vgmax$_2$ is the gate voltage corresponding with the transconductance second peak. When the devices with and without strain are evaluated, it was noticed that all kind of devices present the same "U" shape behavior as shown in figure 2. For strained devices the Vgmax$_2$ is lower due to the band gap variation. For tensive devices, the polysilicon gate material is compressed and it results in an increase of the hole valence band tunneling current. Meanwhile, this HVB increase is responsible for the higher body potential variation and causes an earlier GIFBE onset.

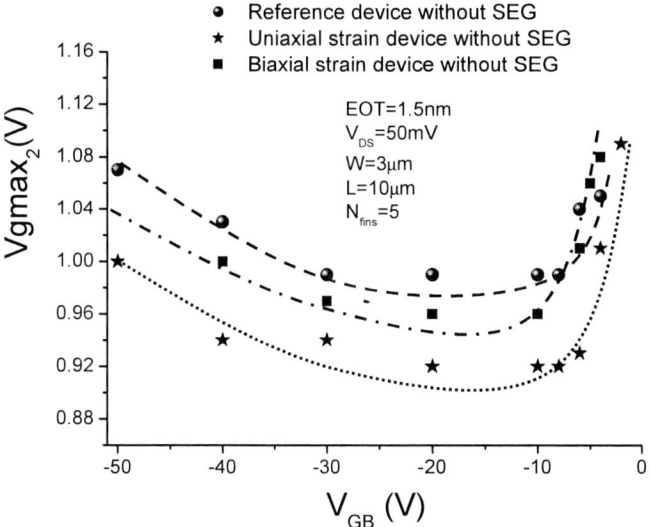

Figure 2: Experimental Vgmax$_2$ as a function of back gate bias for different FinFETs devices.

Although the "U" shape behavior is obtained for all measured devices, a comparison between devices with and without SEG (figure 3) points out that for devices without SEG the gm second peak occurs later (higher values) than for devices with SEG. It is known that GIFBE lowering results from the competition between the threshold voltage reduction and the apparent mobility degradation increase, and it is also known that the series resistance contributes with the effective mobility degradation factor increase [5]. Although SEG has been created in order to reduce the series resistance, the smaller serie resistance reduces the apparent mobility degradation and as a consequence aggravates GIFBE.

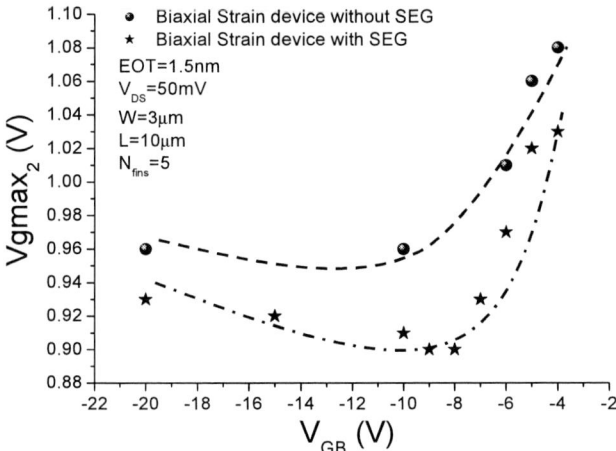

Figure 3: Experimental Vt_2 as a function of back gate bias for biaxial strained devices with and without SEG

Conclusion

The experimental "U" shape was observed in FD FinFET devices independently of the SEG presence and independently of whether or not the channel is submitted to uniaxial or biaxial stress.

For strained devices the GIFBE ($Vgmax_2$) is lower due to the band gap variation and the consequently HVB increase.

When the focus is the devices without SEG, the gm second peak occurs later than for devices with SEG due to the higher series resistance that results in apparent mobility degradation increase which minimizes GIFBE.

Acknowledgments

Paula Ghedini Der Agopian and João Antonio Martino thank CNPq for the financial support for execution of this work.

References

1. J-P Colinge, Solid State Electronics, vol.48, no 6, p. 897-905, 2004.
2. K.F Schuegraf. and Chenming Hu, IEEE Transaction on Electron Devices, Vol.41. p.761- 767, May (1994).
3. P.G.D. Agopian, J.A. Martino, E. Simoen, C. Claeys, Microelectronics Journal, Volume 38, p 114-119, January (2007),
4. W. Xiong, C. Rinn Cleavelin, R. Wise, S. Yu, M. Pas, R.J. Zaman, M. Gostkowisk, K. Mathews, C. Maleville, P. Patruno, T-J King and J-P Colinge, Electronics Letters, vol. 41, no. 8, p. 504-506, (2005).
5. P.G.D. Agopian, J.A. Martino, E. Simoen, C. Claeys, Microelectronics Journal, Volume 37, Issue 8, p. 681-685, August (2006).

QUANTIZATION EFFECT IN CAPACITANCE BEHAVIOR OF NANOSCALE
SILICON MULTIGATE MOSFETS

Aryan Afzalian, Chi-Woo Lee, Ran Yan, Nima Dehdashti Akhavan, Cindy Colinge and
Jean-Pierre Colinge

Tyndall National Institute, Prospect Row, Cork, Ireland

An unusual bump in the gate capacitance characteristics of Si
nanoscale MuGFETs is presented and explained here through 3D
NEGF quantum simulations. The oscillations in the gate
capacitance are shown to be related to the onset of occupation of
subbands: due to the particular 1D density of state and through the
migration of the charge centroid and quantum confinement. We
also present a modeled gate capacitance that allows one to better
understand the nature of the oscillation and show that it is mostly
due to the variation of the 1D DOS. However, the migration of the
charge centroid contributes to a few percents in the total variation
of the gate capacitance. Also, the gate/oxide capacitance ratio
remains low even at high gate voltage and worsen for decreasing
oxide thickness.

Introduction

In a continuous effort to increase current drive and better control short-channel effects,
device dimensions are shrunk. As well, silicon-on-insulator MOS transistors have
evolved from classical, planar, single-gate devices into three-dimensional devices with a
multigate structure. For ultra-scaled devices with cross-section dimensions smaller than
10nm and channel length, L, below 20 nanometers, quantum effects are playing a crucial
role on device performances and parameters. Here the impact of these effects on the gate
capacitance of nanoMOSFETs, which is a major figure of merit to evaluate their
performances, is studied through 3D NEGF quantum simulations [1].

Simulation Results and Interpretation

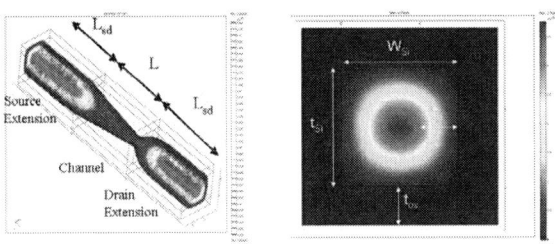

Figure 1. 3D electron concentration (n) at threshold of a 3×3 [100] oriented gate-all-around
(GAA) nanowire. Vd=0.4V. L=L$_{sd}$=10nm (Left). n in the cross-section in the middle of the
channel (x=L/2) (Right).

The gate capacitance, C_g, behavior of ultra scaled devices is expected to be strongly influenced by quantum effect [2,3]. Indeed, C_g, is a series combination of three capacitances: C_{ox}, the constant gate oxide capacitance of thickness t_{ox} and permittivity ε_{ox}, C_d, the depletion capacitance of the dark space as the carrier centroid in the channel is now at a distance t_d from the silicon/oxide interface due to quantum confinement (Fig.1), and C_S, the semiconductor capacitance, which is defined as the differential of the charge in the channel (Q_{ch}) with respect to the surface potential (ϕ_s) and is proportional to the density of states (DOS) at Fermi level, E_F [3]:

$$C_g = \frac{\partial Q_g}{\partial V_g} = -\frac{\partial Q_{ch}}{\partial V_g} = (C_{ox}^{-1} + C_d^{-1} + C_S^{-1})^{-1} = (\frac{\varepsilon_{ox} \cdot A}{t_{ox}}^{-1} + \frac{\varepsilon_{Si} \cdot A}{t_d}^{-1} + \frac{\partial Q_{ch}}{\partial \phi_s}^{-1})^{-1} . \quad [1]$$

where A is the total gate Area and ε_{Si} is the permittivity of silicon. We have extracted the gate capacitance from our 3D NEGF quantum simulations directly from the gate charge Q_g obtained using the integral of the electric field on the gate electrode (Gauss law), and from Q_{ch} given by the NEGF algorithm. Both methods give capacitance value in very good agreement. Fig. 2.Left shows the simulated gate capacitance behavior at low and high drain voltage of a $3x3nm^2$ cross-section GAA (four gates) transistor (with a transversal effective mass of $0.19 \times m_{e0}$ (ISO). Since the cross-section is square and the transversal effective mass is isotropic the second and third energy level are degenerated). These capacitance curves present variations that can be correlated to slope variations, or threshold voltages, in the I_d-V_g curves (Fig. 2.Center). These threshold voltages are themselves related to the occupation onset of the different subbands by electrons in the channel (Fig.2.Left). We can also see that unlike in a classical transistor, C_g remains significantly lower than C_{ox}.

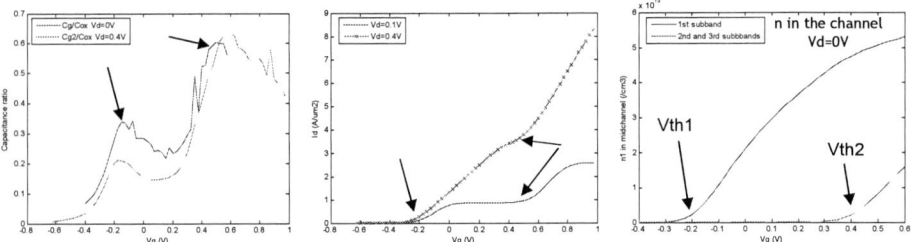

Figure 2. Gate capacitance behavior at V_d=0 (plain) and 0.4V (dashed) (Left) and I_d-V_g curves at V_d=0.1 and 0.4V (Center) and Occupation of the 1st (plain) and 2nd and 3rd (dashed) subbands in the middle of the channel by electrons (V_d=0.1V) (Right).

Oscillations related to the populating of different subbands can arise trough a non-monotonic increase of C_S: Owing to the particular 1D subband structure and density of state (DOS), the DOS at E=E_F can indeed decrease locally for increasing V_g, due to the unavailability of states around the Fermi level between two subbands (Fig. 3.Left). It can also arise through a variation of C_d due to a geometrical effect triggered each time a new subband is populated: Higher order subbands not only increase the DOS and the number of electrons in the channel, but they also move the channel centroid(s) closer to the surface which further increases C_d and the coupling between the gate and the channel and thus C_g as the charge centroid(s) is moved towards the surface (Fig. 3.Right). In order to quantify the impact of each capacitance on the reduction of C_g and on these oscillations, we have also extracted from our simulation C_S and C_d.

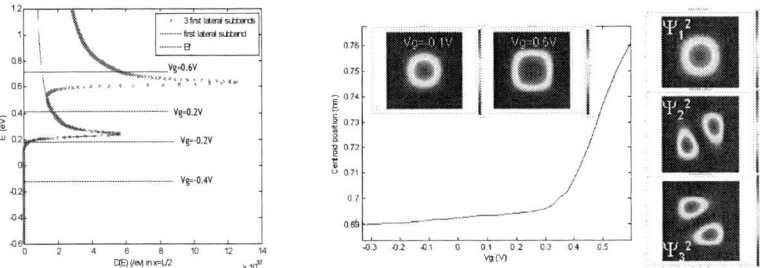

Figure 3. Left: DOS vs. Energy. The relative position of E_F compared to the DOS, as V_g moves the DOS, is shown by the plain horizontal lines for different Vg. Right: Position of the mass centroid vs. Vg and electron concentration in the cross-section in the middle of the channel for Vg=-0.1V (Resp. Vg=0.6V) (Inset) when only the first subband (Resp. the 3 first subbands) is (are) occupied.

In order to model C_S, we need to define a suitable ϕ_S since the charge is not located anymore at the Si-SiO$_2$ interface and the potential in different parts of the cross section varies differently with V_g (Fig. 4.Left). Wherever the charge concentration in the channel is negligible, the potential in the channel follows V_g. At low gate voltage, the channel potential varies as V_g everywhere $(\Delta\Phi(x,y))=\Delta V_g)$. At higher V_g, however, electrons start to populate the first subband in the center of the channel, thereby screening V_c, the potential in the center of the channel, from V_g. As a result, the potential at the center of the wire varies slower with V_g, while it increases faster with V_g close to the surface where the electron concentration is lower.

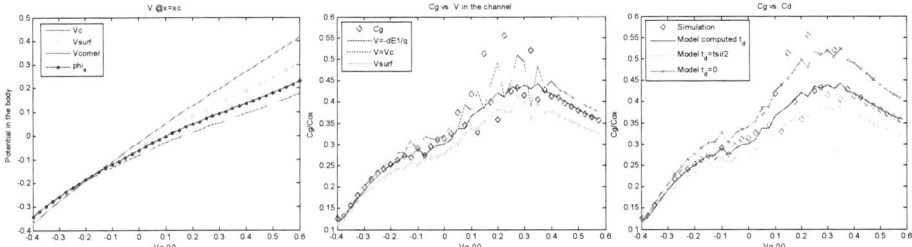

Figure 4. Variation of the potential $(\Phi(x,y))$ in the cross-section in the middle of the channel (x=x$_c$) vs. V_g in the center (V_c) and at the surface in the center, V_{surf}, and in the corner, V_{corner}, and comparison with our proposed surface potential φ_s (eq.2) (Left). Comparison of simulated C_g with modeled C_g for different choices of the surface potential (Center) and for different choice of t$_d$ (Right). 4×4nm^2 GAA nanowire.

Posing $\phi_S=V_c$ leads to overestimate of C_S, while equating ϕ_S to the potential at the surface leads to underestimate ϕ_S. Therefore, we propose to use the following averaged potential in the cross section at $x=x_c$ in the center of the channel that gives very good agreement with the simulated C_g (Fig. 4.Center):

$$\varphi_S(x_c) = \oint_{y,z} V(x_c, y, z) |\psi_1(x_c, y, z)|^2 \, dydz \qquad \frac{\partial \varphi_S}{\partial V_g} = -\frac{\partial E_1/q}{\partial V_g} \qquad [2]$$

This is the integral of the potential normalized by the square of the first wafunction. The variation of this potential with V_g is equal to the variation of the first energy level E_1 with V_g. To model C_d, the simulator computes the position of the centroid of the electron concentration from the center of the cross-section vs. V_g, which yields the value of t_d. Fig. 4.Right compares the value of C_g from the simulation to the value of C_g modeled using three different values of t_d: the actual result from numerical simulation, C_g estimated using $t_d = t_{si}/2$ (charge centroid in the center), and C_g estimated using $t_d = 0$ (charge centroid at the surface). Our modeled t_d gives the best fit.

The modeled gate capacitance resulting from the series combination of C_{ox}, C_d and C_S shows very good agreement with the one directly extracted from our simulations in a wide range of cross-section sizes, oxide thicknesses and crystal orientations. This modeling allows one to better understand the nature of the oscillation and show that it is mostly due to the variation of the 1D DOS. However, the migration of the centroid contributes to a few percents (typically less than 5%) of the value of gate capacitance variation. Several observations can be made:

- Below threshold, the first subband is barely occupied and thus the electron concentration, and therefore, C_S are very low. When V_g is increased, however, the electron concentration, and thus C_S, first increases monotonically. As a result, C_g increases monotonically with V_g until it reaches a first maximum as the peak of the first subband of the DOS coincide with the Fermi level for a V_g value of approximately - 0.1V (Fig. 2. and 3.Left).

- When V_g is further increased above this first threshold, the first subband eventually becomes full, and any further increase in V_g no longer results in an increase of the electron concentration: the DOS(E) now decreases and the electron concentration saturates, which yields a reduction and then a saturation of the gate capacitance. The current also saturates, especially at low drain voltage (Fig. 2). As the gate voltage is further increased, however, higher order subbands become populated (Fig. 3) and the current and capacitance values start to increase again as higher order subbands increase the electron concentration.

- The centroid of charge, and therefore C_d, stays constant as long as only the first subband is occupied since only the number of electrons increases with V_g but the shape of the channel remains unchanged. As higher order subbands become populated, however, they move the channel closer to the Si-SiO$_2$ interfaces (Fig. 3.Right), which further increases C_d and the coupling between the gate and the channel and thus C_g.

The shape of the channel and the peaks of electron concentration impact C_g. Both depend on the number of occupied subbands vs. V_g. Subband energy is mainly related to the product of the cross-section dimensions by the effective masses in the lateral directions:

$$E_{k,l}(t_{si}, W_{si}) = \frac{\hbar^2}{2}\left(\left(\frac{\pi.k}{m_y^* t_{si}}\right)^2 + \left(\frac{\pi.l}{m_z^* W_{si}}\right)^2\right)$$ [3]

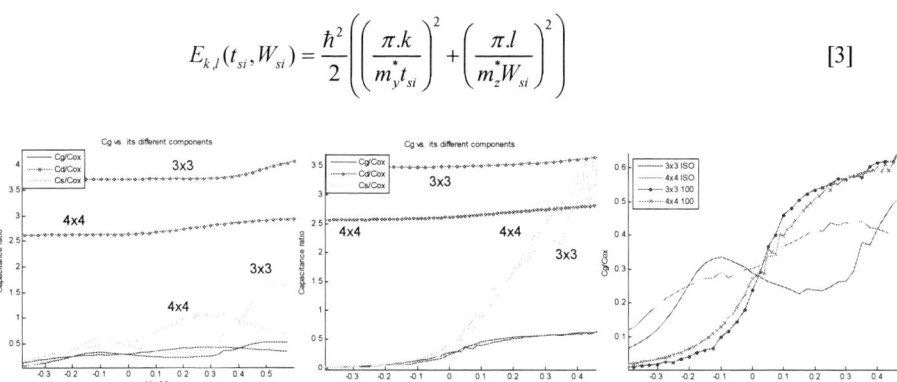

Figure 5. Evolution of the gate capacitance ratio and its different components at V_d=0.1V for a $3\times3\text{nm}^2$ and $4\times4\text{nm}^2$ GAA nanowire, ISO (Left) and [100] Si crystal orientation (Center) and gate capacitance ratio for the 4 different cases (Right). t_{ox}=1nm. ε_{ox} =3.9×ε_0.

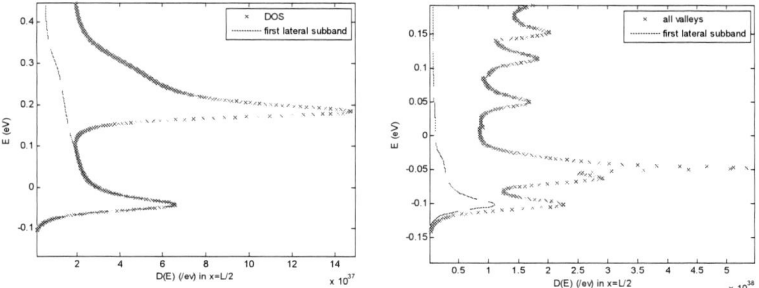

Figure 6. DOS *vs.* Energy for a $4\times4\text{nm}^2$ GAA nanowire, ISO (Left) and [100] (Right).

If this product increases due to an increase of the effective mass and/or the cross-section dimensions, the energy separation between subbands becomes so small that the oscillations become harder and harder to distinguish and eventually disappear completely. This can be observed when comparing the simulation results with an isotropic effective mass in devices with 3×3 and $4\times4\text{nm}^2$ cross sections (Fig. 5.Left and Right). When the full band structure of silicon is taken into account it is more difficult to see the oscillations (Fig. 5.Center and Right). Indeed, for the [100] channel orientation, the DOS is the sum of the three pairs of valleys: there is one pair of valley having the same lateral effective masses than in the isotropic case and 2 pairs of valleys having one of their lateral effective mass nearly 5 times larger. This drastically reduces the spacing in between the subbands and therefore makes the electron concentration increase faster with V_g (Fig. 6). This also explains the higher value of C_S, and, therefore, higher C_g value at high gate voltage. Small oscillations start to be visible however in cross-section around 4x4nm and below and that for different crystal orientations (Fig 7.Left). Note also that as pointed in [3], changing the crystal orientation does not affect much the value of the capacitance. By reducing the cross-section, oscillations become more visible. However, for smaller cross-section, smaller than approximatively between 3x3 and $2\text{x}2\text{nm}^2$, the energy separation between subbands is so large that in the usual V_g range only the first subband is occupied. This explains why no bump was observed in [2] for Si in a [100] 2×2 nm^2 nanowire as confirmed also by our own 1.5×1.5 and 2×2 nm^2 simulation results.

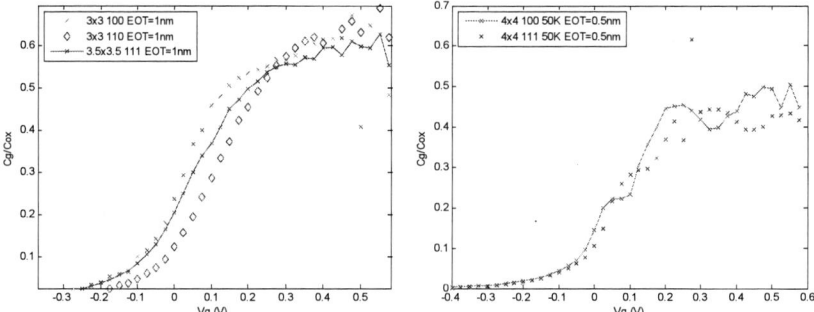

Figure 7. Gate capacitance ratio C_g/C_{ox} vs. V_g at T=300K for the 3 main different Si channel orientations (Left) and at T=50K for a [100] and [111] 4×4nm^2 nanowire (Right).

The oscillations become much more visible when temperature is reduced (Fig. 7.Right). This is attributed to the sharper distribution of the electron around Fermi at lower temperature. C_S is indeed related to the increase of the charge in the device which can be seen as the convolution of the DOS with the Fermi-Dirac distribution around the Fermi level. By reducing the temperature, the Fermi-Dirac distribution around Fermi becomes sharper, which makes C_S becomes a sharper image of the DOS at Fermi and the peaks can better be discriminated. These numerical simulation results corroborate previously reported analytical modelling [4] and experimental [5] results that show oscillations of I_d-V_g curves in nanowire transistors at low temperature, and the vanishing of these oscillations when temperature or drain voltage is increased.

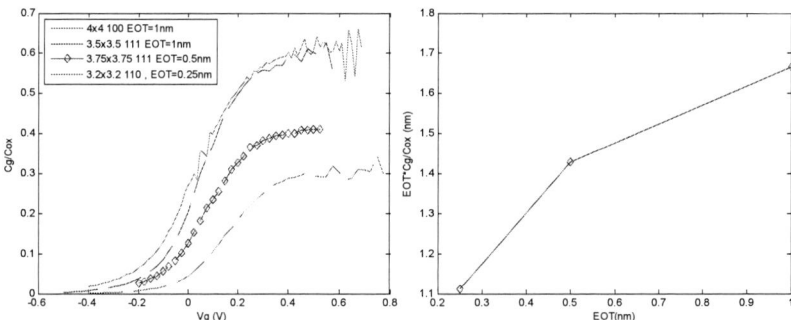

Figure 8. C_g/C_{ox} vs. V_g at T=300K for different orientations and cross-section vs. EOT (Left) and actual oxide thickness vs. EOT (Right).

Finally, in Fig.8, we have changed the equivalent oxide thickness (EOT) by keeping t_{ox} equal to 1nm but increasing ε_{ox}. As it can be seen, by reducing EOT, the ratio C_g/C_{ox} is reduced as t_d is decreased, and therefore, C_d does not scale with t_{ox}. C_S increases because of the better electrostatic control which helps to populate the DOS faster when V_g is increased, but the ratio C_S/C_{ox} decreases. As a result, C_g does not scale down with EOT.

The product of EOT by the value of the capacitance ratio at high gate voltage gives us the effective oxide thickness in term of gate capacitance value.

CONCLUSIONS

An unusual bump in the gate capacitance characteristics of Si nanoscale MuGFETs of a few nm^2 cross sections is presented and explained here through 3D NEGF quantum simulations. Oscillations that are related to the onset of occupation of subbands can arise trough C_S, the semiconductor capacitance, due to the particular 1D subband structure and density of state (DOS), the DOS at E_F can indeed decrease locally for increasing V_g, due to the unavailability of states around the Fermi level in between two subbands. It can also arise through C_d, the dark space capacitance related to quantum confinement, due to a geometrical effect triggered each time a new subband is populated: higher order subbands move the channel closer to the surface. Thus C_g increases as the charge centroid is moved towards the surface. The oscillations have also been shown to increase with decreasing temperatures. We have also presented a modeled gate capacitance resulting from the series combination of C_{ox}, C_d and C_S that shows very good agreement with the capacitance directly extracted from numerical simulations over a wide range of cross-section sizes, oxide thicknesses and crystal orientations. Owing this modeling, we have better understood the nature of the oscillation and show that it is mostly due to the variation of the 1D DOS. However, the migration of the centroid contributes to a few percents in the total variation of the gate capacitance and help to improve the ratio of C_g/C_{ox}. This ratio remains, however, low and worsen with the decrease of the oxide thickness below 1nm, as the gate capacitance does not scale with C_{ox}.

ACKNOWLEDGEMENTS

This material is based upon works supported by Science Foundation Ireland under Grant 05/IN/I888.

REFERENCES

[1] Aryan Afzalian, Dimitri Lederer, Chi-Woo Lee, Ran Yan and Jean-Pierre Colinge, « Ultra Scaled MultiGate SOI MOSFETs: Accumulation-Mode vs. Inversion-Mode », Procedding of EuroSOI 2008, Jan. 2008.
[2] A. Marchi, E. Gnani, S. Reggiani, M. Rudan, G. Baccarani, "Investigating the performances limits of Silicon-nanowire and carbon-nanotube FETs", Solid-State Electronics 50, pp. 78-85, 2006.
[3] N. Neophytou, A. Paul, M.S. Lundstrom, G. Klimeck, "Bandstructure Effects in Silicon Nanowire Electron Transport", IEEE TED, Vol. 55, 2008.
[4] R. Kim and M. Lundstrom, "Characteristic Features of 1-D Ballistic Transport in Nanowire MOSFETs", IEEE Trans. Nanotech. 7-6, pp. 787-794, 2009.
[5] J.-P. Colinge, A.J. Quinn, L. Floyd, G. Redmond, J.C. Alderman, W. Xiong, C.R. Cleavelin, T. Schulz, K. Schruefer, G. Knoblinger, P. Patruno, "Low-Temperature Electron Mobility in Trigate SOI MOSFETs", IEEE Electron Device Letters, Vol. 27, no. 2 , pp. 120- 122, 2006.

ECS Transactions, 19 (4) 329-334 (2009)
10.1149/1.3117426 ©The Electrochemical Society

**Pseudo-MOSFET Analysis of
Proton Irradiated and Annealed SOI Wafer**

Sung-Hoon Jung[1], Jae-Min Kim[1], Lak-Myung Jung[1],
Yong-Hyun Lee[1], Sorin Cristoloveanu[2], and Young-Ho Bae[3*]

[1]School of Electrical Engineering and Computer Science, KNU, Daegu, 702-701, Korea
[2]IMEP-LAHC, INPG, Grenoble, 38016, France
[3]Division of Electronics Engineering, Uiduk University, Gyeongju, 780-713, Korea
e-mail : yhbae@uu.ac.kr

The total dose irradiation effect on the SOI wafer is analyzed
at the material level by pseudo-MOSFET technique. The proton
irradiation induces positively charged traps in the buried oxide
(BOX) and amphoterically charged traps at the film-BOX interface.
The amphoteric interface trap charges contribute to the shift in
threshold and flatband voltages by modifying the effect of the
positive fixed charge in the BOX. The inherent ambipolar pseudo-
MOSFET characteristics reveal both NMOS and PMOS properties,
making it possible to identify and separate the charges that
contribute to the shift of the turn-on voltage. The negatively
charged acceptor-like states are located in the upper part of
bandgap whereas the positively charged donor-like states are
situated in the lower part of bandgap. These interface trap states
can be removed by low-temperature annealing, so that only the
oxide trapped charges continue to govern the turn-on voltage shift.

Introduction

SOI structures are traditional materials for radiation hardened semiconductor
devices, utilized for example in aerospace applications. SOI has advantage of outstanding
immunity to single event upsets due to its structure composed of a thin silicon layer on
top of the buried insulator (BOX). Still, SOI has drawbacks such as trapped charge
generation in buried oxide which gives rise to total dose irradiation effects (1).
Understanding the behavior of irradiation-induced interface traps and fixed charges in the
BOX is important. Although it has been studied for a long time, still it is not clear what
the polarity of the interface trapped charge is and where the charge states are located in
the bandgap (2-3). In general, characterization of irradiation effects on SOI is carried out
at the device level. Pseudo-MOSFET is a method to evaluate SOI at the wafer level. Two
pressure-adjustable point probes contacted on the film surface act as source and drain.
Buried oxide acts as gate insulator and the substrate is the gate electrode. This
measurement technique is very useful because it needs minimum device fabrication,
hence the change in material properties during device processing can be avoided (4). In
this paper, protons are irradiated into SOI wafers under various dose conditions and
annealed to study the total dose irradiation effect on the buried oxide.

329

Experimental

The SOI wafer used for this work was standard Unibond with thickness of 65nm and 145nm for silicon film and buried oxide, respectively. Large silicon islands (5mm x 5mm) were isolated to prevent the leakage current through the possible pinholes of buried oxide and wafer edges during pseudo-MOSFET measurements. Wet etching was used to minimize the process-induced defect generation during the island isolation (5). Proton irradiation was carried out with doses of 100 krad, 500 krad, 1 Mrad, 2 Mrad(Si). The dose rate and energy of proton irradiation were fixed at 334 rad/sec and 36.5 MeV. The wafers were annealed at 300°C for 1 hour in nitrogen ambient to stabilize the irradiation-induced defects. The electrical properties, such as turn-on voltage shift and interface trap density, were characterized by pseudo-MOSFET technique before and after annealing.

Results and Discussions

Figure 1 shows drain current-gate voltage curves for several doses. These curves illustrate the behavior of holes (for $V_G < 0$) and electrons (for $V_G > 0$). As the dose increases, the I-V curves are shifted to negative direction. This means that a net positive oxide charge is generated during proton irradiation. The amount of curve shift on the hole side is greater than that on the electron side, as shown in Figure 1(a). Therefore, as-irradiated I-V curves are 'asymmetrical'. The key point is that, after low-temperature annealing, the amount of I-V curve shift changes from asymmetrical to symmetrical (*i.e.*, equal shifts for electrons and holes), as demonstrated in Figure 1(b). This change is the interesting aspect which motivates our work.

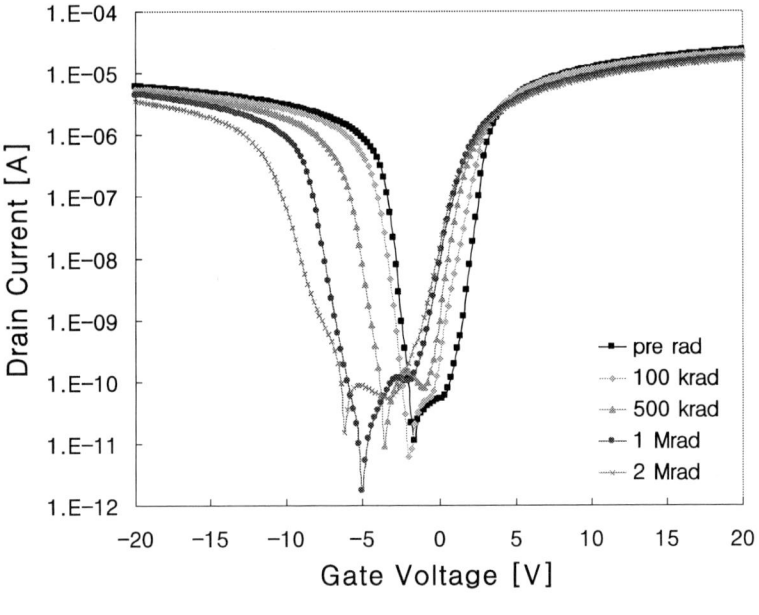

(a)

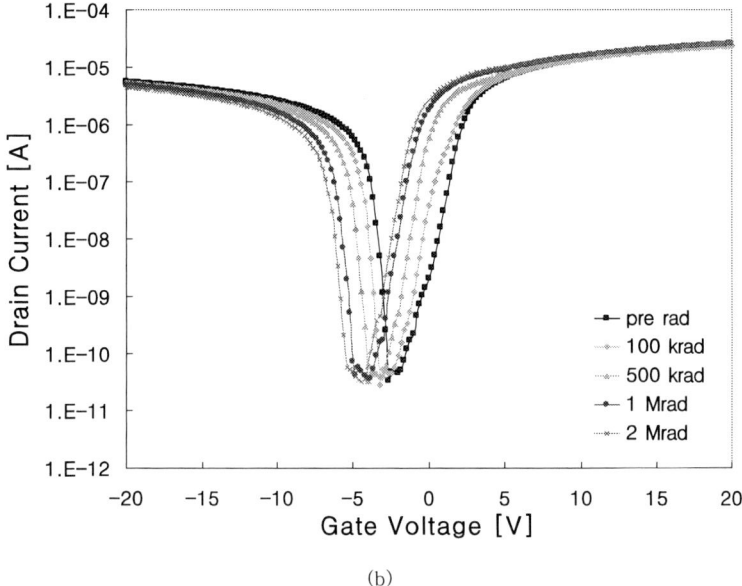

(b)

Figure 1. Drain current versus gate voltage I_D-V_G characteristics of Pseudo-MOSFET for increased total dose radiation. (a) after irradiation, (b) after subsequent annealing.

It is known that the radiation induces positive oxide trapped charges and amphoteric interface traps which can be positively charged (donor-like) or negatively charged (acceptor-like) (2). The position of each charge state in the energy gap is still under debate. It has been showed that acceptor-like states are located in the upper part of the bandgap and also that donor-like states may be situated in the upper part of bandgap. Our aim is to clarify this issue.

Using the I_D/sqrt(g_m) vs V_G curve, the turn-on voltages of both electron and hole channels are extracted and plotted in Figure 2. They correspond to the flatband voltage for hole channel and to the threshold voltage for the electron channel. Comparing the shift of turn-on voltage before annealing, the amount of flatband voltage shift with irradiation is larger than that of the threshold voltage. However, after annealing, the flatband and threshold voltage shifts become quite similar, as the I-V curves in Figure 1 implied. Before annealing, the amphoteric interface trap charge can shift the turn-on voltage in negative or positive directions depending on its charge polarity and location in the bandgap. For the threshold voltage, the amount of shift is smaller because of the compensating effect of negative interface traps on the electron inversion channel. In this case, the surface band bending is downward, and the interface traps are occupied by electrons (acceptor-like negative traps). This is why after annealing the radiation-induced negative shift increases more rapidly with dose.

For the hole channel, the amount of flatband voltage shift is larger before annealing, meaning that there is more positive charge at the interface. In this case, the band bending due to negative gate bias is upward. So, the interface states emit electrons and are positively charged (donor-like states).

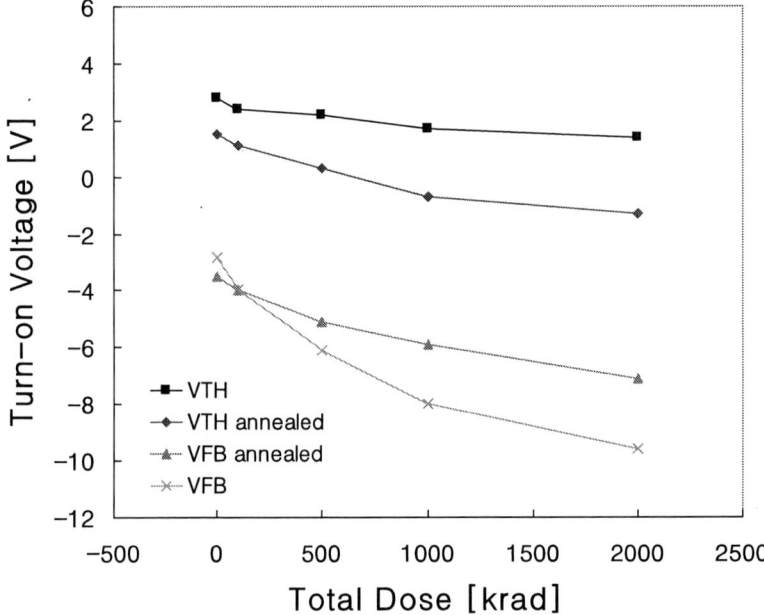

Figure 2. Comparison of threshold and flatband voltage shifts with total dose, before and after annealing.

It is concluded that acceptor-like interface states are in the upper part of bandgap and donor-like interface states are in the lower part of bandgap. After annealing, these amphoteric interface charges are removed and only the positive charges trapped inside the BOX subsist. Hence, the amount of turn-on voltage shift with dose increases equally for both polarities (electron and holes).

Additional information is obtained by extracting the interface trap density D_{it} from the subthreshold region

$$S = 2.3 \frac{kT}{q} \left(1 + \frac{C_{si} + qD_{it}}{C_{ox}} \right)$$ [1]

where S is the subthreshold swing, $dV_G/dlogI_D$.

Figure 3 shows the variation of interface trap density with total dose. Before annealing, D_{it} steadily increases with dose. But after low-temperature annealing, the trap density is reduced and becomes rather independent of total dose. Note that this D_{it} value is slightly higher than that before irradiation. This small increase is attributed to the residual traps, which have not been totally cured by annealing, and also to the top surface, exposed to air in pseudo-MOSFET measurements. Indeed, in thin SOI structures the channel-to-surface coupling cannot be ignored (6). It is reasonable to assume that the surface condition (traps and charges) changes during low-temperature annealing and modifies the channel characteristics. Nevertheless, the radiation-induced traps at the film-BOX interface are removed during annealing process, and then only the oxide trapped charge in the BOX continues to govern the turn-on voltage shift. This is why the two curves in Fig. 2 become parallel. The radiation-induced interface trapped charge is more easily annealed out than the fixed oxide trapped charge.

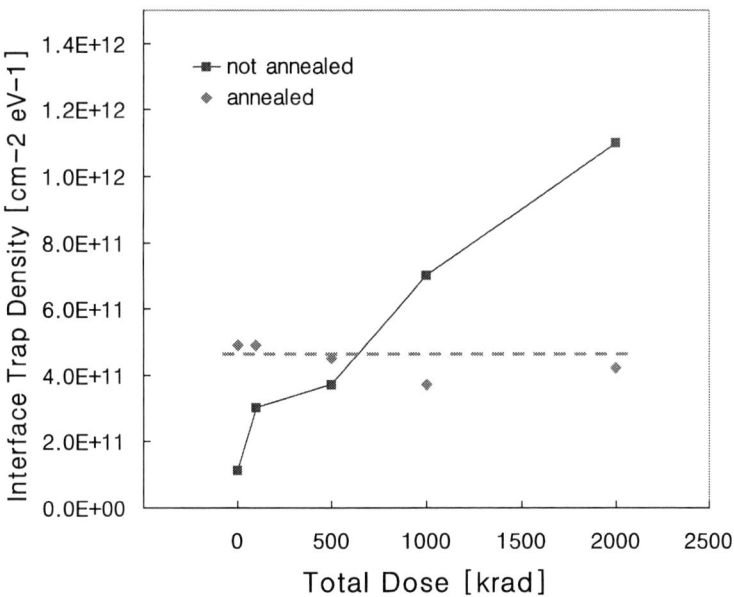

Figure 3. Density of interface trap with total dose before and after annealing.

Conclusion

The irradiation effect on SOI wafers, investigated by pseudo-MOSFET technique, shows that proton irradiation induces positively charged oxide traps and amphoteric interface traps. The negatively charged acceptor-like states, located in the upper part of the bandgap, compensate the reduction of the electron channel threshold voltage imposed by the positive BOX charge. The positively charged interface states located in the lower part of the bandgap enhance the negative shift of flatband voltage (hole channel). These radiation-induced interface trapped charges can be removed easily by low-temperature annealing process, leading to similar shifts in turn-on voltage for electron and hole channels.

References

1. J.R. Schwank, V. Ferlet-Cavrois, M.R. Shaneyfelt, P. Paillet, and P.E. Dodd, *IEEE. Trans. Nuclear Science,* **50**, 522 (2003).
2. P.J. McWhorter, P.S. Winokur, and R.A. Pastorek, *IEEE Transaction on Nuclear Science*, **35**, 1154 (1988).
3. M.L. Alles, D.R. Ball, R.D. Schrimpf, D.M. Fleetwood, R.A. Reed, and B. Jun, in *Silicon-On-Insulator Technology and Devices XII*, G. Celler, S. Cristoloveanu, J. Fossum, F. Gamiz, K. Izumi, Editors, PV 2005-03, p.87, The Electrochemical Society Proceedings Series, Pennington, NJ (2005)
4. S. Cristoloveanu, D. Munteanu, M.S.T. Liu, *IEEE. Trans. Electron Devices,* **47**, 1018 (2000).
5. Y.H. Bae, K.W. Kwon, J.H. Lee, J.H. Lee, H.J. Woo, S. Cristoloveanu, in *Silicon-On-Insulator Technology and Devices XII*, G. Celler, S. Cristoloveanu, J. Fossum, F. Gamiz, K. Izumi, Editors, PV 2005-03, p.295, The Electrochemical Society Proceedings Series, Pennington, NJ (2005).
6. G. Hamaide, F. Allibert, H. Hovel, S. Cristoloveanu, *Journal of Applied Physics*, **101**, 114513 (2007).

High-K Ta$_2$O$_5$ Polyoxide Deposited on Polycrystalline with NH$_3$ Plasma Treatment

Chyuan. Haur. Kao, K. S. Chen, J. S. Chiu, C. S. Chen, T. C. Chan,
W. S. Luo, Y. T. Chung,

Department of Electronic Engineering,
Chang Gung University,
259, Wen-Hwa 1st Road, Kwei-Shan, Tao-Yuan, Taiwan
Tel: +886-3-2118800 ext 5783 Fax: +886-3-2118507
e-mail: chkao@mail.cgu.edu.tw

The high-k Ta$_2$O$_5$ in our study is used to apply as gate dielectric in polyoxide and low-temperture polycrystalline silicon thin-film transistor (LTPS-TFT) for performance improvements. Therefore, we employ the NH$_3$-plasma treatment to improve the electrical characteristics of the high-k Ta$_2$O$_5$ polyoxide dielectric such as higher breakdown voltage and lower leakage current. It's seen that the nitrogen can pile up at the high-k Ta$_2$O$_5$ dielectric and polysilicon interface to form strong Si-N bondings from the SIMS and XPS analysis.

Introduction

The quality of polysilicon oxide (polyoxide) plays an important role as the gate dielectric of polysilicon TFTs [1]. It's well known that different varieties of high-k gate dielectrics have been suggested as possible alternatives for the SiO$_2$ gate dielectrics. The dielectric constant of Ta$_2$O$_5$ is about 26 and its band gap is about 4.5. In previous studies, the tantalum pentoxide (Ta$_2$O$_5$) is become more promising high-κ gate dielectric in MOSFETs due to its outstanding dielectric properties, including high dielectric constant (k~26), lower gate-leakage current, and superior reliability. Therefore, the high-k Ta$_2$O$_5$ can be used to apply as gate dielectric in polyoxide capacitors and low-temperture polycrystalline silicon thin-film transistor (LTPS-TFT).

Besides, hydrogenation leads to tie up the grain boundary dangling bonds with hydrogen, thereby significantly improve the characteristics of the poly-Si TFT's. [2] However, it can be found that the characteristics of the poly-Si TFT's after hydrogenation suffer low hot-carrier endurance and low thermal stability [3], [4]. Furthermore, previous studies show that the quality of oxide is improved by nitrogen incorporated at oxide-silicon interface [5]-[6]. Since the nitrogen incorporated at the gate oxide/silicon interface can avoid hot-carrier damages and prevent boron penetration and improve interface endurance to Fowler-Nordheim (F-N) stress. Theses improvements were suspected to be due to the strain relaxation of the interfaces. The pile-up nitrogen at the interface could form oxynitride to promote the plasma passivation. Therefore, a simpler technique, NH$_3$-plasma post-treatment was employed to enhance the electrical characteristics of the poly-Si TFT's [7].

In our paper, the high-k Ta$_2$O$_5$ as gate polyoxide dielectric combined with NH$_3$ plasma post-treatment was studied and showed excellent characteristics such as higher breakdown voltage and lower leakage current.

Experiment

At first, the p-type Si wafers were thermally oxidized to have an oxidized layer of 550nm. Then, a polysilicon film of 200nm was deposited at 625℃ by low pressure chemical vapor deposition (LPCVD) system. It was implanted with phosphorous at a 5×10^{15} cm^{-2} dosage and 30KeV energy following activated for 10sec in an N_2 ambient at 950℃ to obtain a sheet resistance of 100~115 Ω/sq. Next, the native oxide was removed by dipping in diluted hydrofluoric acid (HF) solution. A 30nm Ta_2O_5 gate dielectric was deposited by RF Sputtering system, and then annealed at 600℃ for 30 min in O_2 ambient to improve thin-film quality. An aluminum gate of a thickness of 300nm was deposited on gate dielectric films. After self-aligned defining aluminum gate and gate dielectric pattern to form polyoxide capacitors by a two step wet etching process. The 300nm Al gate and 30nm Ta_2O_5 gate dielectric layer were etched by Al etching solution and using wet etching, respectively. The high-k Ta_2O_5 dielectric capacitor deposited on polysilicon was fabricated in this study.

Polyoxide thickness was determined by high-frequency (100 kHz) capacitance-voltage (C-V) measurement. The amounts of nitrogen in the polyoxide were measured using secondary ion mass spectroscopy (SIMS) with Cs+ as the primary ions. The current density versus voltage (J-V) characteristics was monitored using a HP4156C semiconductor parameter analyzer.

Results and Discussion

Figure 1 shows the typical J-V characteristics of the high-k Ta_2O_5 dielectric with post NH_3 plasma for 30min, 60min treatment and without NH_3 plasma under both positive and negative Vg gate bias, respectively. It is clearly that the high-k dielectric with NH_3 post plasma have lower leakage currents and higher breakdown electric fields than those of the non-implanted polyoxide for both cases of positive gate bias (electron injection from the bottom polysilicon interface) and negative gate bias (electron injection from the top polysilicon interface) and the breakdown voltage is increased as increasing NH_3 plasma time. The improvements of the electrical characteristics are believed to be due to that the incorporated nitrogen can passivate the dangling bonds and break the strained bonds to form Si-N bonds in the polysilicon grain boundaries and high-k Ta_2O_5 dielectric /polysilicon interface. So, the local stress can be relaxed in the dielectric network. Since the stress between the polyoxide and the polysilicon affects the polyoxide breakdown strength, and the lower oxide stress, the higher oxide breakdown strength.

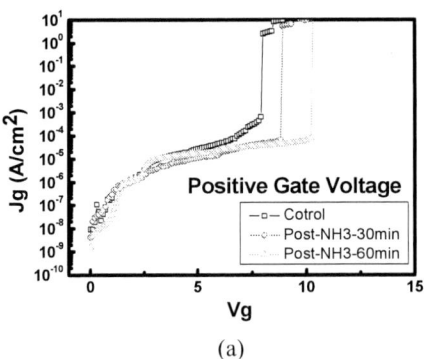

(a)

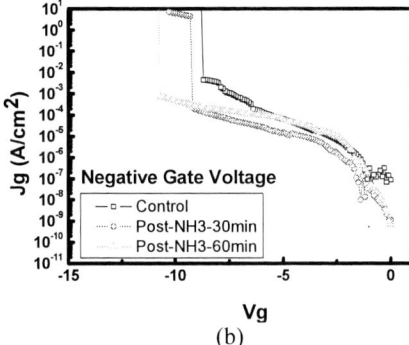

(b)

Figure 1. The J-V characteristics of the high-k Ta_2O_5 dielectrics with post NH_3 plasma for 30min, 60min treatment and without NH_3 plasma (control) under(a) positive and (b) negative gate bias.

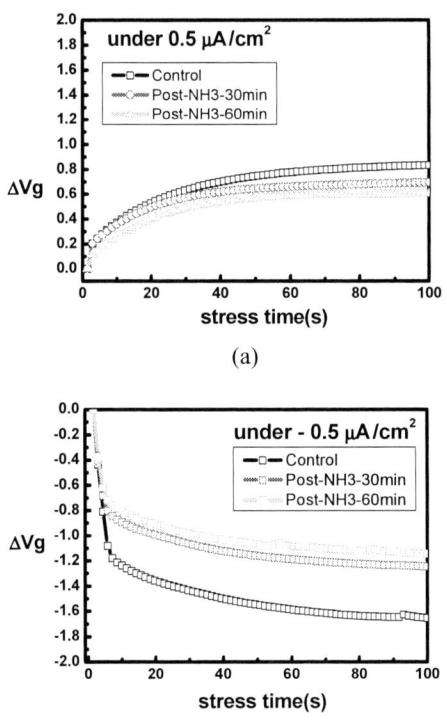

(b)

Figure 2. The curves of gate voltage shifts (ΔVg) versus stress time of the high-k Ta_2O_5 dielectrics without(control), and with post-NH_3 plasma for 30 min and 60 min treatment under the top gate applied with (a) positive gate constant current (under $0.5\mu A/cm^2$) stress and (b) negative gate constant current (under $-0.5\mu A/cm^2$) stress.

Figure 2 shows the gate voltage shift ($\square Vg$) versus stress time under $\pm0.5\mu A/cm^2$ constant current stress for the above samples. It can be seen that those samples with NH_3 plasma treatment had smaller gate voltage shifts in comparison with the control sample, and the gate voltage shift was decreased by increasing plasma time both for positive and negative bias. This means that the longer the NH_3 plasma time, the fewer the trapped electron. This is also due to the NH_3 plasma post-treatment can passivate and reduce trap states at the interface between the high-k dielectric/polysilicon. Since there are many strained bonds within the high-k Ta_2O_5 polyoxide, and these strained bonds are easy to be broken by high field stress and cause dielectric breakdown. When the nitrogen was piled up at the high-k Ta_2O_5 dielectric/polysilicon interface, the nitrogen rich layer formed not only strengthens the high-k Ta_2O_5 dielectric structure and thereby improves the high-k Ta_2O_5 polyoxide quality and performance. Figure 3 shows the weibull plots of charge to breakdown for the above samples. It's also seen that the sample with NH_3 plasma post for 60min treatment had the largest Q_{bd} and uniform Q_{bd} under both positive and negative $\pm0.5\mu A/cm^2$ constant current stress. These improvements were attributed to be due to not only the hydrogen passivation of dangling bonds, but also the nitrogen can pile-up at the high-k Ta_2O_5 dielectric/polysilicon interface to form strong Si-N bonding to terminate dangling bonds and traps in the high-k dielectric and the interface between the high-k Ta_2O_5 dielectric and polysilicon.

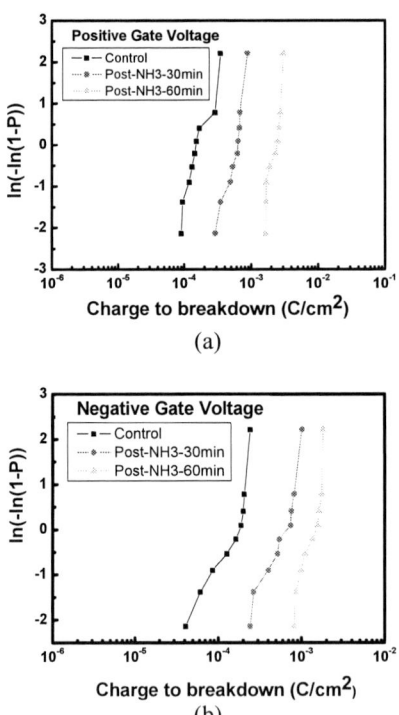

Figure 3. The Weibull plots of charge to breakdown (Q_{bd}) distribution for the high-k Ta_2O_5 dielectrics without(control), and with post-NH_3 plasma for 30 min and 60 min treatment under the top gate applied with (a) positive gate constant current (under $0.5\mu A/cm^2$) stress and (b) negative gate constant current (under $-0.5\mu A/cm^2$) stress.

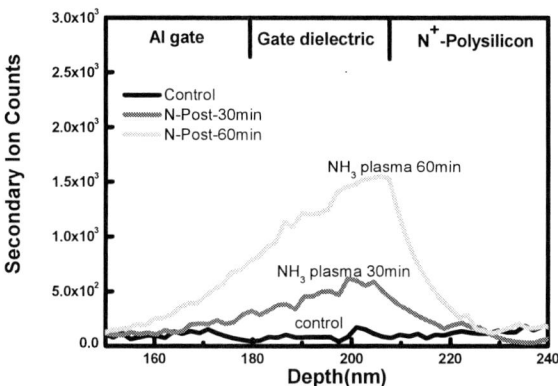

Figure 4. The secondary ion mass spectroscopy (SIMS) profile analysis of nitrogen for the high-k Ta_2O_5 dielectrics without (control), and with post-NH_3 plasma for 30 min and 60 min treatment.

Figure 4 shows the secondary ion mass spectroscopy (SIMS) profile of nitrogen ion for the above samples. It's clearly seen that the nitrogen accumulated and piled up at the interface between the high-k Ta_2O_5 dielectric and polysilicon and the longer the NH_3 plasma time, the higher the nitrogen peak value. There are also some amounts of nitrogen existing especially within the high-k Ta_2O_5 dielectrics.

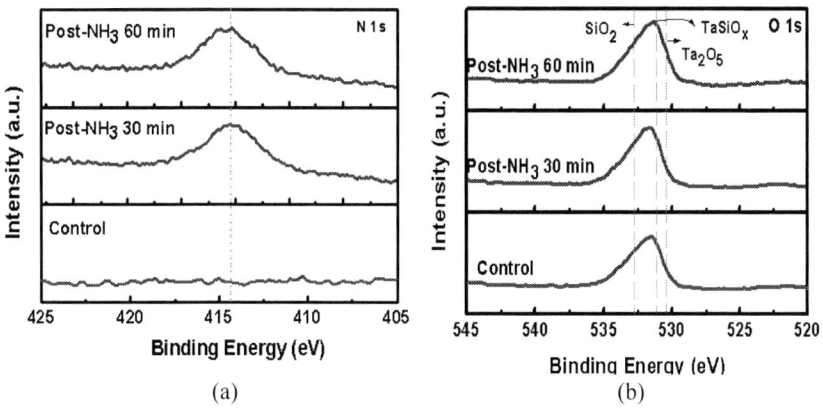

Figure 5. (a) N 1s and (b) O 1s XPS spectrum for the high-k Ta_2O_5 dielectrics without (control), and with post-NH_3 plasma for 30 min and 60 min treatment.

Figure 5 shows the nitrogen (N 1s) and oxygen (O 1s) XPS spectrum for the high-k Ta_2O_5 dielectrics without (control), and with post-NH_3 plasma for 30 min and 60 min treatment. It can be seen that the Si-N bonding is found obviously for the sample with post-NH_3 plasma in comparison with the control sample from the N 1s level analysis. Besides, the intensity of the Ta_2O_5 structure for the post-NH_3 60 min sample becomes slightly higher than that in control sample from the O 1s level analysis. This means that the NH_3 plasma post-treatment can passivate and reduce trap states at the interface between the high-k dielectric/polysilicon. The incorporated nitrogen can form strong Si-N bonding to terminate dangling bonds and traps in the interface between the high-k Ta_2O_5 dielectric and polysilicon, thereby strengthens and improves the high-k Ta_2O_5 polyoxide quality and performance.

Summary

In this study, the electrical characteristics of high-k polyoxide capacitors combined with NH_3 plasma post treatment have been improved (i.e.,lower leakage current, higher electrical breakdown filed and lower electron trapping rate, larger Q_{bd}).

These improvements were attributed to be due to the nitrogen can pile-up at the high-k Ta_2O_5 dielectric/polysilicon interface to form strong Si-N bonding to terminate dangling bonds and traps in the interface between the high-k Ta_2O_5 dielectric and polysilicon. By the way, the quality and integrity of the high-k Ta_2O_5 polyoxide is also improved by the NH_3 plasma post-treatment.

Acknowledgments

This work was supported by the National Science Council of Taiwan, R.O.C., under Contract NSC 97-2221-E-182-006.

References

1. M. K. Hatalis, J. H. Kung, J. Kanicki, and A. A. Bright, Mater. Res. Soc. Symp. Proc, 182, 357 (1990).
2. B. A. Khan and R. Pandya, IEEE Electron. Device. 37 (1990) 1727.
3. S. Banerjee, R. Sundaresan, H. Schichijo, and S. Malhi, IEEE Electron. Device. 35 (1988) 152.
4. M. Hack etal , IEEE Electron. Device. 40 (1993) 897.
5. S. Hadda, and M. S. Liang, IEEE Electron Device Lett, 8 (1987) 58.
6. Mirabedini, M.R.; Kamath, A.; and Yeh, W.C, IEEE Electron Device Lett, 24 (2003) 301.
7. C. K. Yang, T. F. Lei, and C. L. Lee, IEEE Electron Device Lett, 15 (1994) 389.

Author Index

Abbadie, A.	79	De Keersgieter, A.	45
Afzalian, A.	229, 321	de Souza, M.	265
Agopian, P. G.	317	Dehdashti Akhavan, N.	229, 321
Andrieu, F.	37, 55	Delaye, V.	37
Armstrong, G.	283, 305	Denis, H.	37
Augendre, E.	145	Denorme, S.	55
		Deresmes, D.	201
Bae, Y.	297, 329	Donetti, L.	235
Balestra, F.	161	Doria, R. T.	289
Barbe, J.	37	Dubois, E.	201
Baudot, S.	37	Duffy, R.	45
Baumgartner, O.	15		
Bawedin, M.	243	Elbuluk, M. E.	181
Bernasconi, S.	93	Endo, K.	273
Biesemans, S.	45	Eymery, J.	37
Boeuf, F.	55		
Bonnaud, O. A.	195	Faynot, O.	37, 55
Breil, N.	201	Fenouillet-Beranger, C.	55
Brevard, L.	37, 55	Ferreira, F. A.	139
Buj, C.	55	Figuet, C.	79
Bulteel, O.	175	Fiori, V.	55
		Flandre, D.	139, 175, 243, 265
Campidelli, Y.	93		
Celler, G. K.	3, 71	Gaillard, T.	195
Cerdeira, A.	139, 289	Gallon, C.	55
Chen, C. S.	335	Gamiz, F.	127, 235
Chen, K. S.	335	Garcia Ruiz, F.	235
Chen, T. C.	335	Ghyselen, B.	145
Cheng, W.	65	Gianesello, F.	257
Chiu, J. S.	335	Gimenez, S. P.	153
Choi, B.	189	Godoy, A.	127, 235
Choi, B.	101	Guegan, G.	257
Chung, Y. T.	335	Guiot, E.	79, 145
Claeys, C.	133, 317		
Clavelier, L.	93, 145	Haendler, S.	257
Cleavelin, C.	297	Halimaoui, A.	201
Colinge, C.	229, 321	Hammoud, A.	181
Colinge, J.	229, 321	Hayashi, O.	121
Collaert, N.	45, 133	Hoffmann, T.	45
Coulon, N.	195		
Cristoloveanu, S.	55, 93, 145, 243, 297, 329	Idrisi, H.	29
		Ishikawa, Y.	273
Damlencourt, J.	93, 145	Jung, H.	101
Davidson, J.	189	Jung, L.	329

Jung, S.	329	Perreau, P.	55
Jurczak, M.	45, 133	Pham Nguyen, L.	55
		Planes, N.	257
Kang, W.	189		
Kao, C. H.	335	Rashmi, R.	283, 305
Kim, J.	329	Raynaud, C.	257
Kolbesen, B.	29	Reckinger, N.	201
Kostrzewa, M.	37	Rieutord, F.	37
Kranti, A.	283, 305	Rodrigues, M.	133
		Rodriguez, N.	235
Lamrani, Y.	37	Romanjek, K.	145
Lander, R. J.	45	Rooyackers, R.	45
Larrieu, G.	201	Ruiz, F. G.	127
Le Royer, C.	93, 145		
Le Vaillant, Y. -M.	79	Sakamoto, K.	273
Lee, C.	229, 321	Schanovsky, F.	15
Lee, J.	101	Selberherr, S.	15
Lee, J.	101	Simoen, E.	133, 317
Lee, J.	297	Skotnicki, T.	55
Lee, Y.	329	Souma, S.	211
LeQuan, X. C.	189	Sverdlov, V.	15
Lin, C.	311		
Liu, Y.	273	Tahara, Y.	113
Luo, W. S.	335	Tang, X.	201
		Teramoto, A.	65
Ma, Z.	71	Tienda-Luna, I.	127
Martineau, B.	257	Tosti, L.	55
Martino, J. A.	133, 289, 317	Toure, H.	195
Masahara, M.	273	Touret, P.	257
Matsukawa, T.	273	Tsuchiya, H.	211
Mercha, A.	133	Tsukada, J.	273
Morand, Y.	93		
		Udrea, F.	243
Na, K.	297	Uzelli, F.	29
Nakano, S.	121		
Nguyen, T.	93	van Dal, M. J.	45
Nirgin, S.	305	Van Den Daele, W.	145
		Veloso, A.	45, 133
Ogawa, M.	211	Vincent, B.	93
Ohmi, T.	65	Vizioz, C.	37
Omura, Y.	113, 121, 221		
Ouchi, S.	273	Wakabayashi, H.	121
		Wang, X.	311
Pang, H.	71	Weber, O.	55
Patruno, P.	297	Wei, X.	311
Patterson, R.	181	Widiez, J.	37
Pavanello, M.	139, 265, 289	Windbacher, T.	15
Pawlak, B. J.	45	Witters, L.	45
Pejic, M.	29	Wu, A.	311

Xiong, W.	297
Yamada, Y.	211
Yamakawa, S.	121
Yamauchi, H.	273
Yan, R.	229, 321
Yarekha, D. A.	201
Zhang, B.	311
Zhang, M.	311